GW01372804

PRACTICAL AND EXPERIMENTAL ROBOTICS

PRACTICAL AND EXPERIMENTAL ROBOTICS

FERAT SAHIN
Rochester Institute of Technology, New York, USA

PUSHKIN KACHROO
Virginia Tech, Blacksburg, USA

CRC Press
Taylor & Francis Group
Boca Raton London New York

CRC Press is an imprint of the
Taylor & Francis Group, an **informa** business

CRC Press
Taylor & Francis Group
6000 Broken Sound Parkway NW, Suite 300
Boca Raton, FL 33487-2742

© 2008 by Taylor & Francis Group, LLC
CRC Press is an imprint of Taylor & Francis Group, an Informa business

No claim to original U.S. Government works
Printed in the United States of America on acid-free paper
10 9 8 7 6 5 4 3 2

International Standard Book Number-13: 978-1-4200-5909-0 (Hardcover)

This book contains information obtained from authentic and highly regarded sources. Reprinted material is quoted with permission, and sources are indicated. A wide variety of references are listed. Reasonable efforts have been made to publish reliable data and information, but the author and the publisher cannot assume responsibility for the validity of all materials or for the consequences of their use.

No part of this book may be reprinted, reproduced, transmitted, or utilized in any form by any electronic, mechanical, or other means, now known or hereafter invented, including photocopying, microfilming, and recording, or in any information storage or retrieval system, without written permission from the publishers.

For permission to photocopy or use material electronically from this work, please access www.copyright.com (http://www.copyright.com/) or contact the Copyright Clearance Center, Inc. (CCC) 222 Rosewood Drive, Danvers, MA 01923, 978-750-8400. CCC is a not-for-profit organization that provides licenses and registration for a variety of users. For organizations that have been granted a photocopy license by the CCC, a separate system of payment has been arranged.

Trademark Notice: Product or corporate names may be trademarks or registered trademarks, and are used only for identification and explanation without intent to infringe.

Library of Congress Cataloging-in-Publication Data

Sahin, Ferat.
   Practical and experimental robotics / Ferat Sahin and Pushkin Kachroo.
      p. cm.
   Includes bibliographical references and index.
   ISBN-13: 978-1-4200-5909-0 (hardcover : alk. paper)
   ISBN-10: 1-4200-5909-2 (hardcover : alk. paper)
   1. Robotics. I. Kachroo, Pushkin. II. Title.
 	
TJ211.S33 2007
629.8'92--dc22   2007014718

Visit the Taylor & Francis Web site at
http://www.taylorandfrancis.com

and the CRC Press Web site at
http://www.crcpress.com

To the joy of books and learning

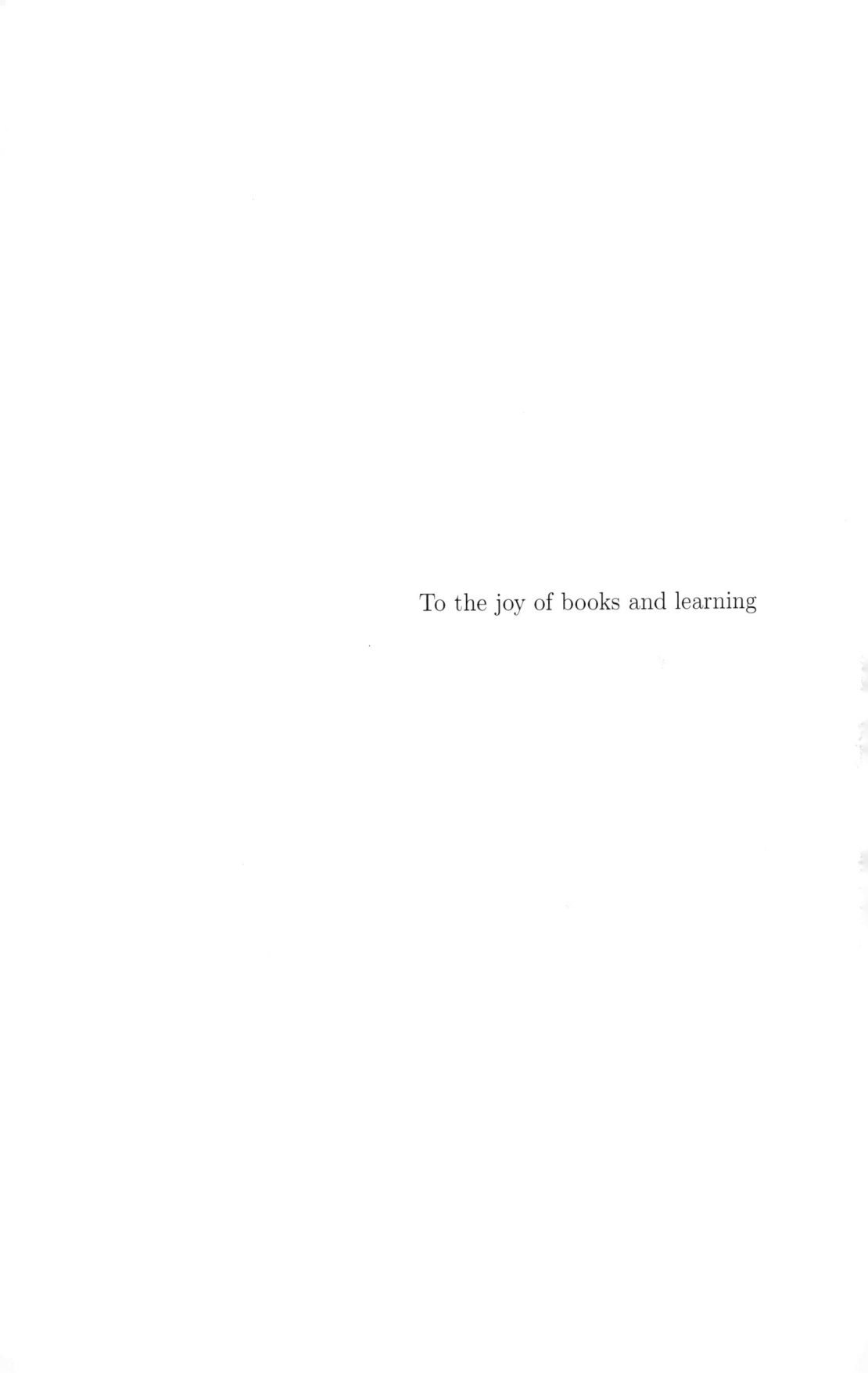

For our parents

Aslan Sahin and Zehra Sahin

&

Dr. P. L. Kachroo and Sadhna Kachroo

Acknowledgments

I have benefited enormously from the love, support, and editorial advice of my family and friends in the course of writing the book. Particularly, I am thankful to my wife, Selhan Garip Sahin, for her editorial help, suggestions, and limitless patience. I am thankful to my co-author, Pushkin Kachroo, for his encouragement and perseverance in publishing this book. The critical reviews of Dr. Wayne Walter and Dr. Mo Jamshidi were tremendously helpful in shaping the technical content of the book. I am also thankful to my students Dr. Ajay Pasupuleti, Archana Devasia, Nathan Pendleton, and Joshua Karpoff for their help in various chapters.

Dr. Ferat Sahin

About the Authors

Ferat Sahin received his B.Sc. in Electronics and Communications Engineering from Istanbul Technical University, Turkey, in 1992 and M.Sc. and Ph.D. degrees from Virginia Polytechnic Institute and State University in 1997 and 2000, respectively. In September 2000, he joined Rochester Institute of Technology, where he is an Associate Professor. He is also the director of Multi Agent Bio-Robotics Laboratory at RIT. He is currently on sabbatical at the University of Texas San Antonio. His current research interests are System of Systems, Robotics, MEMS Materials Modeling, Distributed Computing, and Structural Bayesian Network Learning. He has about seventy publications including journals. He is a member of the IEEE Systems, Man, and Cybernetics Society, Robotics and Automation Society, and Computational Intelligence Society. Locally, he has served as Secretary (2003), section Vice-chair (2004 and 2005) in the IEEE Rochester Section, and the faculty adviser for IEEE Student Chapter at RIT in 2001 and 2002. He has served as the Student Activities chair (2001 - 2003) and the Secretary of the IEEE SMC society since 2003. He has received an "Outstanding Contribution Award" for his service as the SMC Society Secretary. He was the publications Co-Chair for the IEEE International Conference on System of Systems Engineering (SOSE 2007). He is an Associate Editor of IEEE Systems Journal and the Deputy Editor in Chief of International Journal of Electrical and Computer Engineering.

Pushkin Kachroo received his Ph.D. in Mechanical Engineering from University of California at Berkeley in 1993, his M.S. in Mechanical Engineering from Rice University in 1990, and his B.Tech. in Civil Engineering from I.I.T Bombay in 1988. He obtained the P.E. license from the State of Ohio in Electrical Engineering in 1995. He obtained M.S. in Mathematics from Virginia Tech in 2004. He is currently an Associate Professor in the Bradley Department of Electrical & Computer Engineering at Virginia Tech. He was a research engineer in the Robotics R&D Laboratory of the Lincoln Electric Co. from 1992 to 1994, after which he was a research scientist at the Center for Transportation Research at Virginia Tech for about three years. He has written four books (Feedback Control Theory for Dynamic Traffic Assignment, Springer-Verlag, 1999, Incident Management in Intelligent Transportation Systems, Artech House, 1999, Feedback Control Theory for Ramp Metering in Intelligent Transportation Systems, Kluwer, 2003, Mobile Robotics Car Design, McGraw Hill, (August 2004)), three edited volumes, and overall more than eighty publications including journal papers. He has been the chairman of ITS and Mobile Robotics sessions of SPIE conference multiple times. He received the award of "The Most Outstanding New Professor" from the College of Engineering at Virginia Tech in 2001, and Deans Teaching Award in 2005.

Preface

In recent years, the robotics market has grown dramatically with the new family of robots which are simple and easy to use. These robots can be used/ explored by a large variety of people ranging from hobbyists to college students. In addition, they can also be used to introduce robotics to K-12 students and increase their attention and interest in engineering and science. The book has chapters on basic fundamentals of electrical and mechanical systems as well as some advanced topics such as forward and inverse kinematics of an arm robot, dynamics of a mobile robot, and vision control for robots. Each chapter starts with basic understanding of the topic covered. Later in the chapters, the advanced topics are explored so that hobbyists and K-12 students can still assimilate the topic covered in the chapter.

The book also presents a variety of robots from arm robots to robotic submarines most of which are available as kits in the market. In the chapters, we first describe basic mechanical construction and electrical control of the robot. Then, we give at least one example on how to use and operate the robot using microcontrollers or software. We present two arm robots, a two-wheel robot, a four wheel robot, a legged robot, flying robots, submarines, and robotic boats. In addition, we present topics which are commonly utilized in robotics.

The following is an overview of what can be found in each chapter, pointing the goal of the chapters.

Fundamentals of Electronics and Mechanics

In this chapter, we present fundamentals of electrical and mechanical systems and components. We first start with basic electrical components: Resistor, Capacitor, and Inductor. Then, we explore semiconductor devices such as diodes, transistors, operational amplifiers, logic components, and circuitries. In diodes, we present different kinds of diodes mostly used in robotics such as zener diodes, light emitting diodes (LED), photodiodes, and their applications. Then, we introduce transistor theory and transistor types mostly used in robotics and their applications. We discuss bipolar transistors (BJT) and field effect transistors (FET). Discussion continues on special electrical components, namely operational amplifiers (OPAMPs). Most common applications of OPAMPS are also discussed. Finally, we discuss digital systems and their basic components such as logic gates, flip-flops, registers, and some circuitry designed with these components.

In the Mechanical systems section we discuss common mechanical components such as gears, pulleys, chains, cams, ratchets and pawl, bearings, belt and chain drives. These components are introduced and examples of robotics related applications are given for each component.

Basic Stamp Microcontroller

In this chapter, we introduce a commonly used microcontroller in robotic kits. It is called Basic Stamp Microcontroller and used in later chapters. It is a microprocessor which can be programmed with BASIC programming language. Basic Stamp Microcontroller has a PIC microcontroller as a core microcontroller and related electronics. These electronics let users program the PIC microcontroller with a BASIC programming language. We present several Basic Stamp microcontrollers: BASIC Stamp I, BASIC Stamp II, BASIC Stamp IIsx, BASIC Stamp IIp, and BASIC Stamp IIe. In addition some evaluation boards used for BASIC Stamp IIe: BASIC Stamp II Carrier Board (Rev. B), BASIC Stamp Super Carrier Board (Rev. A), Board of Education (Rev. B), and BASIC Stamp Activity Board (Rev. C). Then, we discuss the BASIC Stamp Editor and how to connect evaluation boards to PC and program them. Finally, we present PBASIC programming fundamentals and give example programs on the topics. In this discussion, we also discuss BASIC Stamp math functionality and format. At the end of the discussion, we present commands needed to control a Hexapod robot which has six two-degrees-of-freedom legs.

PC Interfacing

In order to be able to program a robot for repetitive tasks or to integrate with sensors like cameras, we need to be able to connect the robot to a controller. We will use a PC as the robot controller for some robots in this book. Therefore, we need to interface the robot with a PC. There are many ways the robot can be connected to a PC. We can control the robot using relays by developing a sensor board that connects to some computer port, such as the parallel port or a USB port or a serial port. This chapter first discusses parallel port interface. How to setup a parallel port for a windows operating system using Microsoft Visual Studio C++ libraries is presented. In addition, a parallel interfacing using a Borland C++ compiler is also discussed with example programs. Hardware signals related to parallel port interfacing are presented with example circuitries such as active-low switch, active-high switch, LED driving with transistors, and driving relays with transistors. A PC interfacing board is introduced and design and construction of the board are discussed. In addition to C++, a Visual Basic access to parallel port is presented with example programs and setup directions. In addition to parallel port interfacing, serial port interfacing and USB interfacing are discussed. In the serial port interfacing, PC-to-PC, and PC-to-microcontroller and PC-to-Device serial communication are discussed and explained with example circuitry and code. Finally, a board for USB interfacing is introduced and related setup and programming information is provided.

Robotic Arm

In this chapter we will study a robotic arm that is built using DC motors and is controlled by switches. The robotic arm we will study is the OWI-007 robotic arm trainer. First, mechanical construction, properties, and components of the arm robot are studied. Electrical control of the arm robot is studied with the example programs to control the robot. Parallel port interface circuitry using relays is explained and a sample C code is provided for the readers. In addition to the parallel port, USB interface using relays is studied and necessary circuitry and sample code are provided. In addition to relays, the robot can be controlled by transistors. Parallel port interface circuit using transistor is presented with sample C code.

Robotic Arm Control

In this chapter, we explore ways to control two arm robots: OWI-007 and another 6 degrees-of-freedom (DOF) arm robot by MCII Robot. In Chapter 4, the arm robot is controlled in a open loop fashion where the programmer turns on a joint motor for a specific time so that the desired angle can be reached. In this chapter, we present some control techniques in a closed loop fashion using encoders and camera. Related hardware components and software code are provided for the user. For the advanced reader, we also present kinematics equation of the OWI-007 and related analysis for more precise control of the robot. The second half of the chapter focuses on the 6-DOF arm robot. The kinematics analysis of the robot is extended to forward and inverse kinematics equations and calculations. Finally, sample programs in MATLAB are provided for the user to explore the inverse and forward kinematics formulations. Using the inverse kinematics equations, the readers can give the desired location of the gripper of the robot to the MATLAB programs and obtain respective angles for the joints. This part of the chapter is intended for the advance readers.

Differential Drive Robot

In this chapter, we present a two-wheel differential drive mobile robot. First, construction and mechanics of the mobile robot are presented in detail. Pulley systems, drive belt, and DC motor dynamics are also revisited. Then, basic robot movements are presented with detailed breadboard connection explanations. Forward, backward, and various turns are explored. In addition, timed movements using R-S flip-flops are presented with a discussion of famous 555 timer IC. Infrared vision based robot design is explored with IR receiver and transmitter circuitries. Some other ways of controlling the robot are explored such as obstacle avoidance, robot ears, robotic pet, sound based robot, music dancer robot, and robot speed control. Finally, robot kinematics is studied with velocity equations.

Four Wheel Drive Robot

In this chapter, we study a four wheel robot with differential drive by Rigel. Construction and mechanics of the robot are presented. The robot is controlled by an OOPIC R microcontroller. Each wheel is independently controlled by the microcontroller. In the electrical control, we discuss how to connect the OOPIC R with Rigel 4WD robot. A short review of OOPIC with objects used for the Rigel robot is presented with sample codes. Objects used for this robot are oButton, oServo, oSonarPL, oIRPD1, and OOPIC object. In the oServo object discussed, a basic operational theory of a servo motor is also presented. Finally, a sample code to drive the Rigel 4WD robot is presented.

Hexapod Robot

In this chapter, we study a six-legged robot, named hexapod. This is one of the legged robots we cover in this book. We study two types of Hexapod using 2-DOF legs or 3-DOF legs. The control of the servos is done by a servo controller and a BASIC Stamp microcontroller. The chapter explores the construction and mechanics of the hexapods in detail. Each leg and the body construction are described studied with figures. Then, electrical control of the hexapods is done with Next Step Carrier Board with BASIC Stamp IIe (BS2e). The control commands of the BASIC Stamp go through a servo controller which actually drives the servos. Basic intro to servo operation, calibration of the servo controllers, and connecting BS2e to the Hexapod are also presented. Finally, programming BS2e for the hexapod is presented with walking schemes and corresponding sample codes.

Biped

In this chapter, we study three biped robots: Bigfoot by Milford Instruments Limited, England, Biped Lynxmotion Inc, and Robosapien. The construction and mechanics of the robots are studied and presented with figures and corresponding equations. Bigfoot is controlled by a BASIC Stamp controller. Example code for symbol definitions and shuffling movements are presented. Repeated shuffling is what makes the robot walk. Lynxmotion Biped robot is controlled by a PIC based servo controller. Setup and connections of the servo controller are explained. The Robosapien mechanical construction is not covered because the robot is sold as a preassembled robot. The robot can be controlled by a special remote control in four modes. Three modes are sensor programs. One is a master program. Chapter also describes how to control the Robosapien by a PC and a camera using USB port of the PC. Finally, chapter presents autonomous Robosapien robots which have better controllers and programmers. A discussion on how to convert a Robosapien into an autonomous Robosapien is also given.

Propeller Based Robots

In this chapter, we study propeller based robots: flying robotic planes, robotic helicopters, robotic boats, and submarines. In these systems, our approach is taking remote controlled systems (RC planes and helicopters) and using a microcontroller to control the actuators of the robots instead of the RC receivers doing the control. In addition to giving information about the robots and their controller, we present theories for wings and propellers. The chapter explores RC planes and gas powered RC planes. To make the RC planes autonomous, reader needs to add a controller to the system such as Micro Pilot (MP) controller. Micro Pilot controller is studied and details on how to integrate with an RC plane is also presented. A controller for an RC helicopter is also studied to convert RC helicopters into an autonomous helicopter. Kinematics analysis for RC planes, RC helicopters, robotic boats, and submarines are provided in the chapter. Some experiments are also provided for the readers.

Ferat Sahin and Pushkin Kachroo

Contents

1 **Fundamentals of Electronics and Mechanics**   1
  1.1 Fundamentals of Electronics 1
      1.1.1 Electrical Components 1
      1.1.2 Semiconductor Devices 8
      1.1.3 Digital Systems 25
      1.1.4 Sequential logic circuits 27
      1.1.5 Common Logic IC Devices 31
  1.2 Practical Electronic Circuits 32
      1.2.1 Power 32
      1.2.2 Fixed Voltage Power Supplies 35
      1.2.3 Adjustable Power Supplies 37
      1.2.4 Infrared Circuits 40
      1.2.5 Motor Control Circuits 42
  1.3 Fundamentals of Machines and Mechanisms 45
      1.3.1 Simple Machines 45
  1.4 Mechanisms 51
      1.4.1 Gears 51
      1.4.2 Chains and Belts 56
      1.4.3 Linkages 56
      1.4.4 Cam 57
      1.4.5 Ratchet Mechanism 57
      1.4.6 Quick Return Mechanism 60
      1.4.7 Intermittent Motion 60
      1.4.8 Springs and Dampers 60
      1.4.9 Brakes 62
      1.4.10 Clutches 62
      1.4.11 Couplers, Bearings, and Other Miscellaneous Items . . 62

2 **BASIC Stamp Microcontroller**   67
  2.1 Different Versions of BASIC Stamp 67
      2.1.1 BASIC Stamp 1 67
      2.1.2 BASIC Stamp 2 67
      2.1.3 BASIC Stamp 2sx 69
      2.1.4 BASIC Stamp 2p 69
      2.1.5 BASIC Stamp 2e 71
  2.2 Development Boards for BS2e 74

		2.2.1	BASIC Stamp 2 Carrier Board (Rev. B)	74
		2.2.2	BASIC Stamp Super Carrier (Rev. A)	75
		2.2.3	Board of Education (Rev. B)	75
		2.2.4	BASIC Stamp Activity Board (Rev. C)	75

2.3 BASIC Stamp Editor 76
    2.3.1 Connecting BS2e to the PC 76
    2.3.2 Installing the BASIC Stamp Editor 76
    2.3.3 Software Interface for Windows 79
    2.3.4 Software Interface for DOS 81

2.4 PBASIC Programming Fundamentals 84
    2.4.1 Declaring Variables 84
    2.4.2 Defining Arrays 85
    2.4.3 Alias 85
    2.4.4 Modifiers 86
    2.4.5 Constants and Expressions 87
    2.4.6 BASIC Stamp Math 88
    2.4.7 Important PBASIC Commands Used while Interfacing the BS2e with the Lynxmotion 12 Servo Hexapod   90
    2.4.8 DEBUG 92
    2.4.9 FOR...NEXT 93
    2.4.10 END 94
    2.4.11 RETURN 94
    2.4.12 PAUSE 94
    2.4.13 GOSUB 95
    2.4.14 SEROUT 95
    2.4.15 SERIN 100

3 PC Interfacing   103

3.1 Parallel Port Interface 103
    3.1.1 Port Access Library for Windows XP 106
    3.1.2 Hardware Signals 109
    3.1.3 PC Interfacing Board 115
    3.1.4 Visual Basic Access to Parallel Port 123
    3.1.5 Breadboarded Output Circuit 125

3.2 Serial Port Interfacing 130
    3.2.1 PC to PC Communication 130
    3.2.2 PC to Microcontroller Serial Communication   133

3.3 USB Interfacing 137

4 Robotic Arm   139

4.1 Construction and Mechanics 141
    4.1.1 DC Motor and Gear Box 141
    4.1.2 Gear Torques and Speed 143
    4.1.3 Gripper Mechanism 145
    4.1.4 Wrist Mechanism 147

	4.1.5	Elbow and Shoulder Mechanism	149
	4.1.6	Robot Base Mechanism	153
4.2	Electrical Control		153
	4.2.1	Robot Programming	155
4.3	Parallel Port Interface Circuit using Relays		158
	4.3.1	Code for PC Robotic Arm Control	161
4.4	USB Interface using Relays		163
4.5	Parallel Port Interface Circuit using Transistors		165
	4.5.1	Code for PC Robotic Arm Control	171

5 Robotic Arm Control — 173

5.1	Programmed Tasks		173
	5.1.1	Encoder Feedback	174
	5.1.2	Potentiometer Feedback	182
	5.1.3	Joint Control	182
5.2	Automatic Control Using a Camera		183
5.3	Robot Kinematics		187
	5.3.1	Velocity Kinematics	192
5.4	Dynamics and Control		193
5.5	Some Experiments		193
5.6	Control of a Six-Degrees-of-Freedom Arm Robot		194
	5.6.1	Mechanical Construction	194
	5.6.2	JM-SSC16 Mini Servomotor Controller Board	198
	5.6.3	Control Software	198
	5.6.4	Forward and Inverse Kinematics	199
5.7	Examples and MATLAB Programs		209
	5.7.1	DH.m: M-File for a Homogenous Transformation of a Row of a DH Table	209
	5.7.2	RobotTF.m: M-File for Calculating the Transformation Matrix of the Robot	211
	5.7.3	RobotSym.m: M-File for Symbolic Analysis of Inverse and Forward Kinematics	212
	5.7.4	InverseKin.m: M-File for Inverse Kinematics Equations	214
	5.7.5	Angle2Servo.m: M-File for Converting Joint Angles to Servomotor Values	215
	5.7.6	Path.m: M-File for Generating Joint Angles to Move the Robot Linearly	216
	5.7.7	A Sample RB File Created by path.m for Mini Servo Explorer	217

6 Differential Drive Robot — 219

6.1	Construction and Mechanics		221
	6.1.1	Robot Base with Breadboard	221
	6.1.2	Traction System	222

		6.1.3	Power System 227
		6.1.4	Relay Board 229
		6.1.5	Basic Robot Movements 229
		6.1.6	Timed Movements 234
		6.1.7	Robot Timed Movements 239
		6.1.8	Infrared Vision-Based Robot (Robot Eyes) 249
		6.1.9	Audio Detection and Response (Robot Ears) 252
		6.1.10	Sound-Based Robot Movements 260
	6.2	Robot Speed Control 262
		6.2.1	PC Control 265
		6.2.2	Feedback Control 266
	6.3	Robot Kinematics 266

7 Four Wheel Drive Robot 269
 7.1 Construction and Mechanics 269
 7.2 Electrical Control 278
   7.2.1 Connecting the OOPic-R with the Rigel 4WD Robot . 278
   7.2.2 A Review of OOP in Reference to the OOPic-R 282
   7.2.3 Important Objects Used While Interfacing the OOPic-R with the Rigel 4WD Robot 285
   7.2.4 Sample Code to Drive the Rigel 4WD Robot Using the OOPic-R 295

8 Hexapod Robot 301
 8.1 Construction and Mechanics 301
   8.1.1 Servomotors 303
   8.1.2 Mechanical Construction of Extreme Hexapod II . . . 303
   8.1.3 Mechanical Construction of Extreme Hexapod III . . . 322
 8.2 Electrical Control 339
   8.2.1 Lynxmotion 12-Servo Hexapod with BS2e 339
   8.2.2 Programming the Hexapod 346
   8.2.3 Adjusting Servomotors to Mid Position 353

9 Biped Robots 359
 9.1 Bigfoot: The Walker 359
   9.1.1 Construction 359
   9.1.2 Programming 365
   9.1.3 Robot Kinematics 369
 9.2 The Lynxmotion Biped 370
   9.2.1 Leg Assembly 370
   9.2.2 Arm Assembly 372
   9.2.3 Torso Assembly 372
   9.2.4 Hand Assembly 375
   9.2.5 Controller 375
 9.3 The Robosapien Biped 378

	9.3.1 Robosapien Motors	382
	9.3.2 Walking	382
	9.3.3 PC Control	382
	9.3.4 Autonomous Robosapien	384

10 Propeller Based Robots 389
- 10.1 Wings . . . 389
- 10.2 Propellers . . . 391
- 10.3 Robotic Planes . . . 394
  - 10.3.1 RC Planes . . . 394
  - 10.3.2 Manual Control for RC Planes . . . 404
  - 10.3.3 Automatic Controllers for RC Planes . . . 404
  - 10.3.4 Kinematics . . . 406
  - 10.3.5 Robotic Experiments . . . 414
- 10.4 Robotic Helicopter . . . 415
  - 10.4.1 Controlling Movements . . . 415
  - 10.4.2 Automatic Controllers for RC Helicopters . . . 423
- 10.5 Robotic Boats . . . 423
  - 10.5.1 Propulsion . . . 424
- 10.6 Robotic Submarines . . . 427

References   431

Index   434

1

Fundamentals of Electronics and Mechanics

In this chapter we will explore the fundamentals of Electronics and Mechanics. The example applications related to electrical and mechanical components are also presented.

1.1 Fundamentals of Electronics

In this section, we explore electrical components, semiconductor devices, Operational Amplifiers (OPAMPs) and their applications, and digital systems components. There are some very good books that cover analysis of electric circuits such as [10] and [2]. There are also some good books on hands on robotics such as [7].

1.1.1 Electrical Components

Resistors, capacitors, and inductors are basic electrical components used in electronic circuits. Some of the electrical components and their symbols are given in Figure 1.1.

1.1.1.1 Resistors

Resistors are components which resist the flow of electronic current. The resistors are mainly used to reduce the voltage applied to other components and to limit the current flowing through other components. The higher the value of the resistance, the lower the current will be. Resistance of a resistor is measured in terms of Ohms (Ω) since the relationship between voltage (V, volts), current (I, Ampere), and resistance (R) is explained by Ohm's law given in 1.1.

$$V = IR \tag{1.1}$$

The most common resistors are made using a carbon rod core with end caps and wire leads. We can categorize resistors into two basic types: fixed and variable resistors (or potentiometers). A fixed resistor is the one which has a fixed resistance value. Variable resistors have variable resistance values. The

COMPONENT	SYMBOL	FUNCTIONALITY
Wire		Connects elements
Connected Wires		Two connected wires
Disconnected Wires		No marks for disconnected wires
Cell		Supplies electrical energy. Larger side is positive
Battery		Supplies electrical energy. It has more than one cell
DC Supply		Direct Current (DC). Supplies electrical energy
AC Supply		Alternating Current (AC). Supplies electrical energy
Fuse		It melts "blow" if the current exceeds the limit.
Transformer		Two coils of wire linked by an iron core. Transformers are used to increase or decrease AC voltage.
Motor		A motor converts electrical energy to kinetic energy.
Resistor		It resists the current passing through.
Capacitor		It stores electrical charge.
Inductor		Solenoid, coil. It creates a magnetic field when current passes through it.
Ground		A connection to earth. This is also 0 volts.
Push Switch		It allows the current pass when it is pushed.
Push-to-Break Switch		It stops the current flow when it is pushed.
On-Off Switch		SPST: Single pole, single throw. It allows current when it is closed (on).
2-way Switch		SPDT: Single pole, double throw. It changes the direction of the current.
Dual On-Off Switch		DPST: Double pole, single throw. It is a dual on-of switch.
Relay	NO / COM / NC	A switch that can be electrically controlled. NO: Normally open. NC: Normally closed. COM: Common. When a voltage is applied to the coil, the switch is pulled by the magnetic force created by the coil and it changes state.

FIGURE 1.1
Common electrical components and their symbols

value of the resistor is often changed by a user by turning a knob or a dial. There are some special resistors designed to change in resistance when heated. They are called Thermistors and are used in temperature measuring circuits. The same idea is also used to design pressure sensors where a membrane is designed to be a resistor. The membrane resistance changes when it is deformed by the pressure in a chamber.

Resistors generate heat and have a wattage rating relating the power level they can handle. The higher the wattage rating the more heat they can dissipate. There are standard wattage ratings such as 1/8, 1/4, 1/2, 1, more watts. In addition to the value and wattage, each resistor has a tolerance regarding their resistance. Standard resistors have 10-20% tolerance but special resistors can have tolerances around 1%. Depending on the application, the proper tolerance rate is chosen. These properties are often marked on the resistors using a color code. Sometimes, they are written on the resistor.

1.1.1.1.1 Resistor Color Code and Standard Resistor Values Fixed value resistors are color coded to indicate their value and tolerance. Some have their value written on them. There are three color coding systems: a 4 Band code, a 5 Band code, and 6 Band code.

The standard color coding method for resistors has 10 colors to represent numbers from 0 to 9: black, brown, red, orange, yellow, green, blue, purple, grey, and white. The first two bands always represent the significant digits on a 4 band resistor. On a 5 and 6 band, the significant digits are the first three bands. The third band is the multiplier or decade which is multiplied by the resulting value of the significant digit color bands. For example, if the first two bands are brown (1) and orange (3) and the third band is red (2), this means 10^2 or 100. Then, this gives a value of 13×100, or 1300 Ohms.

For the decade band, the gold and silver colors are used to divide by a power of 10 and 100 respectively, allowing for values below 10 Ohms. The tolerance of the resistor is represented by the next band. Four colors are used for the tolerance band: brown (+/-1%), red (+/-2%), gold (+/-5%), and silver (+/-10%). For example, if the tolerance band is silver, the true value of the resistor can be 10% more or less than 1300 Ohms. Thus, the actual value of the resistor can be from 1170 to 1430 Ohms. The sixth band on a 6 band resistor reveals the temperature coefficient of the resistor, measured parts per million per degree Centigrade (PPM/C). Seven colors are used for the temperature coefficient: white (1), purple (5), blue (10), orange (15), yellow (25), red (50), and brown (100). The most popular color is brown (100 PPM/C) and will work for normal temperature conditions. The other colors are used for temperature critical applications. Table 1.1 represents all the colors and their meaning depending on their location on resistors.

Figure 1.2 is a 6 band resistor with 27 KOhms, 10% tolerance, and the temperature coefficient of 50 PPM/C.

Since the sizes of the electronic components are shrinking or changing in

TABLE 1.1
Resistor color codes

Color	Band 1	Band 2	Band 3	Decade	Tolerance	Temp. Coefficient
Black	0	0	0	10^0		
Brown	1	1	1	10^1	+/- 1%	100
Red	2	2	2	10^2	+/- 2%	50
Orange	3	3	3	10^3		15
Yellow	4	4	4	10^4		25
Green	5	5	5	10^5	+/- 0.5%	
Blue	6	6	6	10^6	+/- 0.25%	10
Purple	7	7	7	10^7	+/- 0.1%	5
Grey	8	8	8		+/- 0.05%	1
White	9	9	9			
Gold				0.1	+/- 5%	
Silver				0.01	+/- 10%	

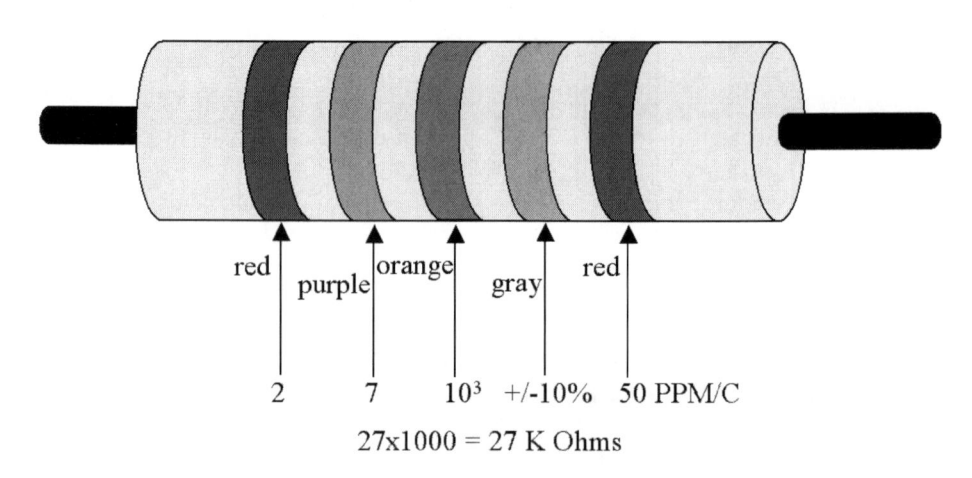

red purple orange gray red
2 7 10^3 +/-10% 50 PPM/C
27x1000 = 27 K Ohms

FIGURE 1.2
An example of a resistor with color codes

Fundamentals of Electronics and Mechanics

shape, it becomes very difficult to put color bands on a resistor. Instead, a simpler alphanumeric coding system is used. This coding system uses three numbers, sometimes followed by a single letter. The numbers play the same role as the first three bands on a 4 band resistor. First two numbers is the significant digits. The third number is the decade. There are five possible letters: M=20%, K=10%, J=5%, G=2%, F=1%. For example, if 473K is written on a resistor array, the 4 and 7 are the significant digits and the 3 is the decade, giving 47 x 1000 or 47000 Ohms. Since the letter is K, the resistor has 10%. The same coding system is also used on the surface mount resistors with SMD package.

Since it could be difficult to see text on some components, the letters K,M and R are used in place of the decimal point. The letter K represents 1000, the letter M represents 1000000, and the letter R represents 0. For example, a 3900 Ohm resistor will have 3K9 on the package and a 7.2 Ohm resistor is represented as 7R2. There are seven standards for resistor values: E3, E6, E12, E24, E48, E96, and E192 based on their tolerance levels 50%, 20%, 10%, 5%, 2%, 1%, and less than 0.5% respectively. E3 standard is no longer used. E6 standard is used very seldom. The most used standards are E12 and E24. In the E12 standard, the resistors take all decades of the following values: 1.0, 1.2, 1.5, 1.8, 2.2, 2.7, 3.3, 3.9, 4.7, 5.6, 6.8 and 8.2 In the E24 standard, the resistors take all decades of the following values: 1.0, 1.1, 1.2, 1.3, 1.5, 1.6, 1.8, 2.0, 2.2, 2.4, 2.7, 3.0, 3.3, 3.6, 3.9, 4.3, 4.7, 5.1, 5.6, 6.2, 6.8, 7.5, 8.2 and 9.1

1.1.1.2 Inductors

An inductor is an electronic component composed of a coil of wire. The magnetic properties of a coil come into effect. When a voltage is applied, a current starts flowing in the coil and a magnetic field is created as shown in Figure 1.3. While the field is building, the coil resists the flow of the current. Once the field is built, current flows normally. When the voltage is removed, the magnetic field around the coil keeps the current flowing until the field collapses. Thus, the inductor can store energy in its magnetic field, and resist any change in the amount of current flowing through it. The unit of inductance is the Henry (H). In order to increase the inductance, we can use core materials like Soft iron, Silicon iron, etc. The most common type of inductor is the Bar Coil type. The others are surface mount inductors, Toroids (ring-shaped core), thin film inductors, and transformers. The choice of inductor depends on the space availability, frequency range of operation, and certainly power requirements.

1.1.1.3 Capacitors

A capacitor is an energy storing device, made up of two parallel conductive plates separated by an insulating material. This insulating material is called a dielectric. It stores a charge because electrons crowd onto the negative

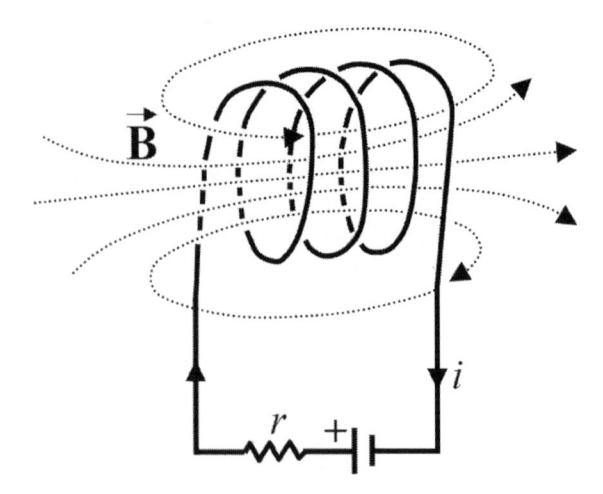

FIGURE 1.3

An inductor and its magnetic field

plate and repel electrons on the positive plate, thereby inducing an equal and opposite charge. The unit of the capacitance is Farad (F). However, practical values of a capacitor are in micro and nano Farad ranges. Figure 1.4 presents an electrolytic capacitor and its symbols.

There are two different types of capacitors: Electrolytic and Non-electrolytic. Non-electrolytic capacitors use mica or polyester as dielectric. Electrolytic capacitors use aluminum metal plates on either side of a sheet of paper soaked in aluminum borate. Ceramic capacitors are used in high frequency applications. These are stable at high frequencies. Tantalum bead capacitors are very small in size, thus commonly used as surface mount components.

Large capacitors have the value printed plainly on them but smaller ones often have just 2 or three numbers on them. It is similar to the resistor codes. The first two are the 1st and 2nd significant digits and the third is a multiplier code. Sometimes, one or two letters are added for tolerance and temperature coefficient. Table 1.2 presents the meaning of the numbers and letters on capacitors.

The values calculated using the digits on a capacitor is in pF (pico Farad). For example, if a capacitor has 105F on it, the capacitor has 10× 100,000 = 1000000 pF = 1000 nF (nano Farad) = 1 μ F (micro Farad) value and 1% tolerance. There are two letters used for temperature coefficient: P (+100) and Z (+80). There are other standards such as EIA (Electronic Industrial Association) where there are more letters for a detailed tolerance and temperature coefficients.

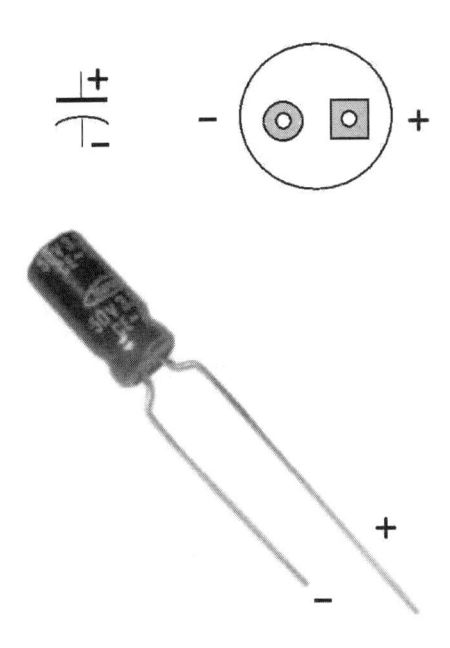

FIGURE 1.4
        An electrolytic capacitor and its symbols

TABLE 1.2
Meaning of the third digit and ketters on capacitors

3rd Digit	Multiplier	Letter 2	Tolerance
0	1	D	0.5 pF
1	10	F	1%
2	100	G	2%
3	1000	H	3%
4	10000	J	5%
5	100000	K	10%
6,7		M	20%
8	0.01	P	+100, -0%
9	0.1	Z	+80, -20%

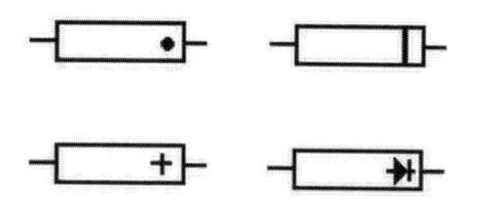

FIGURE 1.5
Standard diode markings

1.1.2 Semiconductor Devices

In this section, we will explore commonly used semiconductor devices: diodes, transistors, and their derivatives.

1.1.2.1 Diodes

Diode is an electrical device allowing current to move through it in one direction with far greater ease than in the other direction. The most common type of diode in modern circuit design is the semiconductor diode. Diodes are polarized, which means that they must be inserted in the correct way. Diodes have two connections: an anode (positive) and a cathode (negative). The cathode is always identified by a dot, ring, or some other mark, shown in Figure 1.5.

Diodes are said to be biased, based on the voltages applied to it. To forward bias a diode, the anode must be more positive than the cathode. To reverse bias a diode, the anode must be less positive than the cathode. When forward biased, the device conducts current, but when reverse biased, it prevents the flow of current. Figure 1.6 shows a circuitry to characterize a standard diode and the corresponding current-voltage (I-V) graph.

Note that diode starts to conduct when the voltage on the diode reaches a certain level (in practice this is about 0.7 Volt). Voltages above this value increase the current going through the diode linearly. On the other hand, if the voltage on the diode is reversed, the diode does not let any current pass through itself. However, if the reverse voltage is increased up to a certain level, the diode can be broken and lets a high current pass through itself. This voltage is called breakdown voltage.

The most common application for diodes is voltage rectification. Rectifiers are devices that convert alternating current (AC) into direct current (DC). There are many specialized diodes like the zener diode for certain applications. A diode designed to emit light is called a light-emitting diode, or LED. Figure 1.7 shows the symbols for commonly diode types used in electronics.

Next, we explore commonly used diodes: Zener diode, LED, Laser Diode, and Photo Diode.

Fundamentals of Electronics and Mechanics

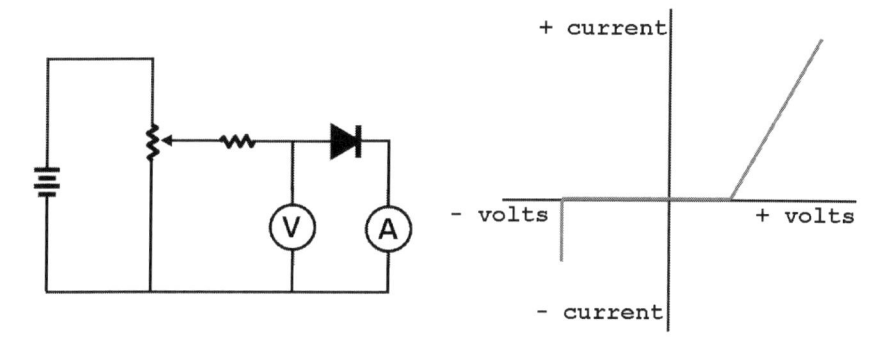

FIGURE 1.6
Standard diode I-V curves and characterization circuitry

FIGURE 1.7
Symbols for commonly used diodes

1.1.2.1.1 Zener Diode The zener diode is operated in reverse bias mode (positive on its cathode). It relies on the reverse breakdown voltage occurring at a specified value. With the application of sufficient reverse voltage, the diode junction will experience a rapid avalanche breakdown and conduct current in the reverse direction. When this process takes place, very small changes in voltage can cause very large changes in current. Zener diodes are available for a wide variety of break down voltages from about 4 volts to several hundred volts. Figure 1.8 is a more accurate I-V curve of a zener diode.

As mentioned before, when the reverse voltage reaches a certain value, a current in reverse direction starts flowing. The small increase on the voltage after this threshold value causes significant current increase. Thus, the voltage on the zener diode is assumed to be constant as long as a healthy current is being passed through the diode. The following applications use this property successfully.

1. As a reference source, where the voltage across itself is compared with another voltage.

2. As a voltage regulator, smoothing out any voltage variations occurring in the supply voltage across the load.

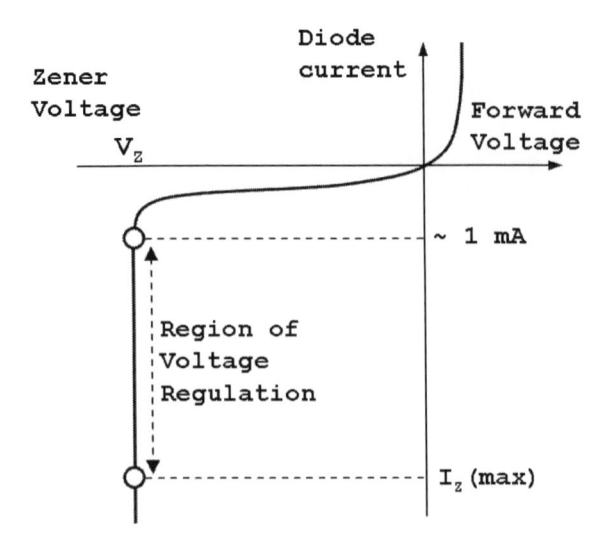

FIGURE 1.8

The V-I curve of a zener diode

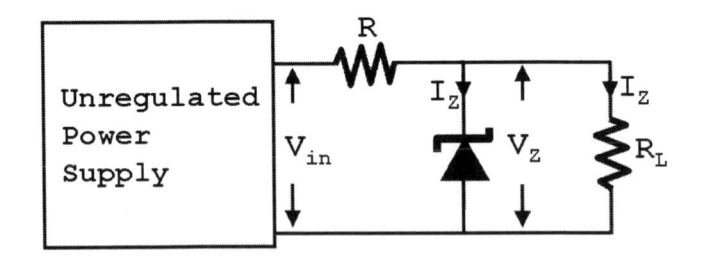

FIGURE 1.9

Voltage regulators with a zener diode

This operation is very useful in the construction of power supplies, voltage regulators, and voltage limiters. Figure 1.9 presents a simple circuitry of a zener diode based voltage regulator. In the circuitry, R_L is the load which needs constant voltage source with a maximum power ($W = V_Z \times I_L$). The voltage V_Z is the breakdown voltage of the zener diode. The current IL is the current passing through the diode. As mentioned above, if the voltage across the zener diode is not controlled, the diode will draw higher currents to keep the voltage constant across itself. This could be harmful for the diode and can burn it. Thus, a resistor R is used to limit the current which passes through the zener diode.

Fundamentals of Electronics and Mechanics 11

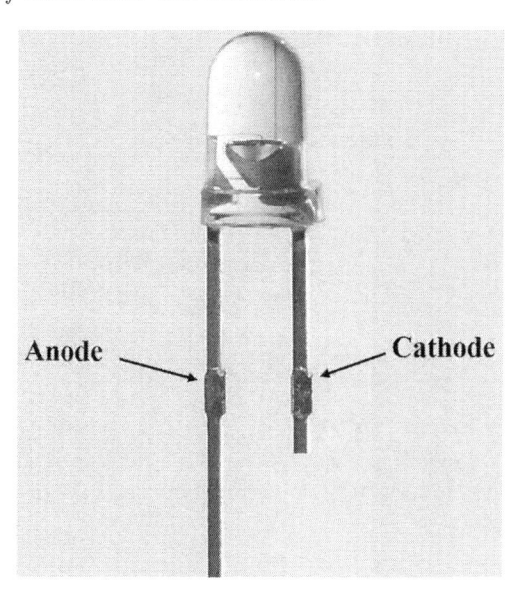

FIGURE 1.10
An example of an LED

1.1.2.1.2 Light Emitting Diode (LED) A light-emitting diode (LED), shown in Figure 1.10, is a semiconductor device that emits incoherent narrow-spectrum light when electrically biased in the forward direction. The longer leg in the figure is the anode, and the shorter one is the cathode. This effect is a form of electro-luminescence. The color of the emitted light depends on the chemical composition of the material used, and can be near-ultraviolet, visible, or infrared.

An LED is a special type of semiconductor diode. Like a standard normal diode, it consists of a chip of semiconductor material impregnated (doped) with impurities to create a structure called a p-n junction. Current flows easily from the p-side (anode) to the n-side (cathode). Charge-carriers (electrons and holes) flow into the junction from electrodes with different voltages. When an electron meets a hole, it falls into a lower energy level, and releases energy in the form of a photon as it does so. The light we see from an LED is created by these photons.

Since there are many LEDs with different packaging and leg types, it may not be easy to determine the anode and the cathode of an LED. The best solution is to test the LED with a resistor and a voltage source. However, the Table 1.3 can help finding the anode and cathode of most of the LED types.

The LEDs are commonly used in electronics circuitry as well as robotics in order to report the operational status of the devices. For example, an LED can be used to show the successful operation of the device or the existence of the voltage source. If you look at your computer, you can see an LED for

TABLE 1.3
How to determine the anode
and cathode of an LED

sign	+	-
polarity:	positive	negative
terminal:	anode	cathode
wiring:	red	black
pinout:	long	short
interior:	small	large
shape:	round	flat
marking:	none	stripe

your power status and an LED for the harddrive operation. Whenever you save something to the harddrive you should see a blinking LED showing the harddrive activity.

1.1.2.1.3 Laser Diode The laser diode is a further development upon the regular light-emitting diode. All semiconductor devices are governed by the principles described in quantum physics. One of these principles is the emission of light energy whenever electrons fall from a higher energy level to a lower energy level as mentioned in the LED discussion. Figure 1.11 is a sketch of a p-n junction of a laser diode showing the light source and other physical properties of the LED.

Laser action can be achieved in a p-n junction formed by two doped gallium arsenide layers. The two ends of the structure need to be optically flat and parallel with one end mirrored and one partially reflective. The length of the junction must be precisely related to the wavelength of the light to be emitted. The junction is forward biased and the recombination process produces light as in the LED (incoherent). Above a certain current threshold the photons moving parallel to the junction can stimulate emission and initiate laser action. Laser Diodes are used in a wide variety of applications. Low power applications are in CD Players (840 nm). Higher power applications are in Laser printers (760 nm). Another important application is in Fiber communications (1300 nm).

1.1.2.1.4 Photodiode A photodiode, shown in Figure 1.12, is special diode which operates when a light in certain frequency is reflected through it. It is also a type of photodetector. Their p-n junction is designed to be responsive to light. In order to pass the light to the sensitive part, photodiodes are provided with either a window or optical fiber connection. If there is no window, they can be used to detect vacuum UV or X-rays.

Photodiodes can be used in two ways: zero bias and reverse bias. In zero bias, a voltage across the device is generated when the light falls on the diode. This leads to a current in the forward bias direction. This is also called the

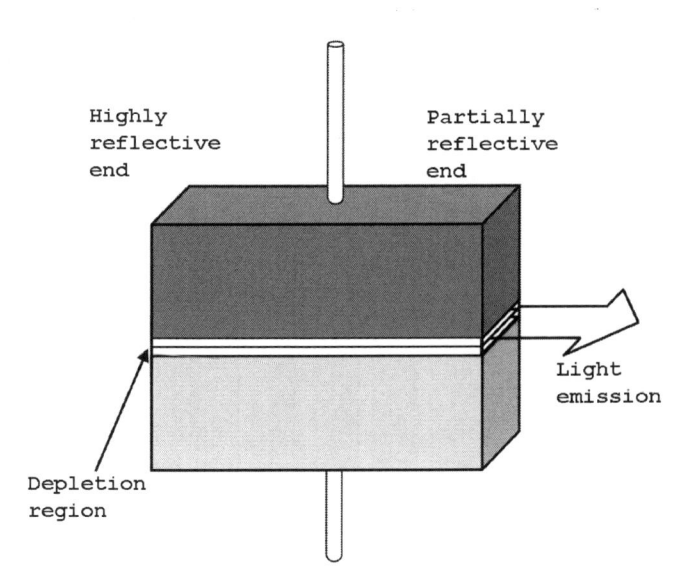

FIGURE 1.11
    Components of a laser diode and generation of light

FIGURE 1.12
    A photodiode

photovoltaic effect and is the basis for solar cells. In fact, a solar cell is just a large number of big and cheap photodiodes.

When a diode is reverse biased, they usually have extremely high resistance. When light of a proper frequency reaches the junction, this resistance is reduced. Thus, a reverse biased diode can be used as a detector by monitoring the current running through it. Circuits based on this effect are more sensitive to light than ones based on the photovoltaic effect.

Photodiodes are mostly used as switches in robotics applications. In addition, they are used as a communication device when the photodiodes are made infrared sensitive. For example, your TV remote control device has an LED which emits light wave in infrared frequency. Respectively, your TV has a photodiode which is activated by this infrared light. In this kind of set up, the transmitter LED sends 0s and 1s as light on and light offs and the receiver detects these 0s and 1s and acts accordingly.

1.1.2.2 Transistors

A transistor can be initially thought of as an "electronically-controlled resistor." Two of the pins act like a normal resistor. The other "control" pin controls the resistance "seen" between the other 2 pins. The "control" pin is called the gate in a Field Effect Transistor (FET) (the other 2 pins are the source and drain). The "control" pin is called the base in a Bipolar Junction Transistor (BJT) (the other 2 pins are the emitter and the collector).

Two electrical quantities can be used to control the resistance between the two terminals - current and voltage. In a FET, the voltage at the gate controls the resistance between source and drain, while in the BJT, the current flowing into the base controls the resistance between the emitter and collector. While often referred to as an amplifier, a transistor does not create a higher voltage or current of its own accord. Like any other device, it obeys the Kirchoff's laws. The resistance of a transistor dynamically changes, hence the term transistor.

One of its popular uses is in building a signal amplifier, but it can also be used as a switch. Today's transistors are mostly found inside ICs. Standalone transistors are used mostly only in high power applications or for power-regulation.

Both the BJT and the FET are popular today (among the FETs, the MOSFET being the most popular form of transistor), each one having certain advantages over the other. BJTs are much faster and high current devices, while FETs are small-sized low-power devices. Understanding the function of a transistor is a key to understanding electronics.

1.1.2.2.1 Bipolar Junction Transistor The current through the collector and emitter terminals of a BJT is controlled by the current through the base terminal. This effect can be used to amplify the input current. BJTs can be thought of as voltage-controlled current sources but are usually char-

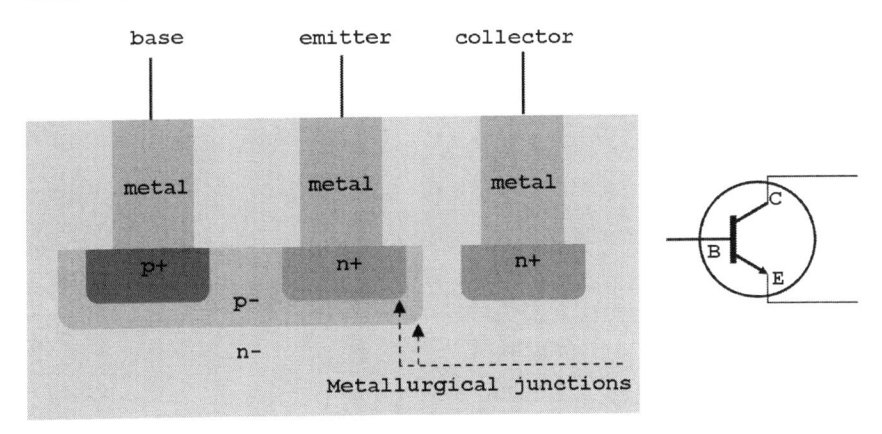

FIGURE 1.13
Cross-section of BJT

acterized as current amplifiers due to the low impedance at the base. Early transistors were made from germanium but most modern BJTs are made from silicon. If one applies the Kirchoff's current law on the device, the current entering the device through all the terminals must add up to zero. Hence IC is not the same as IE.

1.1.2.2.2 BJT Construction A lightly doped region called base is sandwiched between two regions called the emitter and collector respectively. The collector handles large quantities of current; hence its dopant concentration is the highest. The emitter's dopant concentration is slightly lesser, but its area is larger to provide for more current than the collector. The collector region should be heavily doped because electron-hole pair recombines in that region, while the emitter is not such a region. Figure 1.13 is the cross-section of a BJT and its electronic symbol.

We can have two varieties in this kind of transistor based on the type of junctions created: NPN and PNP transistors. Here a lightly doped p-type semiconductor (semiconductor with more holes than electrons) is sandwiched between two well-doped n-type regions. It is like two P-N junctions facing away. An IEEE symbol for the NPN transistor is shown in Figure 1.14. The arrow between the base and emitter is in the same direction as current flowing between the base-emitter junctions. Power dissipated in the transistor is P = $V_{CE} \times I_C$, where V_{CE} is the voltage between the collector and the emitter and I_C is the collector current.

For a PNP transistor, everything is opposite that of NPN. This one is more like two P-N junctions facing each other. Its symbol is shown in Figure 1.15. Again, note the direction of the arrow.

FIGURE 1.14

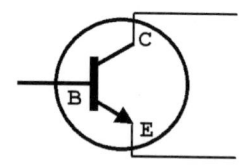

   IEEE symbol for an NPN transistor

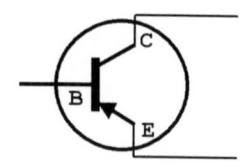

FIGURE 1.15

   IEEE symbol for a PNP transistor

1.1.2.2.3 Operation An NPN bipolar transistor can be considered as two diodes connected anode to anode. In normal operation, the emitter-base junction is forward biased and the base-collector junction is reverse biased. In an NPN type transistor, electrons from the emitter "diffuse" into the base. These electrons in the base are in the minority and there are plenty of holes with which to recombine. The base is always made very thin so that most of the electrons diffuse over to the collector before they recombine with holes. The collector-base junction is reverse biased to prevent the flow of holes, but electrons are swept into the collector by the electric field around the junction. The proportion of electrons able to penetrate the base and reach the collector is approximately constant in most conditions. However, the heavy doping (low resistivity) of the emitter region and light doping (high resistivity) of the base region mean that many more electrons are injected into the base, and therefore reach the collector, than there are holes injected into the emitter. The base current is the sum of the holes injected into the emitter and the electrons that recombine in the base - both small proportions of the total current. Hence, a small change of the base current can translate to a large change in electron flow between emitter and collector. It is important to keep the base region as thin and as free from defects as possible, in order to minimize recombination losses of the minority carriers.

1.1.2.2.4 Field Effect Transistor The FET is simpler in concept than the BJT and can be constructed from a wide range of materials. The most common transistors today are FETs.

Fundamentals of Electronics and Mechanics 17

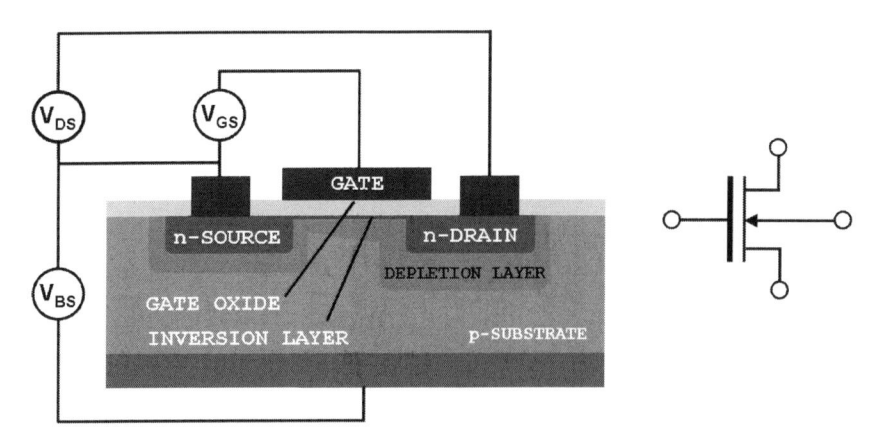

FIGURE 1.16
        Cross-section of a FET and its symbol

1.1.2.2.5 FET Construction The terminals in FET are called gate, drain, and source. Figure 1.16 shows a cross-section of a FET and its symbol. The voltage applied between the gate and source terminals modulates the current between source and drain. These transistors are characterized as having a conductance between source and drain dependent on the voltage applied between the gate and the source terminals. The dependence is linear if the gate to drain voltage is also high along with the gate to source voltage.

One of the issues that come up in circuit design is that as chips get smaller the insulator gets thinner and as a result the insulator starts acting like a conductor. This is known as leakage current. One solution is to replace the insulator by a material with a higher dielectric coefficient.

1.1.2.2.6 FET Operation The shape of the conducting channel in a FET is altered when a potential is applied to the Gate terminal (potential relative to either Source or Drain.) In an N-channel device, a negative Gate potential causes an insulating depletion zone to expand in size and encroach on the channel from the side, narrowing the channel. If the Depletion Zone pinches the channel closed, the resistance of the channel becomes very large, and the FET is turned off entirely. At low voltages, the channel width remains large, and small changes to the Gate potential will alter the channel resistance. This is the "Variable Resistance" mode of FET operation. This mode has uses but is not employed in conventional amplifier circuits.

If a larger potential difference is applied between the Source and Drain terminals, this creates a significant current in the channel and produces a smooth gradient in potential distributed along the channel. This also causes the shape of the Depletion Zone to become asymmetrical, and one part of the channel becomes narrow while another part widens. If the voltage is large enough, the Depletion Zone begins to close the channel entirely. Something unusual then happens: negative feedback arises, since a closed channel would lead to flat potential gradient and a symmetrical Depletion Zone, which would open the channel. Rather than closing entirely, the Depletion Zone shapes itself to produce an extremely narrow channel of variable length. Any attempted increases to the channel current will change the shape of the Depletion Zone, lengthening the channel. This increases the channel resistance and prevents the value of current from increasing. This mode of operation is called "Pinchoff mode." In this mode of operation, the channel behaves as a constant-current source rather than as a resistor. The value of channel current is relatively independent of the voltage applied between Source and Drain. The value of Gate voltage determines the value of the constant current in the channel. The different types of field-effect transistors can be distinguished by the method of isolation between channel and gate:

- MOSFET (Metal-Oxide-Semiconductor FET): standard FET

- JFET (Junction Fet): When voltage is applied between the source and drain current flows. Current only stops flowing when a voltage is applied to the gate.

- MESFET (MEtal-Semiconductor FET): p-n junction is replaced with Schottky junction. Not made with Silicon.

- HEMT (High Electron Mobility Transfer): A MESFET

- PHEMT (Pseudomorphic HEMT).

1.1.2.2.7 Applications

1. The most common use of MOSFET transistors today is the CMOS (complementary metallic oxide semiconductor) integrated circuit, which is the basis for most digital electronic devices.

2. FETs can switch signals of either polarity, if their amplitude is significantly less than the gate swing, as the devices are basically symmetrical. This means that FETs are the most suitable type for analog multiplexing.

1.1.2.2.8 Phototransistor The Phototransistor, shown in Figure 1.17, works in a similar manner as a usual transistor but it is controlled by light. In this case the light striking the base is amplified instead of the voltage applied to the base. Commonly used phototransistor is the NPN type with an exposed base. In the symbol of the photo transistor, there is no base. Thus, it may be possible to confuse the photo transistors with some photo diodes. The main difference between the photodiode and photo transistor is the amplification property of the transistors. Thus, photo transistors are more suitable for high current applications and switching applications.

Fundamentals of Electronics and Mechanics

FIGURE 1.17

A phototransistor

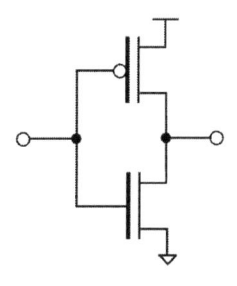

FIGURE 1.18

CMOS inverter circuit

1.1.2.2.9 Complementary Metal Oxide Semiconductor (CMOS) CMOS is not a type of transistor. It is a logic family, based on MOS transistors. Figure 1.18 shows a CMOS inverter.

1.1.2.2.10 Construction and Operation These use a totem-pole arrangement where one transistor (either the pull-up or the pull-down) is on while the other is off. Hence, there is no DC drain, except during the transition from one state to the other, which is very short. As mentioned, the gates are capacitive, and the charging and discharging of the gates each time a transistor switches states is the primary cause of power drain.

The C in CMOS stands for "complementary." The pull-up is a P-channel device and the pull-down is N-channel. This allows busing of control terminals, but limits the speed of the circuit to that of the slower P device (in silicon devices).

Complementary Metal Oxide Semiconductor (CMOS) is made of two FETs blocking the positive and negative voltages. Since only one FET can be on at a time, CMOS consumes negligible power during any of the logic states. But when a transition between states occurs, power is consumed by the device. This power consumed is of two types.

Short-circuit power: For a very short duration, both transistors are on and a very huge current flows through the device for that duration. This current accounts for about 10% of the total power consumed by the CMOS.

Dynamic power: This is due to charge stored on the parasitic capacitance of the output node of the device. This parasitic capacitance depends on the wire's area, and closeness to other layers of metal in the IC, besides the relative permittivity of the quartz layer separating consecutive metal layers. It also depends (to a much smaller extent) upon the input capacitance of the next logic gate. This capacitance delays the rise in the output voltage and hence the rise or fall in the output of a gate is more like that of a resistor-capacitor (RC) network.

1.1.2.2.11 Operational Amplifiers One can clearly generalize and say that the field of electronics largely depends upon manipulating with the input signals such as voltage or current to produce desired output. These manipulations include (not limited to) mathematical operations, such as addition, subtraction, integration, and differentiation. In the analog domain, the most common device that is used to perform the above listed operations is the operational amplifiers or op amps. Current applications of operational amplifiers go far beyond simple mathematical operations. Op-amps are used in many control and instrumentation systems to perform various tasks such as voltage regulators, oscillators, logarithmic amplifiers, peak detectors, and voltage comparators.

Operational amplifiers have special characteristics due to which they are widely used as predictable building blocks in many circuit designs. Some of these characteristics are as follows: very high gain (10,000 to million), high input resistance (10^3 to 10^{15} ohms), small size, low power consumption, good reliability and stability, and last but not the least, low cost of manufacturing. Figure 1.19 illustrates the standard symbol of an operational amplifier. It consists of two input terminals and one output terminal. The input terminal indicated with minus sign is called the inverting terminal and the other input terminal is called the non-inverting terminal. A signal applied at the inverting terminal and ground appears at the output with a $180°$ phase shift. Likewise a signal applied at the non-inverting terminal and ground appears at the output with a $0°$ phase shift.

In the analysis and design of circuits employing op-amps, a simplified circuit model known as ideal op-amp is often used that has the following characteristics.

- Infinite open loop voltage

- Infinite input resistance

- The amplifier draws zero current

- Output resistance is negligible

- The gain is constant and independent of frequency

Fundamentals of Electronics and Mechanics

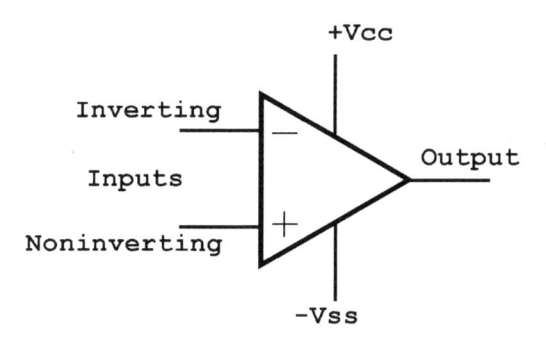

FIGURE 1.19
Standard representation of operational amplifier

As described above, the applications of Op-amps are almost endless and numerous books deal with this subject in detail. However, in the following section some of the applications such as integrators, differentiators, instrumentation amplifiers, and inverting and non-inverting amplifiers shall be illustrated.

1.1.2.2.12 Integrator circuit As the name suggests, op-amps in this application produce an output voltage that is the integral of the input voltage. Figure 1.20 illustrates the circuit that operates as an integrator. The working of this device is based on the impedance of the capacitor Cf and the resistance Rin. The output of the circuitry is expressed with equation 1.2.

$$V_{out} = -\frac{1}{R_{in}C_f} \int V_{in}\, dt \qquad (1.2)$$

For the case of a DC voltage input, equation 1.3 represents the output voltage, where V_{co} is initial voltage of the output.

$$V_{out} = -\frac{V_{in}}{R_{in}C_f} t + V_{co} \qquad (1.3)$$

1.1.2.2.13 Differentiator circuit The schematic circuit that produces an output that represents a differentiation of the input is illustrated in Figure 1.21.

The output voltage of this circuit in the time domain is given in equation 1.4.

$$V_{out} = -R_f C_{in} \frac{dV_{in}}{dt} \qquad (1.4)$$

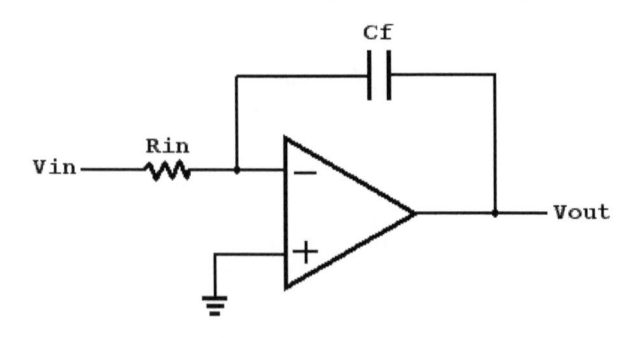

FIGURE 1.20

Integrator circuit

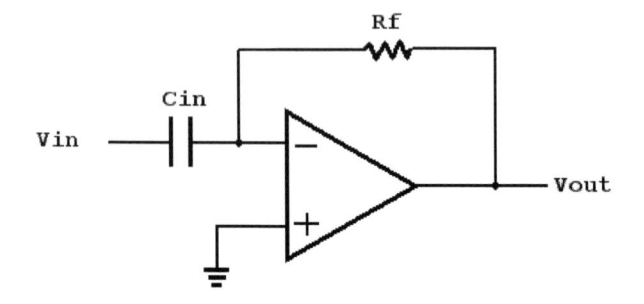

FIGURE 1.21

Differentiator using an inverting Opamp

Fundamentals of Electronics and Mechanics

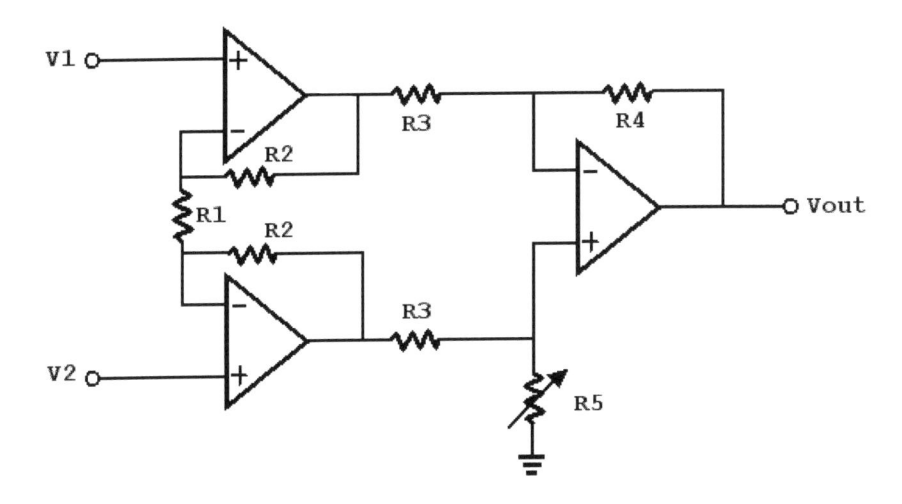

FIGURE 1.22

Instrumentation amplifier

1.1.2.2.14 Instrumentation Amplifier These amplifiers are often used as signal conditioning of low-level dc signals buried in large amounts of noise. The circuitry for this feature is very similar to the differential amplifier with extremely high input impedance. Figure 1.22 illustrates the circuit diagram of an instrument amplifier. As shown in this figure, the circuit consists of two stages. The input stage is the differential amplifier and the output stage is a difference amplifier. The output voltage of this amplifier is given in equation 1.5 assuming $R_4 = R_5$.

$$V_{out} = -(V_1 - V_2)\left(1 + \frac{2R_2}{R_1}\right)\left(\frac{R_4}{R_3}\right) \quad (1.5)$$

1.1.2.2.15 Unity Gain In this circuit any change in the input voltage reflects an equal amount of change in the output voltage. As a result this circuit is also known as the voltage follower circuit. Advantage of such circuit is that it provides high input impedance, low output impedance, and unity gain. Figure 1.23 illustrates the circuit diagram of this amplifier.

1.1.2.2.16 Inverse Gain As the name suggests, this amplifier is capable of inverting and amplifying the input signal. It consists of two resistors that are arranged as shown in Figure 1.24. Using Kirchoff's law and Ohm's law the input-output relationship can be obtained and is given in equation 1.6.

$$\frac{V_{out}}{V_{in}} = -\frac{R_2}{R_1} \quad (1.6)$$

FIGURE 1.23

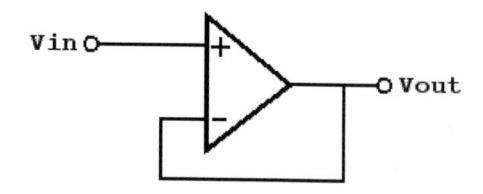

Circuit diagram of a unity gain amplifier

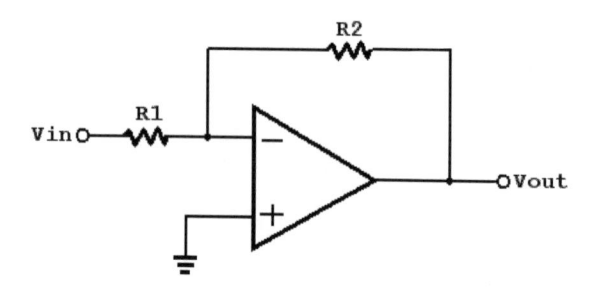

FIGURE 1.24

Inverter amplifier circuit

From the above equation it is clear that the output voltage is always the inverse of the input voltage and the amplification is purely obtained by increasing the ratio of the resistors.

1.1.2.2.17 Summing Amplifier Operational amplifiers are also used to sum/mix signals in practical circuits. A summing amplifier is given in Figure 1.25. The output voltage is proportional to the sum of the signals and given in equation 1.7.

$$V_{out} = -\frac{R_3}{R_1}V_1 - \frac{R_3}{R_2}V_2 \qquad (1.7)$$

If R_1 and R_2 are equal to each other, the output voltage can be expressed as in equation 1.8.

$$V_{out} = -\frac{R_3}{R_1}(V_1 + V_2) \qquad (1.8)$$

More detailed discussions on operational amplifiers and circuitry designed by operational amplifiers can be found in many mechatronics books [20] [21].

Fundamentals of Electronics and Mechanics

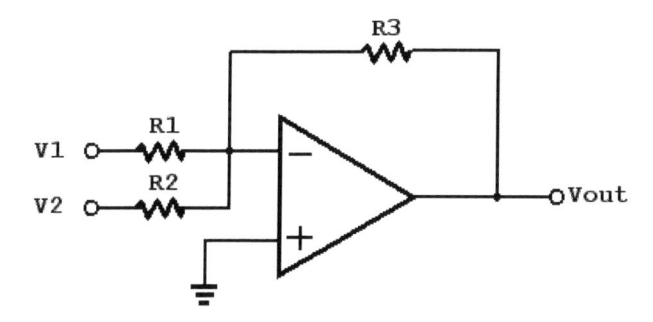

FIGURE 1.25
                    Summing amplifier

1.1.3 Digital Systems

Physical data obtained via sensors is often analog in nature. This data is converted to digital information which is later analyzed by electronic circuits. Circuits that perform these functions are often called digital logic circuits or logic gates.

In digital systems, three main views are followed to realize a product from a conceptual idea to design and development. These views describe the behavior models, structural models, and physical models. In the behavior models, the general behavior of the circuit is described without any details on the actual implementation. On the other hand, in the structural and physical models the actual implementations as well as the physical connections are realized. The views are complemented with four levels of abstraction that include the transistors, gates, registers, and processors. Table 1.4 illustrates the interaction between the components and their corresponding representations in various domains.

Table 1.4 clearly illustrates the general functions of the four main components of digital systems. In the following section gates and registers are described in detail.

1.1.3.1 Logic states

Digital circuits primarily perform binary logic functions. They use two binary variables or states: zero (low) or one (high). In the circuit level, these two states are often represented by two distinct voltages with V_H for high and V_L for low state. However, in practical circuits, it is very difficult to obtain a crisp signal that matches the VH or VL levels. This is due to the presence of noise, temperature sensitive components, and other factors. As a result each of the states are defined in a range of voltages.

TABLE 1.4
Representation and levels of abstraction [22]

Levels	Behavior Forms	Structural Components	Physical Objects
Transistor	Differential equations, current-voltage diagrams	Transistors, resistors, capacitors	Analog and digital cells
Gate	Boolean equations, finite-state machines	Gates, flip-flops	Modules or units
Register	Algorithms, flowcharts, instruction sets, generalized FSM	Adders, comparators, registers, counters, register files, queues	Microchips
Processor	Executable specification, programs	Processors, controllers, memories, ASICs	Printer-circuit boards or multi-chip modules

TABLE 1.5
Truth tables for common logic gates

A	B	A AND B	A OR B	A NAND B	A NOR B	A XOR B	A NOT B
0	0	0	0	1	1	0	1
0	1	0	1	1	0	1	1
1	0	0	1	1	0	1	0
1	1	1	1	0	0	0	0

1.1.3.2 Logic gates

As described in the previous section, logic gates describe the behavior and structural aspects in the design of digital systems. These gates usually combine one or more inputs and produce an output. The logic gates that comprise of the fundamental building blocks are NOT, AND, OR, NAND, NOR, and XOR gates. Table 1.5 illustrates the truth table of these gates.

Figure 1.26 describes the symbols of these gates.

Boolean expressions that describe the behavior of digital systems can be represented using these fundamental gates. Complex systems can be represented using these expressions, as there is no restriction on the number of variables and gates used in the design process. Various techniques are available to minimize the number of states in a representation. Among the shopping list of techniques, Karnaugh maps (K-maps) technique is the most popular. This

Fundamentals of Electronics and Mechanics 27

AND	
OR	
NOT	
NAND	
NOR	
XOR	

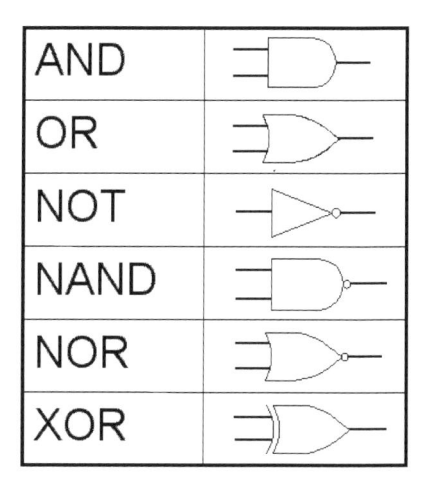

FIGURE 1.26

Symbols for common logic gates

process of minimization is important as it eliminates redundant and unnecessary states thereby reducing the number of physical components used in the design.

1.1.4 Sequential logic circuits

The above described gates are very widely used in digital systems that are static and do not depend upon past responses. However, in applications that require the output to depend upon the history of the inputs and the present inputs, sequential circuits are needed. The history dependence is typically obtained by introducing a memory element in the Boolean expression as well as feeding a part of the output into the input. As a result, the future state of the system is represented as a function of past state and the present state. When the possible number of states is finite then the system is called a Finite State Machine (FSM). Simplest memory units that can introduce a time delay are flip-flops.

1.1.4.1 Flip-Flops

These units are memory units that remember the most recent input by introducing a time delay caused by the process speed limitations of logic devices. The most common flip-flop is the set-reset. Figure 1.27 illustrates the schematic of this flip-flop. The set-reset flip-flop is a sequential circuit that contains two input lines, S and R, connected to two NOR gates and the Q and Q_P states the outputs of the gates. Although an extensive analysis on the working of this flip-flop can be performed with the help of truth tables,

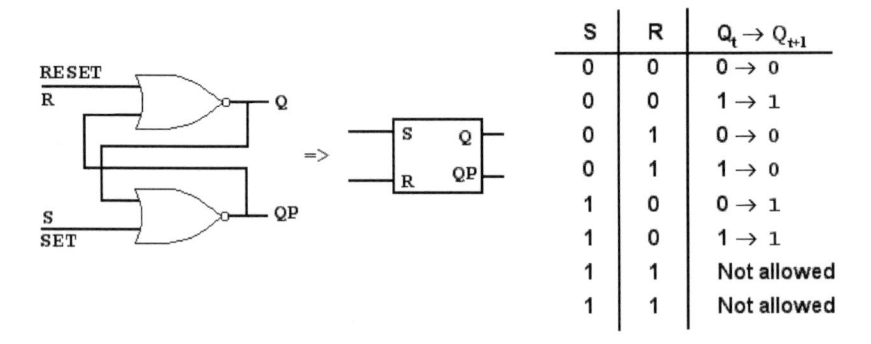

S	R	$Q_t \to Q_{t+1}$
0	0	0 → 0
0	0	1 → 1
0	1	0 → 0
0	1	1 → 0
1	0	0 → 1
1	0	1 → 1
1	1	Not allowed
1	1	Not allowed

FIGURE 1.27
Schematic diagram, symbol, and truth table of a set-reset flip-flop

an intuitive analysis is that the new states of Q_P and Q toggle only when the corresponding input states S and R toggle between 0 and 1.

In practical circuits, these flip-flops also find their applications in control elements to the external circuits in order to perform sequential operations. A combination of these flip-flops is called a register and these can be used in the design of general processors. These are discussed below. Flip-flops can be used to create simple and useful circuits. Figure 1.28 presents an alarm circuitry designed by a flip-flop and a phototransistor [23].

In the alarm circuitry given in Figure 1.28, the alarm is activated when a light beam is interrupted. The alarm continues even if the beam is no longer interrupted. When there is light, the transistor is on, thus S and R inputs of the SR flip-flop are both zero. When R and S are zeros, the output of an SR flip-flop stays the same. Since initially, the Q output of the SR flip-flop is zero, the alarm is not activated. When the light beam is interrupted, the transistor turns off. Thus, the S input becomes logic 1 (5 volts). When S is logic 1 and R is logic zero, the output of the flip-flop is flopped. Thus, the Q output becomes logic 1 and activates the alarm. After this moment, if light beam comes back again, the Q output stays the same because R and S become zeros.

Flip-flops are generally clocked so that they can be used synchronously. Figure 1.29 is a clocked SR flip-flop.

There are other types of flip-flops such as D and JK flip-flops. Figure 1.30 shows their symbols and their truth tables.

Clocked flip-flops are triggered either during the rising and falling edge of the clock cycle. The rising edge triggered flip-flops are also called positive edge triggered flip-flops. Similarly, falling edge triggered flip-flops are also called negative edge triggered flip-flops. Figure 1.31 shows positive and negative edge triggered D flip-flops and their corresponding timing diagrams.

Using edge triggered D flip-flop and some logic gates we can design an alarm

Fundamentals of Electronics and Mechanics   29

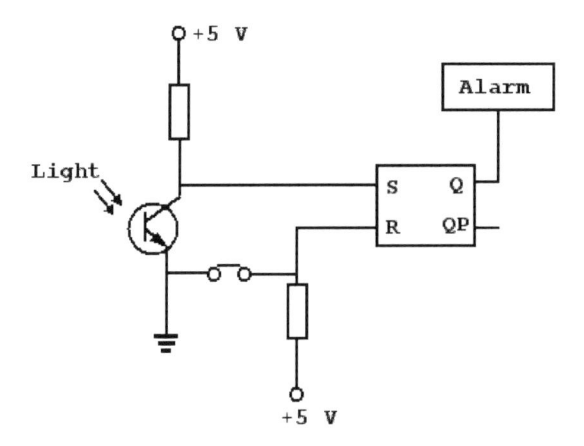

FIGURE 1.28
Alarm circuitry designed by a flip-flop and a phototransistor [23].

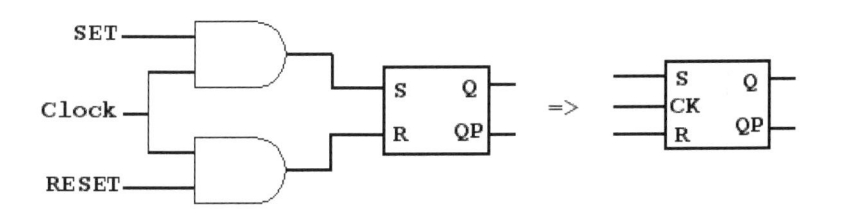

FIGURE 1.29
Clocked SR flip-flop design and its symbol

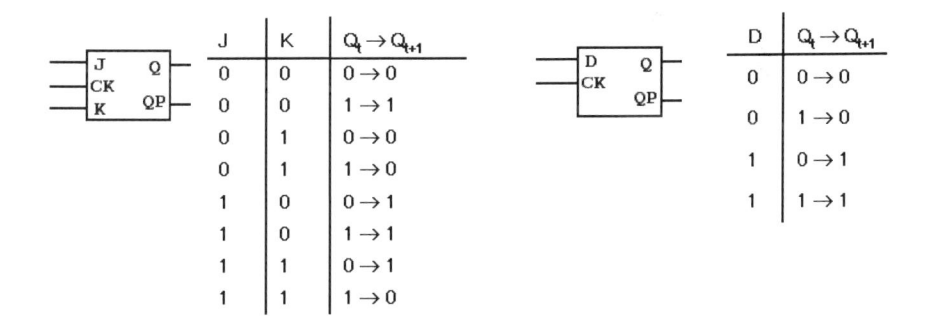

J	K	$Q_t \to Q_{t+1}$
0	0	$0 \to 0$
0	0	$1 \to 1$
0	1	$0 \to 0$
0	1	$1 \to 0$
1	0	$0 \to 1$
1	0	$1 \to 1$
1	1	$0 \to 1$
1	1	$1 \to 0$

D	$Q_t \to Q_{t+1}$
0	$0 \to 0$
0	$1 \to 0$
1	$0 \to 1$
1	$1 \to 1$

FIGURE 1.30
D and JK flip-flop symbols and their truth tables

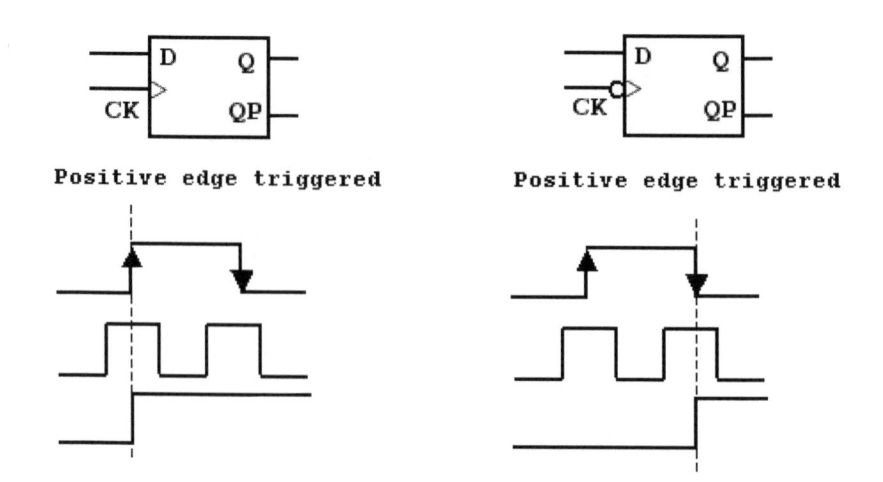

FIGURE 1.31
Positive and negative edge triggered flip-flops and their timing diagrams

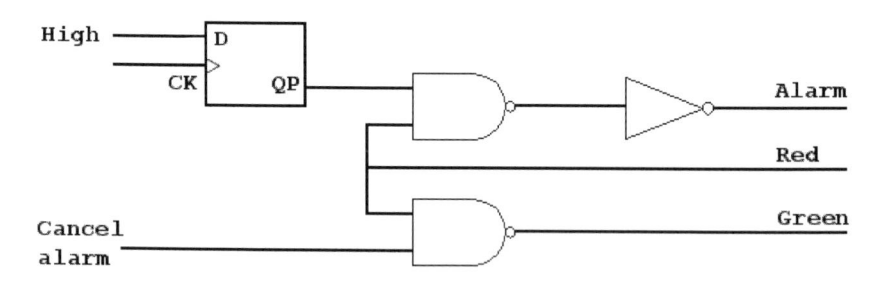

FIGURE 1.32
Alarm circuitry designed by a D flip-flop and logic gates [23].

circuitry shown in Figure 1.32. The system shows a green light when the sensor input is low. When it becomes high, the system shows a red light and an alarm is activated. The red light stays on as long as sensor input is high but the alarm can be switched off by triggering the D flip-flop through its clock. This system can be a monitoring system where the sensor input represents the activity being monitored. This could be a temperature, pressure, or the level of a liquid in a container.

Flip-flops can also be used to design synchronous logic circuits. We will cover one of these circuits, registers, next.

Fundamentals of Electronics and Mechanics 31

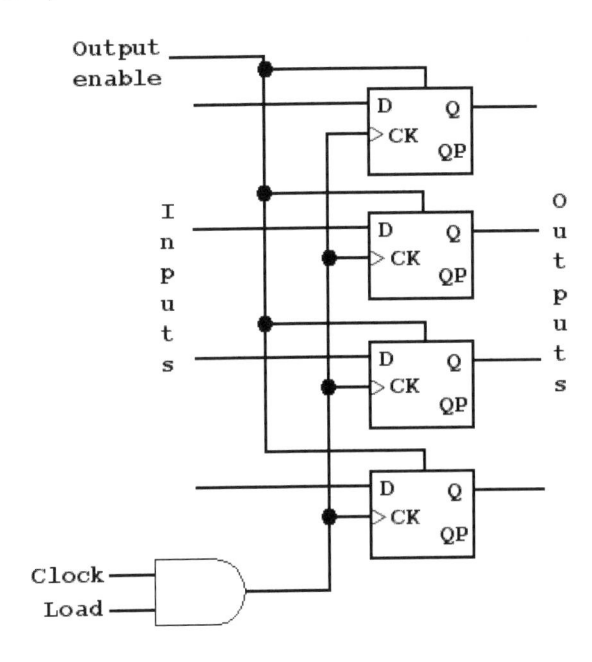

FIGURE 1.33
A register designed with D flip-flops

1.1.4.2 Registers

These components are used to store and/or manipulate information. Figure 1.33 illustrates the schematic of a register designed by D flip-flops. They can also be designed by other flip-flops. The main purpose of this component is to record and release stored information on demand so that the value can be used by other parts of the system. As described in the figure, the record phase is controlled by the Load voltage and the output is controlled by the Output enable signal.

A collection of such registers forms the core of the central processing unit in the computers when they are connected using data buses to the processor along with clock generation systems.

Extensive discussion on logic design and their usage in mechatronics and robotics can be found in [23], [21], [20], and [22].

1.1.5 Common Logic IC Devices

In this section we will introduce commonly used IC devices containing logic gates and components. There are two main IC families for logic gates: 40xx and 74xx series. 40xx series ICs are CMOS and suitable for high voltage systems. They can work with supply voltage of up to 15 volts. They are older

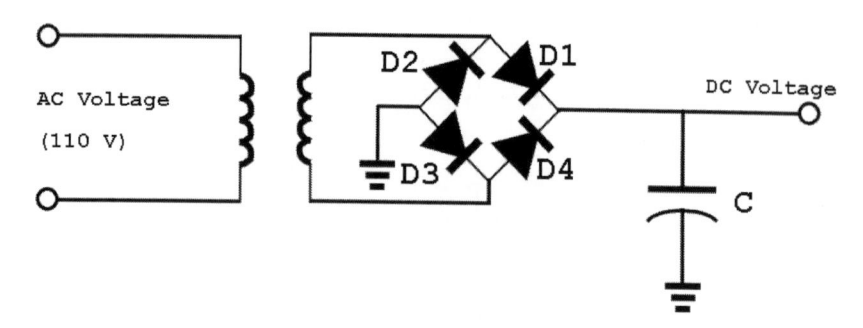

FIGURE 1.34
  AC to DC voltage converter using diodes and a capacitor

ICs and can work in low frequencies, 1 MHz. 74xx series have both CMOS (74HC and 74 HCT) and TTL (74LS) low power versions. HCT and LS series require 5 volts but HC series can work with very low supply voltage, 2 volts. Except 74LS, other series are low power ICs (a few Watt) and require drivers to control devices. 74LS series are relatively high power (a few mW). Table 1.6 gives a summary of the ICs and their components.

1.2 Practical Electronic Circuits

In this section, we will explore some practical and fun circuits commonly used in robotics related projects. We present circuits in power, control, IR communication, motors, and digital systems.

1.2.1 Power

Most electronic circuits need a Direct Current (DC) power supply in order to operate successfully. Before a regulated voltage can be obtained, Alternate Current (AC) voltage is reduced to a desired level by a transformer. Then, the reduced AC voltage is converted to DC voltage by the help of four diodes and a capacitor, as shown in Figure 1.34.

In Figure 1.34, diodes take the absolute value of the AC signal. In the positive cycle of the AC signal D_1 and D_3 are on. Alternatively, in the negative cycle of the AC signal, D_2 and D_4 are on. As a result, only positive components go through D_1 and D_4 resulting in positive signals at the output. This positive changing signal is then smoothened by the capacitor so that it becomes a positive signal with small ripples. These signals are shown in Figure 1.35.

Fundamentals of Electronics and Mechanics 33

TABLE 1.6
Common ICs used in digital systems

Device Type	Common Type	ICs
Gates	Quad 2-input gates	7400 quad 2-input NAND 7402 quad 2-input NOR 7403 quad 2-input NAND with open collector outputs 7408 quad 2-input AND 7409 quad 2-input AND with open collector outputs 7432 quad 2-input OR 7486 quad 2-input EX-OR 74132 quad 2-input NAND with Schmitt trigger inputs
	Triple 3-input gates	7410 triple 3-input NAND 7411 triple 3-input AND 7412 triple 3-input NAND with open collector outputs 7427 triple 3-input NOR
	Dual 4-input gates	7420 dual 4-input NAND 7421 dual 4-input AND
	8-input NAND gate	7430 8-input NAND gate
	Hex NOT gates	7404 hex NOT 7405 hex NOT with open collector outputs 7414 hex NOT with Schmitt trigger inputs
Counters	Ripple Counters	7490 decade (0-9) ripple counter 7493 4-bit (0-15) ripple counter 74390 dual decade (0-9) ripple counter 74393 dual 4-bit (0-15) ripple counter
	Synchronous Counters	74160 synchronous decade counter (standard reset) 74161 synchronous 4-bit counter (standard reset) 74162 synchronous decade counter (synchronous reset) 74163 synchronous 4-bit counter (synchronous reset)
	Up/down Counters	74192 up/down decade (0-9) counter 74193 up/down 4-bit (0-15) counter
Decoders		7442 BCD to decimal (1 of 10) decoder

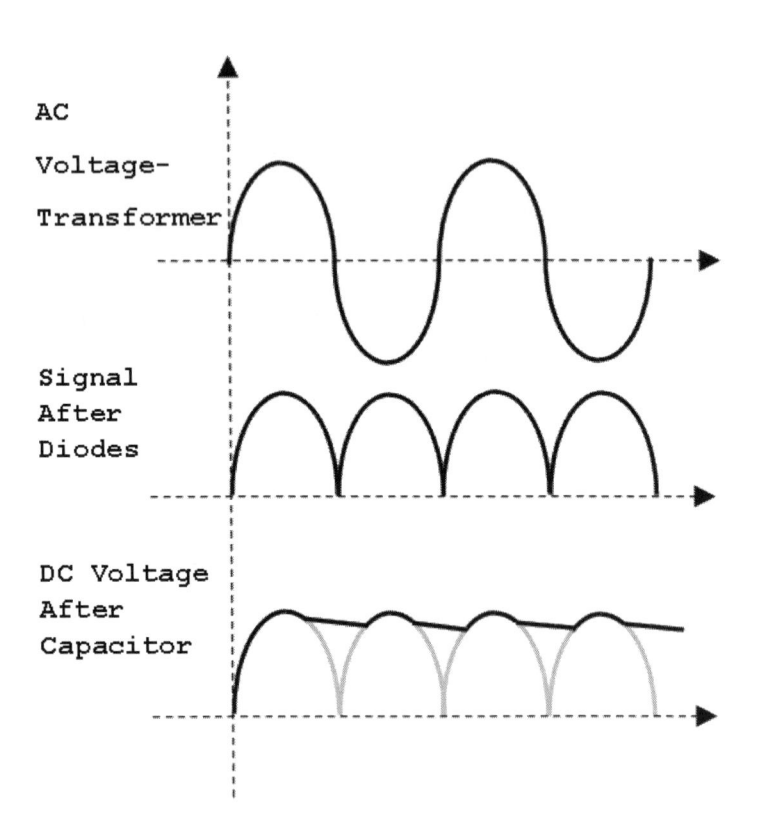

FIGURE 1.35
Converting AC voltage to DC voltage using diodes and a capacitor

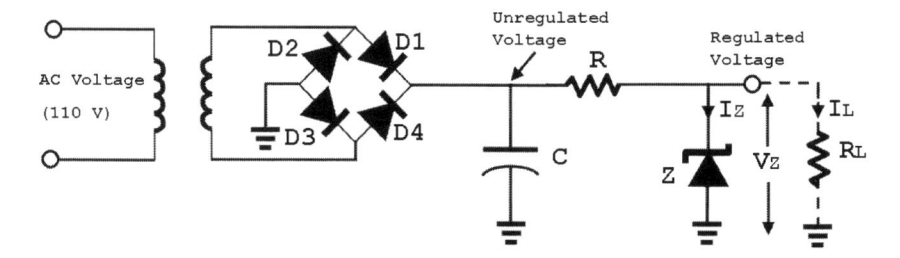

FIGURE 1.36
A fixed voltage power supply design with a zener diode

After creating DC voltage using the circuitry shown in Figure 1.34, we now can do the voltage regulation so that we can have a true DC voltage supply. We can either design a fixed voltage power supply or adjustable voltage power supply.

1.2.2 Fixed Voltage Power Supplies

Two most common ways to design fixed voltage power supplies are using a zener diode and using a regulator IC such as 78xx series. Figure 1.36 presents a fixed voltage power supply designed by a zener diode.

In Figure 1.36, the resistor R_L represents a load, i.e. circuitry or component you are powering. The resistor R protects the zener diode when there is no load. Unregulated voltage needs to be larger than the zener diode voltage so that enough amount of current can be supplied to the load. However, if the unregulated voltage is too large, then the voltage on resistor R (unregulated voltage - zener voltage) becomes so large that the efficiency of the regulator goes down and we waste power on the resistor R. If this voltage is so large and we want larger currents from our regulator, then the resistor R needs to be high power resistor. In practice, the resistor R needs to be as small as possible. In addition, the unregulated voltage (V_{un}) needs to be close to the zener (regulated) voltage so that the efficiency of the power supply can be improved. There are several resources online and offline on how to choose the resistor R, R_L, and V_Z values for required I_Z and I_L. For example, if you have an unregulated voltage of 11 DC +/- 1 volt and you would like to have 5 V output with a current range 0 to 200 mA, then you need to choose resistor R to be 24 Ohms for the safety of the regulator. The equation to calculate the resistor R is given in equation 1.9.

$$R = \frac{min(V_{un} - V_Z)}{min(I_Z) + max(I_L)} \qquad (1.9)$$

If you have load resistance of 100 Ohms, your power supply will provide 50 mA successfully. However, you need to be careful about not allowing load

FIGURE 1.37

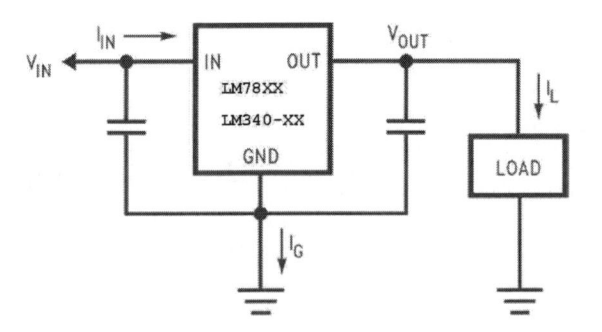

Usage of a regulator IC

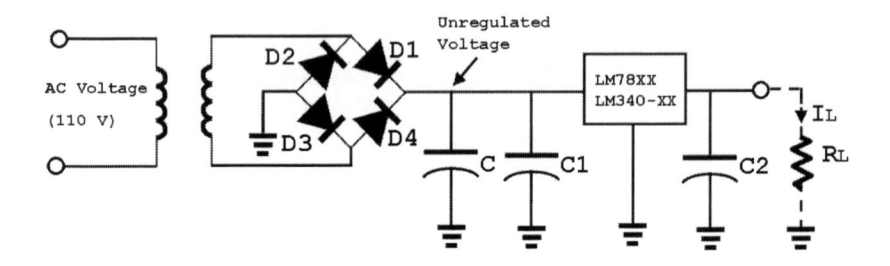

FIGURE 1.38

Fixed voltage regulator using LM78XX series regulator ICs

resistances which will require higher current values than the maximum load current. If you exceed the maximum current limit, the circuitry will not function currently.

Another way of designing fixed voltage power supplies is to use integrated circuits (ICs) designed as regulators such as 78XX series. In these types of regulators, all we need to provide is an unregulated voltage larger than the output voltage of the regulator IC. Figure 1.37 shows how these regulator ICs can be used. In the Figure 1.37, V_{IN} represents the unregulated input voltage; I_{IN} represents the input current; V_{OUT} represents the output voltage, which is specified by the IC's name, i.e., 7805 outputs 5 volts; I_L is the load current which is drawn by the device that you are powering. Finally, I_G represents the current that goes to the ground from the IC. This current needs to be as small as possible. Generally, it is related to the voltage difference between V_{IN} and V_{OUT}.

Figure 1.38 shows a regulator circuitry using an unregulated DC supply we have discussed before.

The values of the capacitors C1 and C2 are 0.22 micro Farad (F) and 0.1 micro F, respectively. Capacitor C1 is needed to help reduce the ripple in the unregulated DC voltage. They are especially needed if the regulator IC is far

Fundamentals of Electronics and Mechanics 37

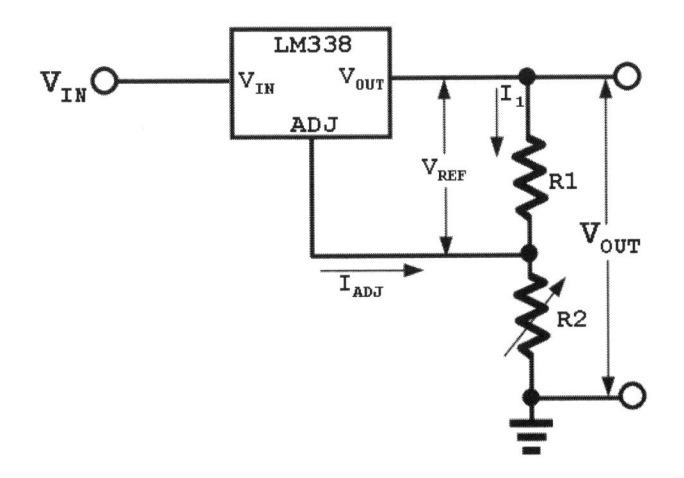

FIGURE 1.39
Adjustable regulator design with a regulator IC

from the unregulated voltage. Capacitor C2 is needed for optimum stability and transient response. Both capacitors should be located as close as possible to the regulator IC.

1.2.3 Adjustable Power Supplies

Most common designs of adjustable regulators are done by using 3-terminal adjustable regulator ICs such as LM117/LM317A/LM317 and LM138/338 (5 Ampere adjustable regulator). The typical adjustable regulator design with these ICs is shown in Figure 1.39.

In regular operation, the LM338 develops a nominal 1.25 reference voltage, VREF, between the output and the adjustment terminals. This voltage is impressed across the resistor R_1, thus I_1 is constant based on the Ohm's Law. The current I_{ADJ} and I_1 together flows through the output set resistor R_2. Based on this circuitry, the output voltage is dependent on the values of resistors R_1 and R_2, reference voltage V_{REF}, and current I_{ADJ}. Thus, the output voltage can be calculated by equation 1.10.

$$V_{OUT} = V_{REF}\left(1 + \frac{R_2}{R_1}\right) + I_{ADJ}R_2 \qquad (1.10)$$

In equation 1.10, I_{ADJ} term represents an error term. Thus, the LM338 is designed to minimize this current and make it very constant with line and load changes. A 0.1 micro F disc or solid tantalum capacitor is recommended as an input bypass capacitor. In addition, the adjustment terminal can also be bypassed to ground so that the ripple in the output voltage is minimized.

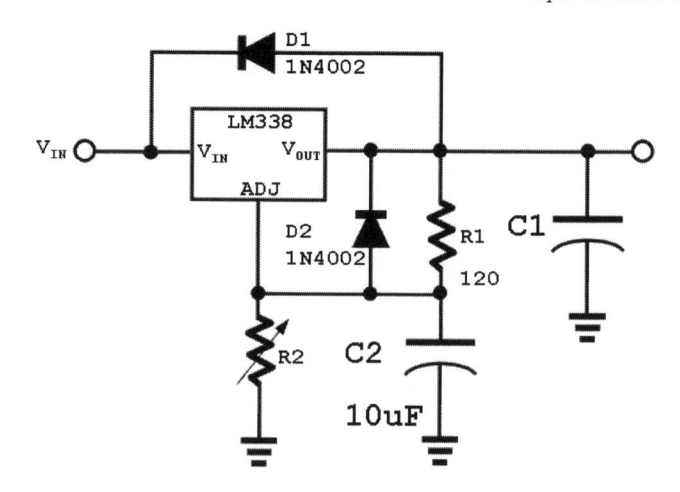

FIGURE 1.40

An LM338 regulator with protection diodes

A 10 micro F capacitor can create enough amount (75 dB) of ripple rejection.

When external capacitors are used with any IC regulator, it is necessary to add protection diodes to prevent the capacitors from discharging through low current points into the regulator. Most 20 micro F capacitors can deliver 20 Ampere spikes when shorted. This amount of current can damage the IC even though it happens in a short period of time. However, for voltages less than 25 volts, the LM 338 has enough internal resistance so that no protection is needed for the capacitors. Figure 1.40 shows an LM338 regulator with protection diodes.

In Figure 1.40, diode D1 protects against C1. Similarly, diode D2 protects against C2. The capacitors discharge through the diodes instead of discharging through the regulator.

Now, we will design an adjustable regulator which can generate voltages from 1.2 volts to 25 volts. Figure 1.41 presents the adjustable voltage regulator without diodes since the maximum output voltage is 25 volts.

In the design shown in Figure 1.41, capacitor C_2 is optional and used for improving transient response. It can be in the range of 1 micro F to 1000 micro F aluminum or tantalum electrolytic capacitors. Capacitor C_1 is needed if the LM338 is more than 6 inches from the filter capacitor (C in Figure 1.34).

The adjustable voltage regulators can also be used for other applications where you need an adjustable voltage. For example, we can design a light controller using an LM338 adjustable voltage regulator by replacing output response resistor (R_2 in Figure 1.41) with a phototransistor. The light controller circuitry is shown in Figure 1.42.

In Figure 1.42, the resistance of the phototransistor is very high when there is no light. In this case, the output voltage will be maximum, thus the lamp

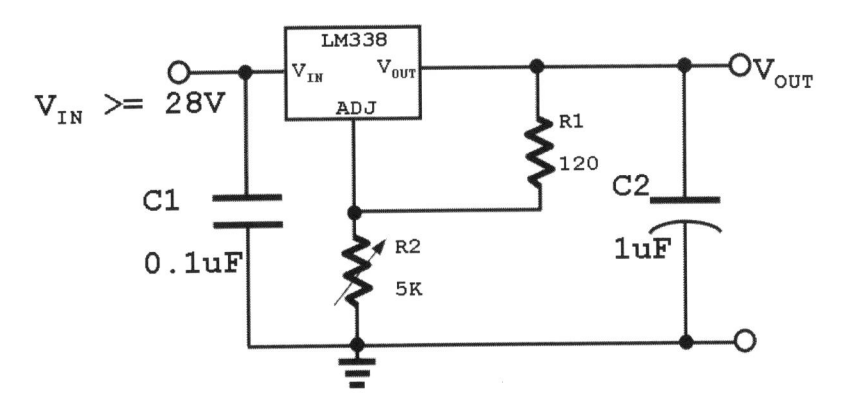

Fundamentals of Electronics and Mechanics 39

FIGURE 1.41
Adjustable voltage regulator design (1.2 volt to 25 volt)

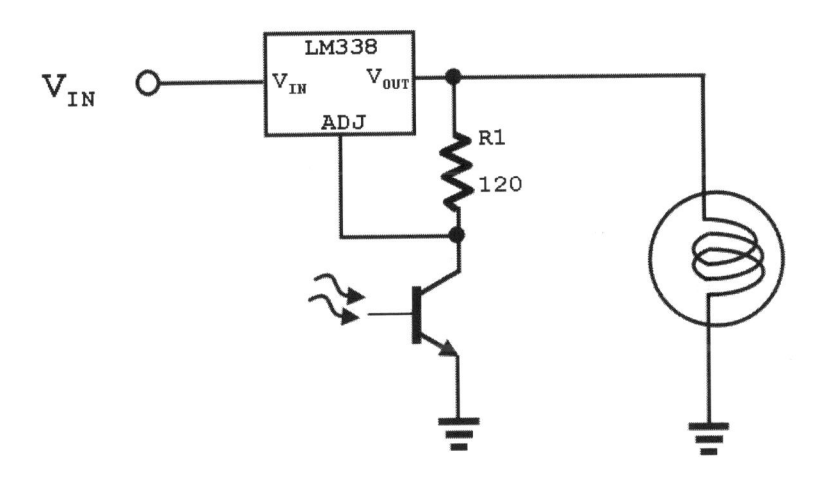

FIGURE 1.42
Light controller design with an LM338 adjustable voltage regulator

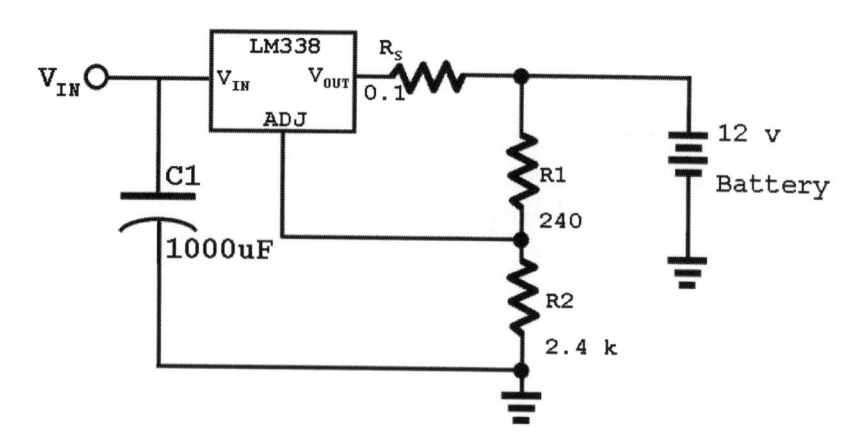

FIGURE 1.43
12 volt battery charger design with an LM338 adjustable voltage regulator

will be on. When there is light, the resistance of the phototransistor becomes very small, thus the output voltage goes down and lamp is not on. So, depending on the outside light condition, the lamp is powered up or down so that there is enough light in the area.

Another useful circuitry you can design with an LM338 is a battery charger. Figure 1.43 presents an example battery charger for a 12 volt battery. In the design, resistor RS allows low charging rates with fully charged battery. The capacitor C_1 is required to filter out transients in the unregulated input voltage. The output impedance of the charger is determined by R_S, as shown in equation 1.11.

$$Z_{OUT} = R_S \left(1 + \frac{R_2}{R_1}\right) \tag{1.11}$$

1.2.4 Infrared Circuits

In this section, we will explore two circuits regarding infrared communication and/or control: Infrared transmitter and infrared receiver.

1.2.4.1 Infrared Transmitter

In order to send data using IR LEDs, you need to modulate the data with about 40 kHz carrier. Using a modulated signal improves the distance the signal can be transmitted successfully. Figure 1.44 presents a simple example of an IR transmitter designed by a 555 Timer.

In the circuitry shown in Figure 1.44, the 555 Timer generates the 30 Hz carrier frequency to modulate the data being sent. When the pin 4 of 555

Fundamentals of Electronics and Mechanics 41

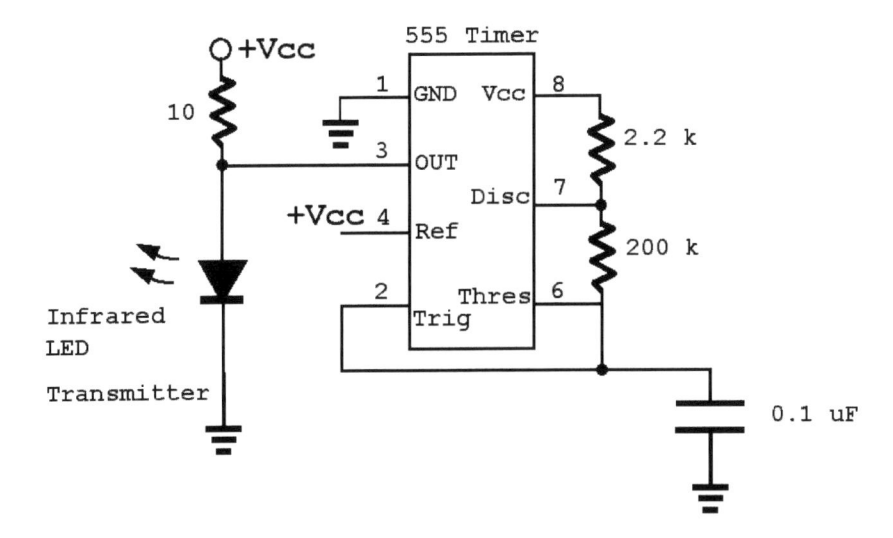

FIGURE 1.44

IR Transmitter designed by 555 Timer

Timer is $+V_{cc}$, the 30 Hz signal is transmitted by the infrared LED. When the pin 4 is 555, the 555 Timer does not create any signal, thus the infrared is not on and not sending any light. Therefore, you can send simple 1 and 0 signal by just changing the Ref input of the 555 Timer.

1.2.4.2 Infrared Receiver

In order to receive the signal sent by the transmitter, you need to have an IR detector and a 30 Hz bandpass filter. If the transmitter is sending the 30 Hz modulated signal (Ref is 1), then the receiver needs to detect this signal and the filter converts it to logic 1 (V_{cc}). If the receiver does not get the modulated IR signal, then the receiver should output logic zero (0 volt). Figure 1.45 presents the IR receiver designed to receive signals sent by the transmitter shown in Figure 1.44.

In the circuitry shown in Figure 1.45, the IR signals sent by the transmitter shown in Figure 11 are detected by the IR detector. When the IR detector detects the IR light, it turns on and the voltage at the input of the filter (before 39 K Ohm resistor) becomes close to zero (0.7 volt - remember that IR detectors are essentially diodes). When there is not light detected, the IR detector is not on, thus the voltage at the input of the filter is almost equal to $+V_{cc}$. So, if the transmitter sends the IR signals at 30 Hz, the detector will provide the same frequency signal to the filter. Since the filter is designed to detect the 30 Hz signals, whenever the IR detector detects this signal, the output of the receiver will be logic 1 (about $+Vcc$). Alternatively, if the signal is not detected, the output of the receiver will be zero. The LM 3900 is

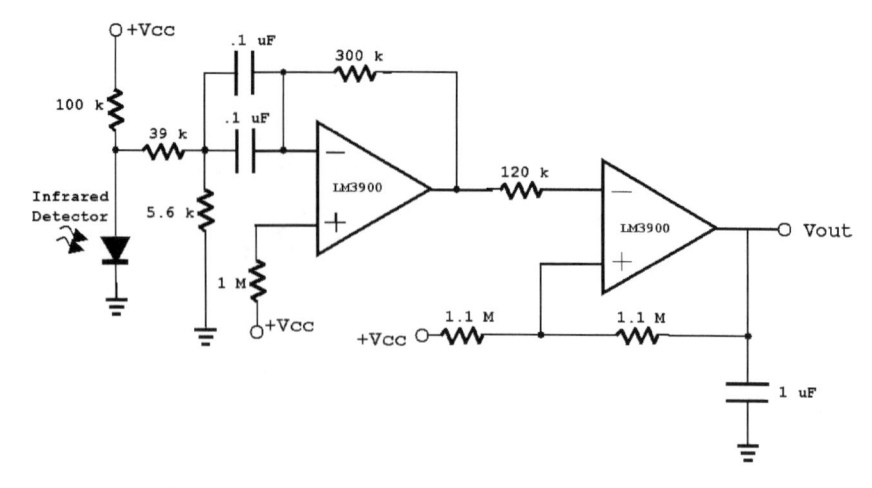

FIGURE 1.45

IR receiver

a quad unipolar Opamp IC. The 0.1 micro F capacitors are used to eliminate the direct light such as sunlight or any other infrared light source.

If you need to have the IR communication in larger distances, you can use an IR remote control receiver such as GP1UD26XK made by HARP electronic. These units have an IR detector and filter to detect the carrier frequencies ranging from 40 kHz to 56.8 kHz. All you need to provide is power. At the transmitter end you use the similar circuitry with higher frequencies. This is possible by changing the resistor values in Figure 1.44. In addition, you can use a Timer of a microcontroller to create the same carrier signal.

1.2.5 Motor Control Circuits

In this section we will cover the circuits used to control DC motors. These circuits are called H-Bridges. You can either design an H-Bridge or buy an IC which has an H-Bridge. We will give example circuits for both cases. Figure 1.46 presents the H-Bridge circuitry designed by switches. Depending on how these switches are implemented your H-Bridge will have different properties but the theory of operation is the same.

In Figure 1.46, there are four switches and a DC motor connected together as a letter H. Depending on which switches are turned on, the DC motor is driven forward or backward and stalled. As can be seen from the figure, when switches Q1 and Q4 are turned on (closed), the positive motor power is connected to positive input of the DC motor and the negative motor power is connected to the negative input of the motor. Thus, the motor is powered correctly and rotates in forward direction. When Q2 and Q3 are closed, then the DC motor is powered reverse. Thus, the DC motor rotates in backward

Fundamentals of Electronics and Mechanics 43

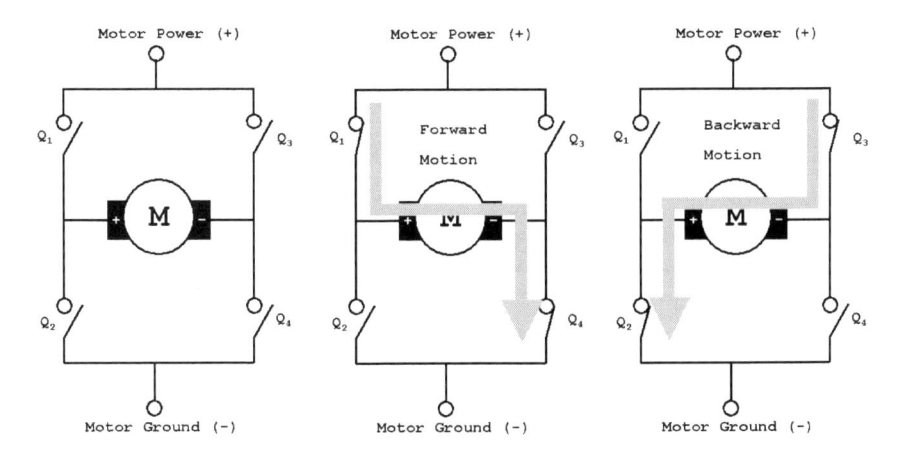

FIGURE 1.46
An H-Bridge and driving a DC motor forward and backward

direction. If Q1 and A3 are closed, both inputs of the motor get positive motor power. This forces the DC motor to stall. If none of the switches are turned on, the motor breaks slowly. As we mentioned above, we can implement these switches in various ways. We will cover only two of them: using transistors and using an H-Bridge IC.

1.2.5.0.1 H-Bridge Design by Transistors In order to design an H-Bridge with bipolar transistors, we need to replace the switches with proper transistors. Depending on how these switches are controlled, we can choose to use PNP or NPN transistors. Figure 1.47 presents an example of an H-Bridge designed by bipolar transistors (TIP102 and TIP107). These transistors can handle larger current levels. In order to control the transistors (switches), we can use Opto-couplers (PS2501) so that control signals are separated from the signals going through the motor. The Opto-couplers can be driven directly by a TTL logic through a 470 Ohm resistor.

In the Figure 1.47, the diodes are used to protect the circuitry from the back emf voltage of the motor. This voltage is created by the DC motor when it is powered on and off. Based on the schematic in Figure 1.47, when Forward (FWD) pin is 1, the Opto-couplers corresponding to Q1 and Q4 switches in Figure 1.46 are turned on. Thus, the DC motor rotates in the forward direction. When Reverse (REW) pin is 1, then the DC motor rotates in reverse direction. In addition, we need to set EN to logic zero so that we can enable the Opto-couplers. If EN is 1, the motor can run only if PWM (Pulse Width Modulation) signal is applied to it. When both FWD and REW are 1, then the DC motor is shorted thus it is stalled. This is also called braking. Finally, when FWD and REW are 0, the motor inputs get 0 volt causing the motor to coast. The motor will eventually stop after a while because of the

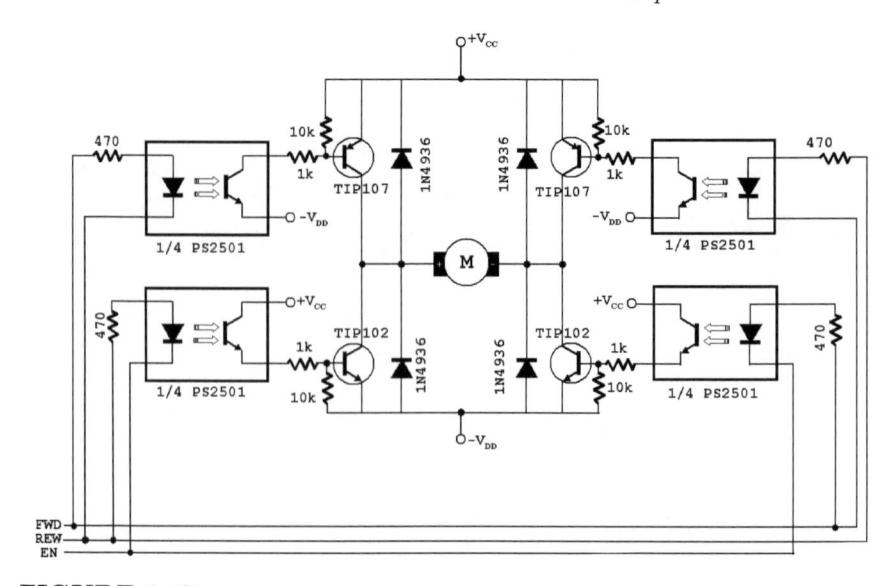

FIGURE 1.47

H-Bridge designed by bipolar transistors

internal and external frictions.

1.2.5.0.2 H-Bridge Design by ICs It is also possible to design H-Bridges using ICs designed specially for this kind of application. Generally, H-Bridge ICs have either high power semiconductor switches or high power Opamps. Figure 1.48 presents an example H-Bridge to control a DC motor using an IC called L298N. This IC has 4 high power controllable Opamps (drivers) enough to control two DC motors. Figure 1.48 also gives the truth table for forward, backward, active braking (fast motor stop), and passive braking (free running motor stop). Four diodes are used to protect the circuitry from back emf voltage during active start and stop. This IC allows you to detect high currents and stop/slow the motor through sensing pins. A current is drawn out of these pins (1, 15) when the motor is rotating. You can put a resistor at this pin (15) so that a voltage should appear on the resistor for the circuitry to notice. When the voltage reaches certain level, the IC shuts down and does not allow excessive current to damage the drivers. Typical value of the sense resistors is 0.5 Ohm. Finally, the diodes used in this circuitry need to be fast diodes in order to react to the quick changes the motor may impose.

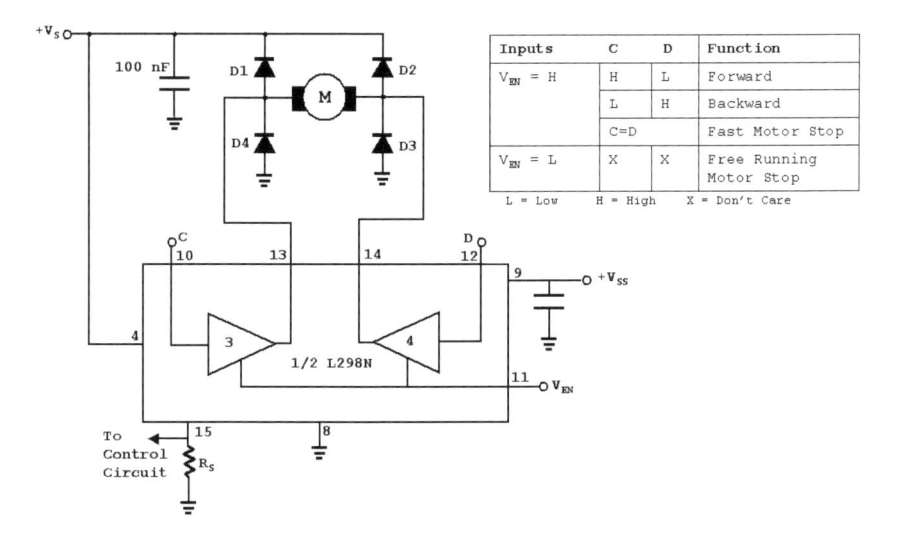

FIGURE 1.48
H-Bridge designed by a specialized IC

1.3 Fundamentals of Machines and Mechanisms

Roboticists need to not only understand basic electronics, but also be able to understand the basic principles of machines and mechanisms. There are some very good books that cover basic machines such as [5] and [15]. There are also some good books that show machine analysis and design from an engineering perspective such as [19].

1.3.1 Simple Machines

Simple machines are the most common tools that can be combined to make compound machines. These transform one motion into another, and make the work easier. In every simple machine, we have the effort force trying to move the load. If the effort required is less than the load, then the movement of the effort is increased compared to that of the load. If the effort required is more, then the movement of the effort required is less than that of the load. In an ideal simple machine, the product of the effort and its distance covered is the same as the product of the load and its distance covered. This is based on the energy conservation principle, since energy is the product of force and distance. This principle can be written as:

$$Effort \times D_{in} = Load \times D_{out} \qquad (1.12)$$

FIGURE 1.49

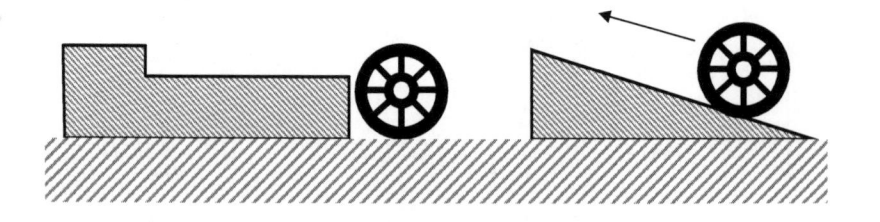

Usefulness of inclined plane

FIGURE 1.50

Wedge

Here D_{in} is the distance traveled by the effort and D_{out} is the distance traveled by the load. The mechanical advantage (M.A.) of a simple machine is defined as the ratio of the load to effort force.

$$M.A. = \frac{Load}{Effort} = \frac{D_{in}}{D_{out}} \qquad (1.13)$$

There are six types of simple machines. These are:

1.3.1.1 Inclined Plane

Inclined plane is a flat surface that is inclined. If the surface was not inclined it would not be useful as a machine. Since it is inclined we can slide or roll objects over the inclined surface. To understand the usefulness of the inclined plane, study Figure 1.49 that shows how a wheel can not roll up a step, but can roll up an inclined plane.

The next two simple machines are in fact derivatives of the inclined plane.

1.3.1.2 Wedge

A wedge (see Figure 1.50) has both surfaces inclined, and the two surfaces help in producing motion that requires less effort for movement.

To understand the usefulness of the wedge, study Figure 1.51 that shows its various applications.

Fundamentals of Electronics and Mechanics 47

Peg   Nail   Zipper

FIGURE 1.51

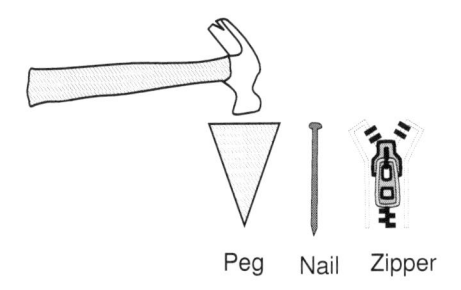

Usefulness of wedge

cylinder

Inclined plane

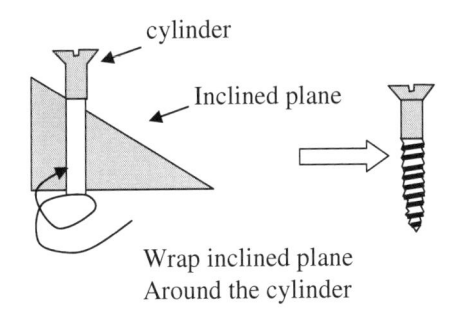

Wrap inclined plane
Around the cylinder

FIGURE 1.52

Screw

1.3.1.3 Screw

A screw 1.52 is an inclined plane wrapped around a cylinder, in such a way that when you turn the screw one time, it moves the amount of distance called pitch.

Screws are used in car jacks (see Figure 1.53), jar lids, light bulbs, etc.

1.3.1.4 Wheel and Axle

Wheel and axle (see 1.54) allows for one circular motion to be transferred to another. Axle is the rod that goes through the wheel.

Wheel and axles can be seen as parts of many compound machines, shown in Figure 1.55. They are also seen in cars, bicycles, gears, doorknobs, pulleys, etc.

1.3.1.5 Lever

A lever has three parts: a fulcrum, load, and effort. There are three types of levers based on the relative positions of these three points. These three types

FIGURE 1.53

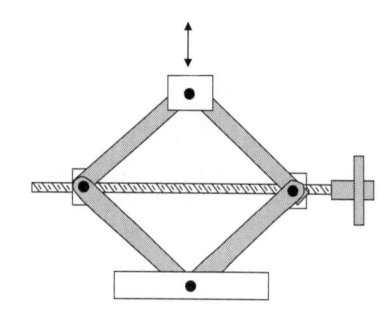

Car jack screw

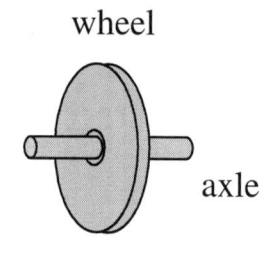

FIGURE 1.54

Wheel and axle

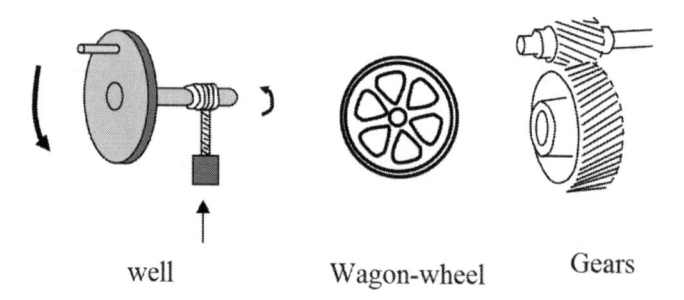

FIGURE 1.55

Some wheel and axle applications

Fundamentals of Electronics and Mechanics 49

FIGURE 1.56

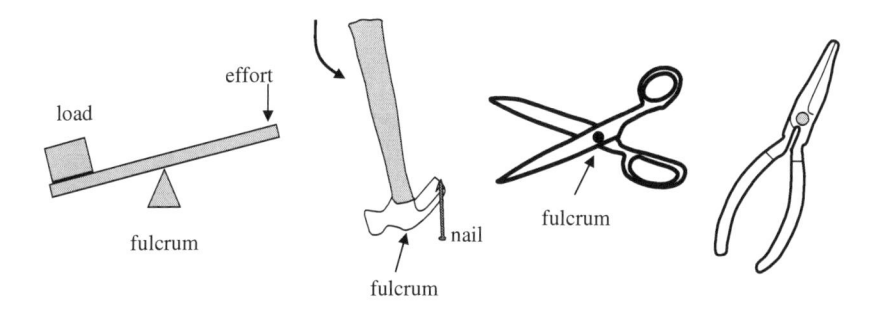

First-class levers

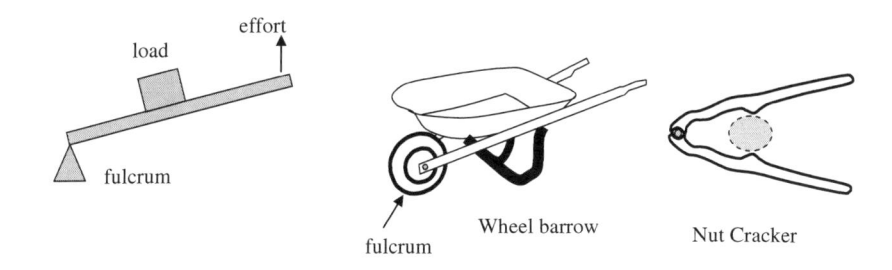

FIGURE 1.57

Second-class levers

of levers are:

1.3.1.5.1 First Class Lever First class levers have the fulcrum in between the load and the effort. Examples of this type of lever are: seesaw, hammer (when used to pull out a nail), and scissors as shown in Figure 1.56.

1.3.1.5.2 Second Class Lever Second class levers have the load in between the fulcrum and the effort. Examples of this type of lever are wheelbarrow and nutcracker as shown in Figure 1.57.

1.3.1.5.3 Third Class Lever Third class levers have the effort in between the load and the fulcrum. Examples of this type of lever are tweezers and tongs as shown in 1.58.

1.3.1.6 Pulley

A pulley consists of a wheel that rotates with a rope going around it. There are different kinds of pulleys: fixed, movable, and compound.

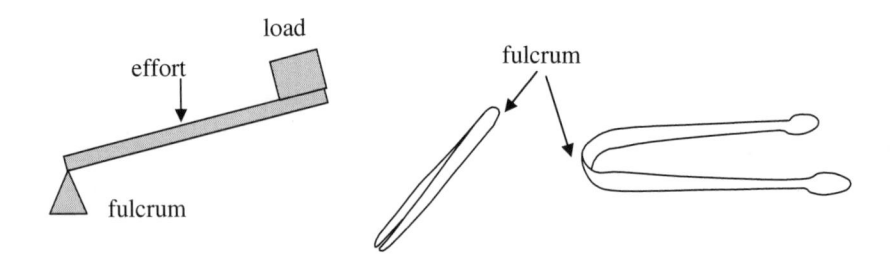

FIGURE 1.58

Third-class levers

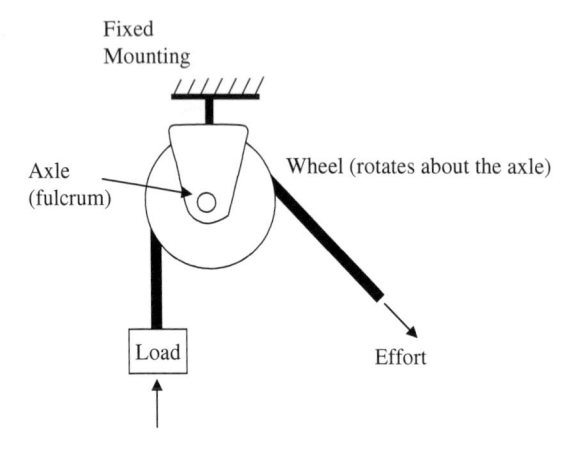

FIGURE 1.59

Fixed pulley

1.3.1.6.1 Fixed Pulley A pulley consists of a wheel that rotates with a rope going around it. There are different kinds of pulleys: fixed, movable, and compound.

In a fixed pulley, the effort pulls on the rope on one side and the load moves on the other as shown in Figure 1.59. The wheel rotates about its axle but remains at the same vertical position. This pulley functions as a direction changer in that the direction of motion of the effort is different than that of the load. The amount of force is the same as the weight of the load.

In a movable pulley, the effort pulls on the rope on one side and the load moves on the other as shown in Figure 1.60. The wheel rotates about its axle and with the load. The amount of force is half of the weight of the load, but the effort has to move double the amount of the movement of the load.

We can change the direction of effort in the movable pulley by adding a

Fundamentals of Electronics and Mechanics

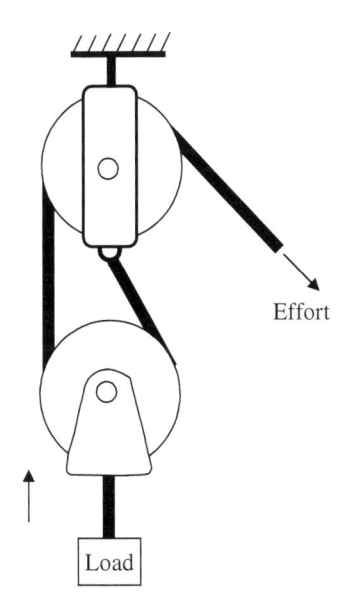

FIGURE 1.60

Movable pulley

fixed pulley to it as shown in Figure 1.61.

1.4 Mechanisms

There are many different types of mechanisms that can be useful in building complex mechanical devices. We will study a few here.

1.4.1 Gears

Gears are essentially levers that work using rotational motion. They transfer one rotational motion into another. The radii of the different gears dictate if the effort required will be greater or less than the load. There are different types of gears, depending on what type of motion has to be transferred. These are:

1.4.1.1 Spur Gears

Spur gears transfer rotational motion in the same plane. An example is shown in Figure 1.62.

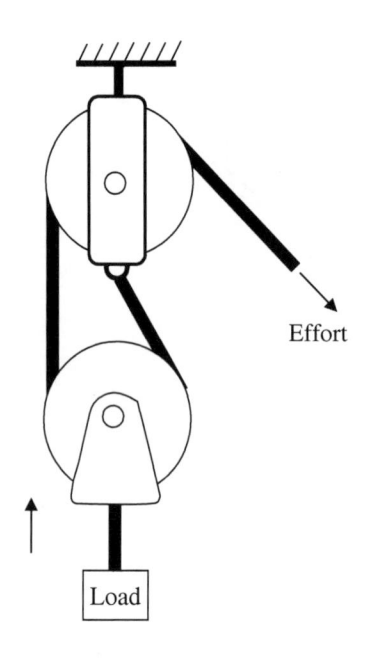

FIGURE 1.61

Compound pulley

FIGURE 1.62

Spur gear (Copyright Emerson Power Transmission Corporation)

Fundamentals of Electronics and Mechanics

FIGURE 1.63
Hellical gear (Copyright Emerson Power Transmission Corporation)

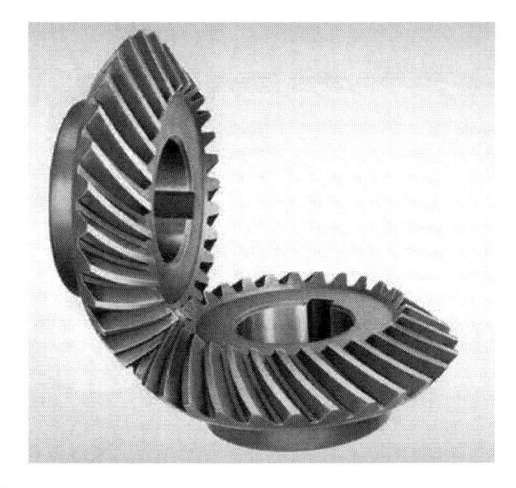

FIGURE 1.64
Bevel gear (Copyright Emerson Power Transmission Corporation)

1.4.1.2 Hellical Gears

Hellical gears also transfer rotational motion in the same plane; however, the gear teeth are designed to make the motion transfer smoother. An example is shown in Figure 1.63.

1.4.1.3 Bevel Gears

Bevel gears help in changing the direction of the rotational motion. An example is shown in Figure 1.64. The shafts in bevel gears do not have to be at right angles. They can be at any other angle such as 45 degrees.

1.4.1.4 Worm Gears

Worm gears also change the direction of the rotational motion, but also create large gear reductions. Worm gear produces motion at right angle to the input gear. An example is shown in Figure 1.65.

FIGURE 1.65
   Worm gear (Copyright Emerson Power Transmission Corporation)

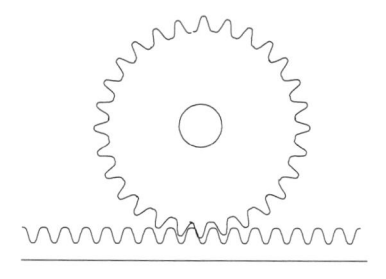

FIGURE 1.66
   Rack and pinion gear

1.4.1.5 Rack and Pinion Gears

Rack and pinion gears convert rotational motion into translational or vice versa. An example is shown in Figure 1.66.

1.4.1.6 Planetary Gears

Planetary gears provide more complex control over rotational motion. An example is shown in Figure 1.67. It has a central gear (sun gear), planet gears, ring gear, and an arm. The input and output shafts can be connected to the ring gear, the sun gear or the arm, giving different angular speeds. One can have different number of planet gears including one.

1.4.1.7 Compound Gears

Gears can be connected to each other to produce very large mechanical advantages. These are called compound gears. An example is shown in 1.68.

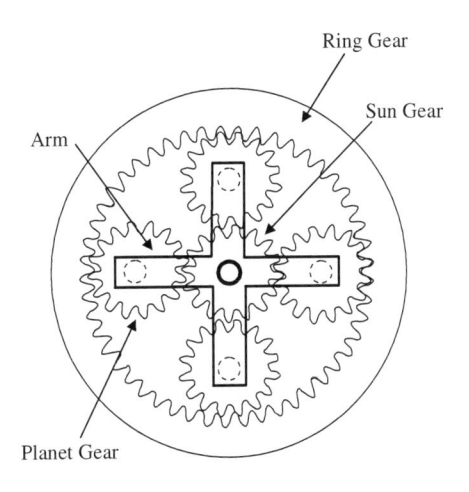

FIGURE 1.67 Planetary Gear

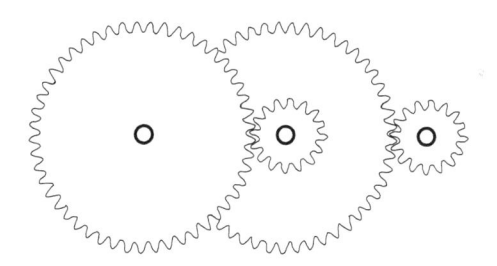

FIGURE 1.68 Compound gear

FIGURE 1.69

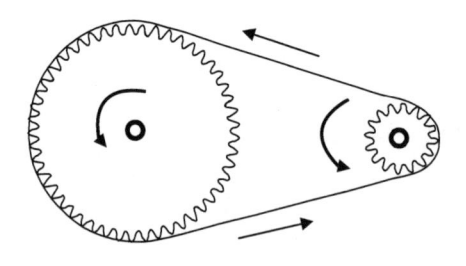

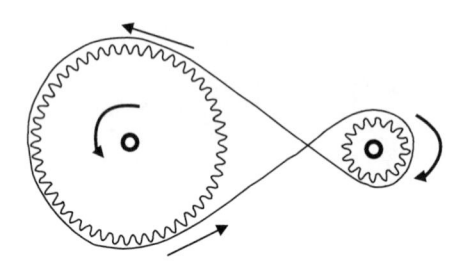

Belt

FIGURE 1.70

Reverse motion belt

1.4.2 Chains and Belts

Instead of using gears, we can also use Chains to connect different gears. One example is shown in Figure 1.69. In the figure, the big wheel is being driven and that makes the chains move as shown, causing the smaller wheel to follow the shown motion.

The direction of motion can be changed by using the belt as shown in 1.70.

Belts can be used with wheels that have grooves on the rims. Chains need gears for their use.

1.4.3 Linkages

There are many different types of linkages that can be designed for the kind of mechanical movement transfer that is desired. We review a few here.

1.4.3.1 Bell Crank

This linkage allows for motion direction to be changed by 90 degrees as shown in Figure 1.71. The motion direction can be changed by any other angle, also by designing the rotating piece to be of a specific sector of a circle.

Fundamentals of Electronics and Mechanics 57

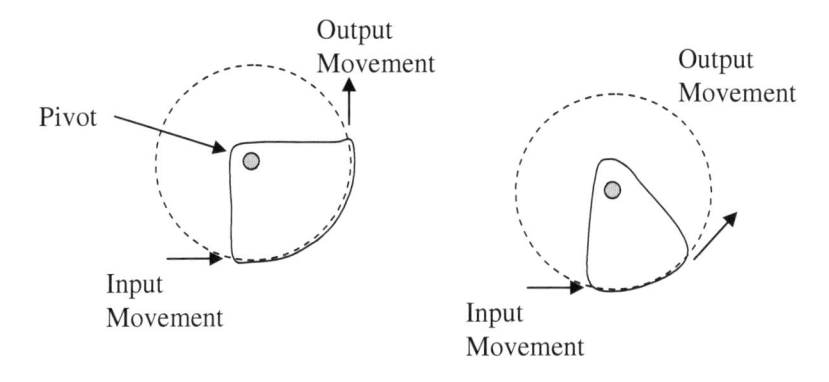

FIGURE 1.71
Bell crank linkage

1.4.3.2 One Bar Linkage

This linkage is the simplest allowing a circular motion of a link about a pivot (see Figure 1.72). There is only one degree of freedom for this linkage.

1.4.3.3 Two Bar Linkage

This linkage has the two links connected to each other and one is free to rotate about a pivot, and the other is free to rotate about the first link (see Figure 1.73). There are two degrees of freedom for this linkage. We can restrict one degree of freedom to obtain a crank and slider linkage.

1.4.3.4 Four Bar Linkage

This linkage allows for parallel motion to take place (see Figure 1.74).

1.4.3.5 Reverse Motion Linkage

This linkage allows for motion direction to be reversed (see Figure 1.75).

1.4.4 Cam

Cams allow rotational motion to be converted into a reciprocating motion of the follower (see Figure 1.76). By changing the shape of the cam, we can achieve different types of oscillations.

1.4.5 Ratchet Mechanism

Ratchet mechanism allows for one directional rotation only (see Figure 1.77). In the figure only counter-clockwise rotation of the wheel is possible.

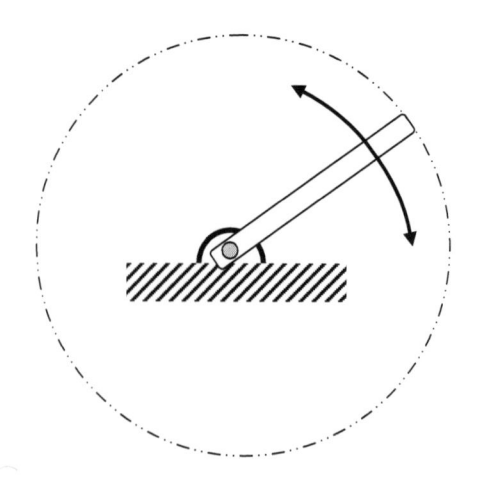

FIGURE 1.72

One bar linkage

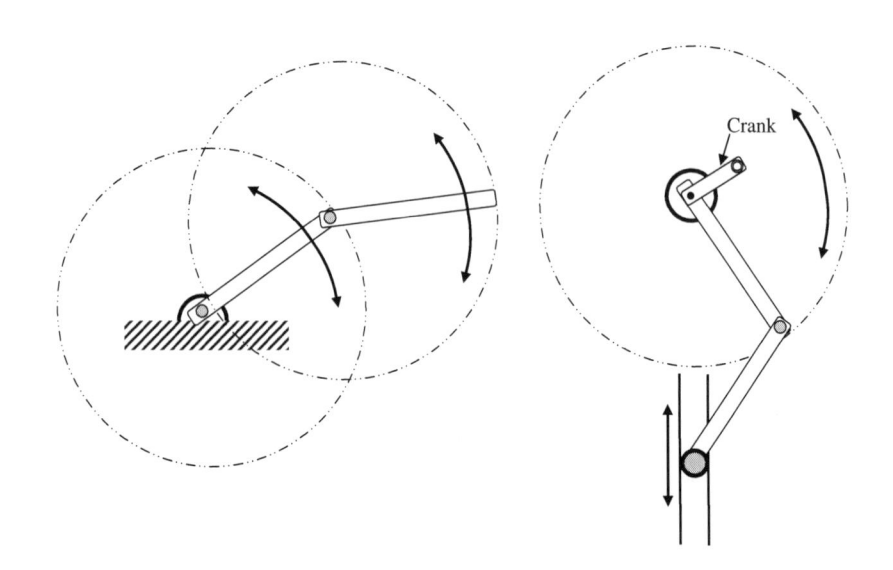

FIGURE 1.73

Two bar linkages

Fundamentals of Electronics and Mechanics

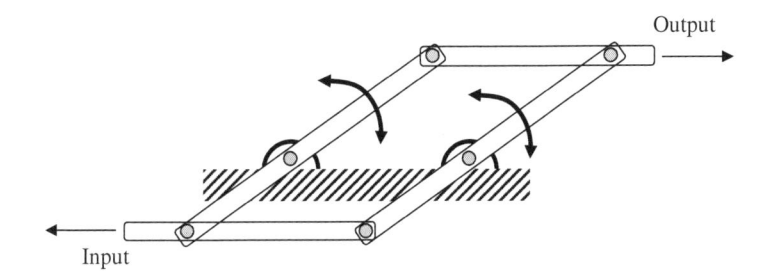

FIGURE 1.74

Four bar linkages

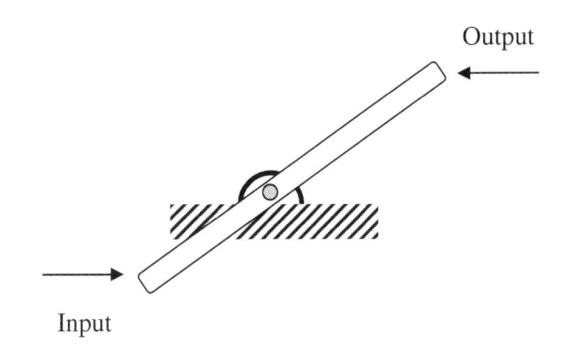

FIGURE 1.75

Reverse motion linkage

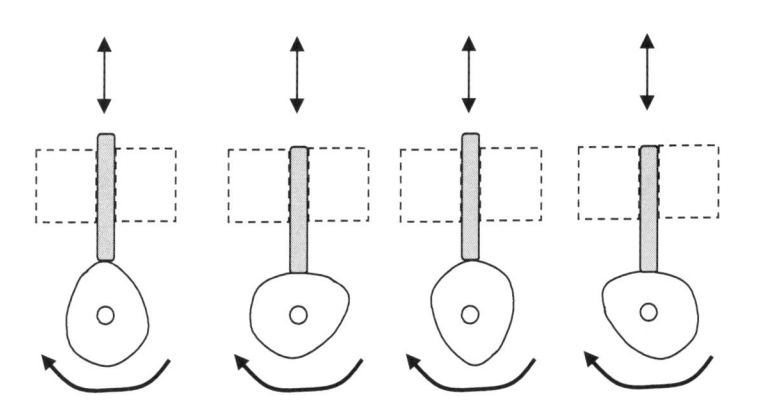

FIGURE 1.76

Cam

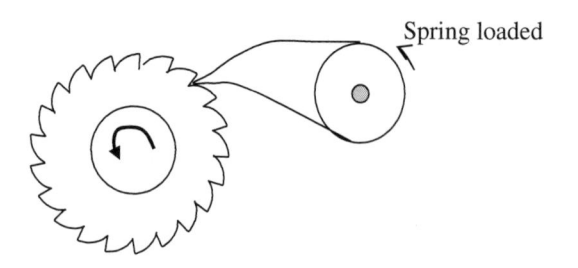

FIGURE 1.77
Ratchet mechanism

1.4.6 Quick Return Mechanism

Quick return mechanism allows for one direction motion of oscillation to happen slowly and then the other direction motion to happen quickly (see Figure 1.78). This happens because of the time the rotating ball spends in the groove going up versus down. The vertical slide oscillates right and left. The circular disk rotates clockwise in the figure.

1.4.7 Intermittent Motion

Intermittent motion can be provided by many ways. One way is to have gears, where the input wheel does not have all teeth (see Figure 1.79).

Another mechanism for this is called the Geneva stop (see Figure 1.80) which accomplishes the same thing without using gears. The wheel turns constantly and the ball on the disk makes the right hand mechanism to rotate intermittently (when it enters the groove, it makes it move till it exits the grove).

1.4.8 Springs and Dampers

Springs and dampers also help in mechanical systems. Springs act against the motion providing a resistive force proportional to motion. Dampers fight movement instead of displacement. They are usually designed by having fluid inside a cylinder so that the piston inside resists any motion against it providing a force proportional to the speed of motion. Extended spring has a resistive force even if there is no motion (the force is independent of speed, and only depends on change in the length of the spring). Dampers produce resistive force only when there is motion (the force is independent of position and only depends on the rate of change of piston position, i.e., speed). The piston head that is moving (Figure 1.81) inside the oil in the damper tube has holes so that the oil can flow through. Many shock absorbers come with

FIGURE 1.78

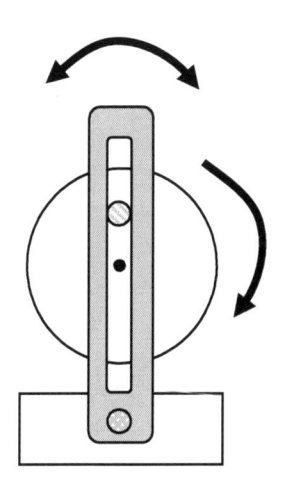

Quick return mechanism

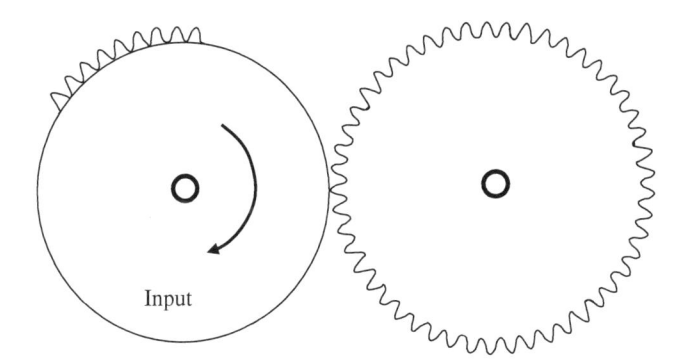

FIGURE 1.79

Intermittent rotation

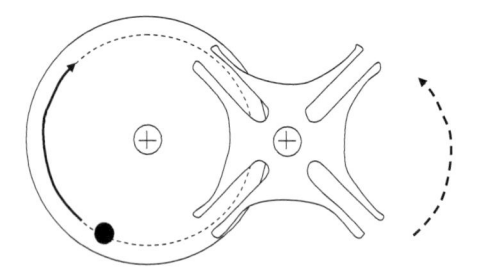

FIGURE 1.80
                Geneva stop mechanism

integrated spring and damper, as shown in Figure 1.81.

1.4.9 Brakes

Brakes allow us to stop moving devices, such as a wheel or a shaft. Brakes work based on friction between moving surfaces. Figure 1.82 shows two different types of bicycle brakes. The left one uses the first class lever principle to make the brake pads press against the wheel, while the right one uses the second-class lever mechanism.

1.4.10 Clutches

To understand how clutches work and what they are, consider Figure 1.83 that shows a motor shaft connected to a rack and pinion gear. Let us say that we want to use the same motor to also move the wheel on the left. We have to switch the motor output gear from the rack on the right to the wheel gear on the left as it is moving. For this to happen, we cannot do this while the motor continues to give rotational power to the gear. Therefore, we need a way to disengage the motor gear from the motor shaft as the motor is still rotating.

To accomplish this we use a clutch. A clutch is just a disk that rotates with another disk due to friction when the two disks are pressed together, and when the two disks are disengaged, and then the motion of one disk does not cause the motion of the other disk. This is shown in Figure 1.84.

1.4.11 Couplers, Bearings, and Other Miscellaneous Items

When two shafts have to be connected together, they can be joined using couplers. If the joint between the shafts has to be flexible, then flexible couplers can be used as shown in Figure 1.85.

When a shaft is rotating and is in contact with another surface, we can

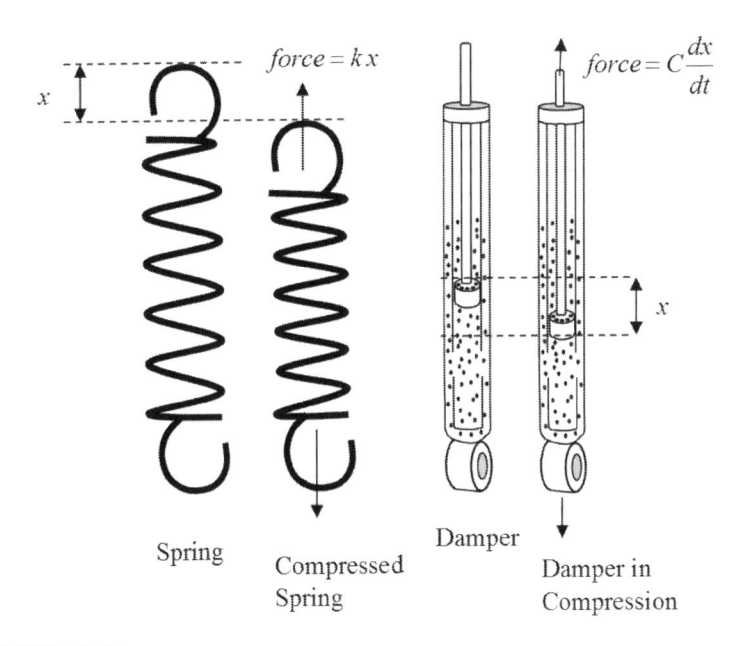

FIGURE 1.81 Spring and damper

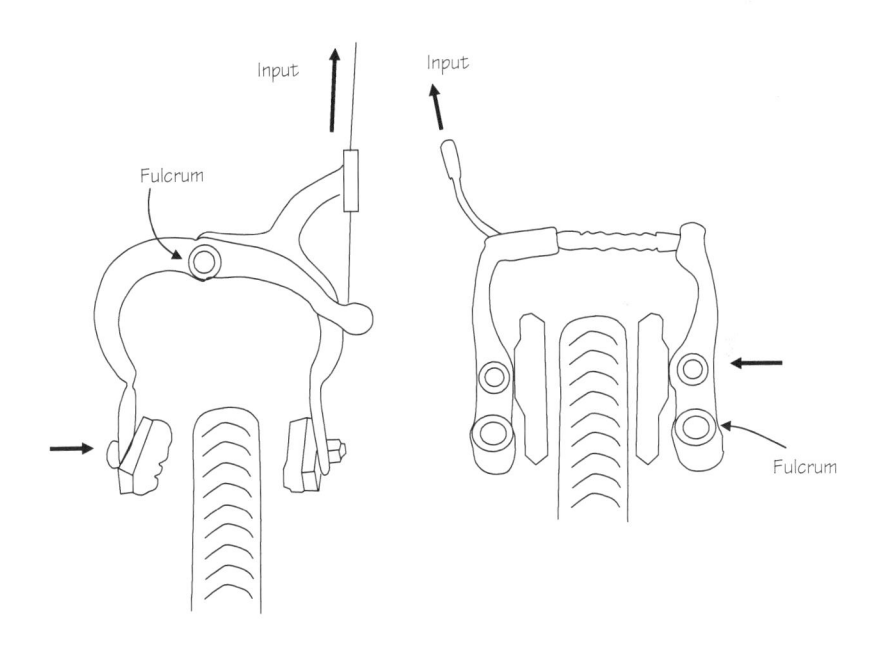

FIGURE 1.82 Clutches and brakes

FIGURE 1.83

Gear change

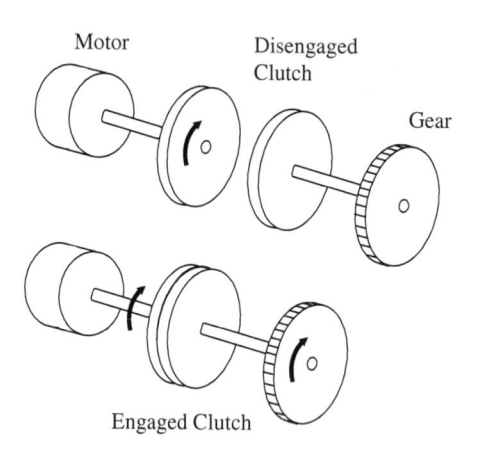

FIGURE 1.84

Clutch mechansim

FIGURE 1.85

Couplings (http://sdp-si.com)

use ball bearings to reduce friction (see Figure 1.86). There are many other mechanical elements such as screws, nuts, bolts, sheet metal, etc. that we do not cover in this book.

FIGURE 1.86
Ball bearings (http://sdp-si.com)

2

BASIC Stamp Microcontroller

The BASIC Stamp BASIC Stamp is a microcontroller developed by Parallax, Inc. It is programmable using a form of BASIC language called PBASIC (Parallax BASIC). It is called Stamp because of its small size. The PBASIC language is a hybrid of the BASIC programming language, and has been designed to exploit all the BASIC Stamp's capabilities. In this chapter, we will explore several kinds of BASIC Stamp controllers used in robotic kits. We will provide an overview of these controllers rather than very detailed explanations of how they are used in robotic kits; specific usage will be explored in later chapters.

2.1 Different Versions of BASIC Stamp

The BASIC Stamp line consists of the BASIC Stamp 1, the BASIC Stamp 2, the BASIC Stamp 2e, BASIC Stamp 2p, and the BASIC Stamp 2sx. These five functional versions of the BASIC Stamp (BS) and 11 physical versions are discussed in the following sections.

2.1.1 BASIC Stamp 1

The BASIC Stamp 1 has three physical versions (package types). The BASIC Stamp 1 Rev. Dx, shown in Figure 2.1, is a through-hole socketed package. The BS1-IC, shown in Figure 2.2, is a 14-pin single inline package (SIP) with surface-mounted components. The OEMBS1 is a 14-pin SIP and features an easier-to-trace layout meant to aid customers who wish to integrate the BASIC Stamp 1 circuit directly into their design (as a lower-cost solution). The OEMBS1 is available in either an assembled form or a kit form. All three packages are functionally equivalent, with the exception that the Rev. Dx does not have a reset pin.

2.1.2 BASIC Stamp 2

The BASIC Stamp 2 is available in two physical versions. The first is the BS2-IC, which is a 24-pin dual inline package (DIP) with surface-mounted

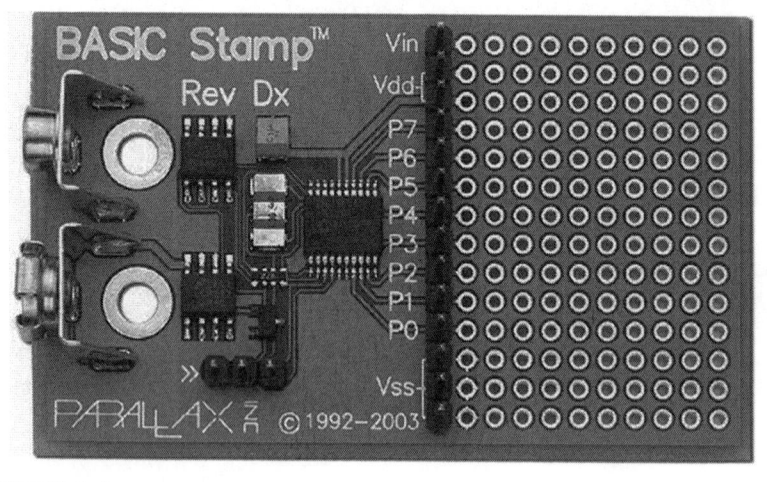

FIGURE 2.1
Commercially available BASIC Stamp 1 Rev. Dx (Copyright 2006 Parallax Inc.)

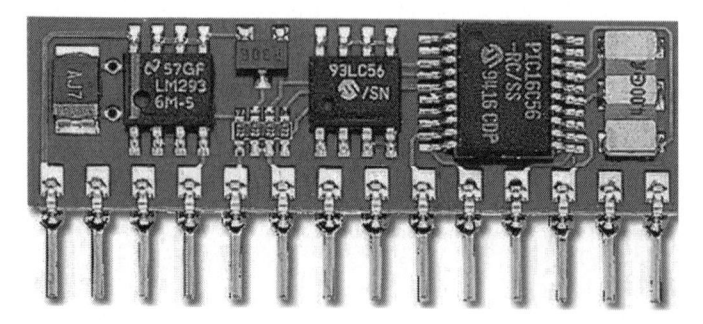

FIGURE 2.2
Commercially available BS1-IC (Copyright 2006 Parallax Inc.)

FIGURE 2.3
Commercially available BS2-IC (Copyright 2006 Parallax Inc.)

components, as shown in Figure 2.3. The second is the OEMBS2, which features an easier-to-trace layout meant to aid customers who wish to integrate the BASIC Stamp 2 circuit directly into their design (as a lower-cost solution). The OEMBS2 is available in either an assembled form or a kit form. BS2-IC and OEMBS2 are functionally equivalent.

2.1.3 BASIC Stamp 2sx

The BASIC Stamp 2sx is available in two physical versions. The first is the BS2sx-IC, which is a 24-pin (DIP) with surface-mounted components, as shown in Figure 2.4. The second is the OEMBS2sx, which, like the other OEM versions, features an easier-to-trace layout for customers who wish to integrate the BASIC Stamp 2 circuit directly into their design (as a lower-cost solution). The OEMBS2sx is available only in assembled form. BS2sx-IC and OEMBS2sx are functionally equivalent.

2.1.4 BASIC Stamp 2p

The BASIC Stamp 2p is available in two package types. The first is the BS2p24-IC, which is a 24-pin dual DIP with surface-mounted components, as shown in Figure 2.5. The second is the BS2p40-IC, which is a 40-pin DIP package, as shown in Figure 2.6. Both packages are functionally equivalent except that the BS2p40 has 32 I/O pins instead of 16 pins.

70   Practical and Experimental Robotics

FIGURE 2.4
   Commercially available BS2sx-IC (Copyright 2006 Parallax Inc.)

FIGURE 2.5
   Commercially available BS2p24-IC (Copyright 2006 Parallax Inc.)

BASIC Stamp Microcontrollers 71

FIGURE 2.6
Commercially available BS2p40-IC, (Copyright 2006 Parallax Inc.)

FIGURE 2.7
Commercially available BS2e-IC, (Copyright 2006 Parallax Inc.)

2.1.5 BASIC Stamp 2e

The BASIC Stamp 2e is available in two physical versions. The first is the BS2e-IC, which is a 24-pin DIP with surface mounted components, and is shown in Figure 2.7. The second is the OEMBS2e, which features an easier-to-trace layout geared toward customers who want to integrate the BASIC Stamp 2 circuit directly into their design (as a lower-cost solution). The OEMBS2e is available only in assembled form. BS2e-IC and OEMBS2e are functionally equivalent.

The pin configuration of BASIC Stamp 2e is shown in Figure 2.8.

Table 2.1 lists the BS2e-IC pins and their functionalities.

Figure 2.9 shows the pin descriptions of DB-25 and DB-9 serial ports.

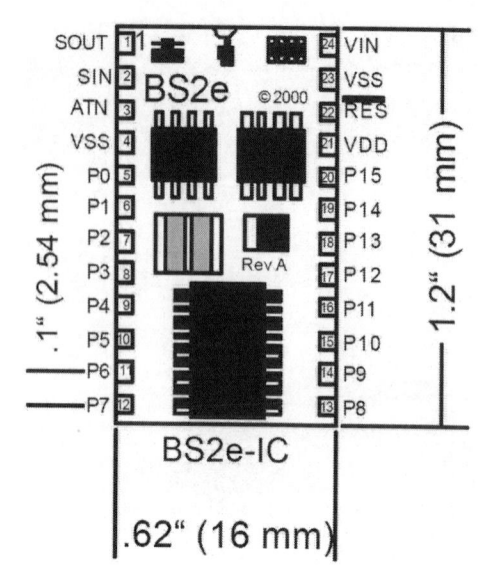

FIGURE 2.8
   Pin configuration of BS2e-IC, (Copyright 2006 Parallax Inc.)

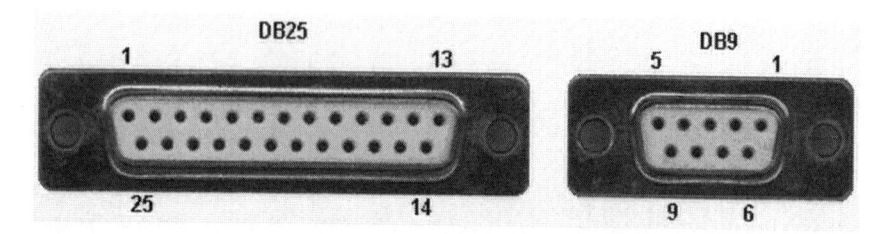

FIGURE 2.9
   Pin configuration for serial port DB-25 (25-pin connector) and DB-9 (9-pin connector

TABLE 2.1
Pin description of BS2e-IC

Pin	Name	Description
1	SOUT	Serial out: connects to the PC serial port Receive (RX) pin (DB9 pin 2 / DB25 pin 3) for programming.
2	SIN	Serial in: connects to the PC serial port Transmit (TX) pin (DB9 pin 3 / DB25 pin 2) for programming.
3	ATN	Attention: connects to the PC serial port Data Terminal Ready (DTR) pin (DB9 pin 4 / DB25 pin 20) for programming.
4	VSS	System ground (same as pin 23): connects to the PC serial port Ground (GND) pin (DB9 pin 5 / DB25 pin 7) for programming.
5-20	P0-P15	General-purpose I/O pins: each can source and sink 30 milli-Ampere (mA). However, the total of all pins should not exceed 75 mA (source or sink) if using the internal 5-volt regulator. The total per 8-pin group (P0-P7 or P8-15) should not exceed 100 mA (source or sink) if using an external 5-volt regulator.
21	VDD	5-volt DC input/output: if an unregulated voltage is applied to the input voltage (VIN) pin, then this pin will output 5 volts. If no voltage is applied to the VIN pin, then a regulated voltage between 4.5 volts and 5.5 volts should be applied to this pin.
22	RES	Reset input/output: goes low when power supply is less than approximately 4.2 volts, causing the BASIC Stamp to reset. Can be driven low to force a reset. This pin is pulled high internally and may be left disconnected if not needed. Do not drive high.
23	VSS	System ground (same as pin 4): connects to power supply's ground (GND) terminal.
24	VIN	Unregulated power in: accepts 5.5-12 volts of direct current, or VDC (7.5 recommended), which is then internally regulated to 5 volts. May be left unconnected if 5 volts are applied to the VDD (+5 V) pin.

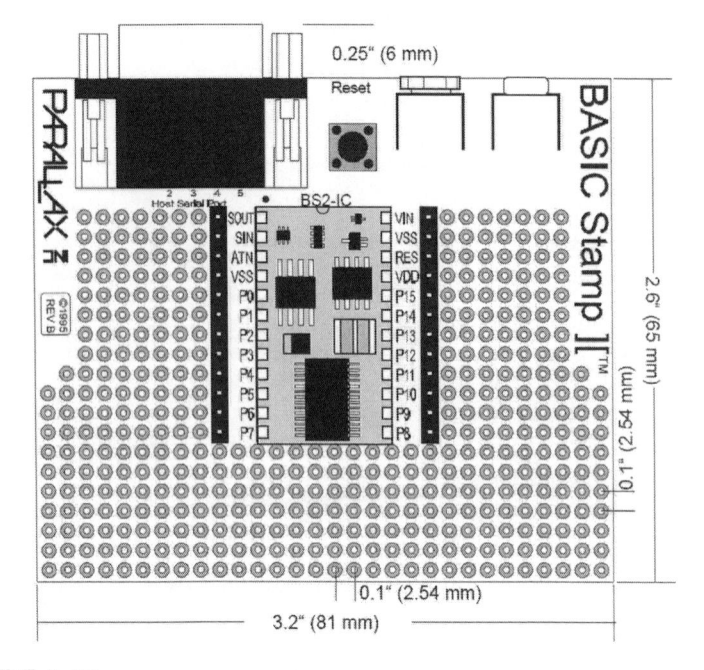

FIGURE 2.10
BASIC Stamp 2 Carrier board (Rev. B) shown with BS2-IC inserted
(Copyright 2006 Parallax Inc.)

2.2 Development Boards for BS2e

Several development boards are used for BS2e. We will discuss four major development boards: BASIC Stamp 2 Carrier Board (Rev. B), BASIC Stamp Super Carrier (Rev. A), Board of Education (Rev. B), and BASIC Stamp Activity Board (Rev. C).

2.2.1 BASIC Stamp 2 Carrier Board (Rev. B)

The BASIC Stamp 2 Carrier Board (also called the BS2 Carrier Board) is designed to accommodate the BS2-IC, BS2e-IC, and BS2sx-IC modules. The BASIC Stamp 2 Carrier Board, shown in Figure 2.10, provides ample prototyping space for simple or moderate circuits. This space can be used to add (solder) any circuitry, depending on the application.

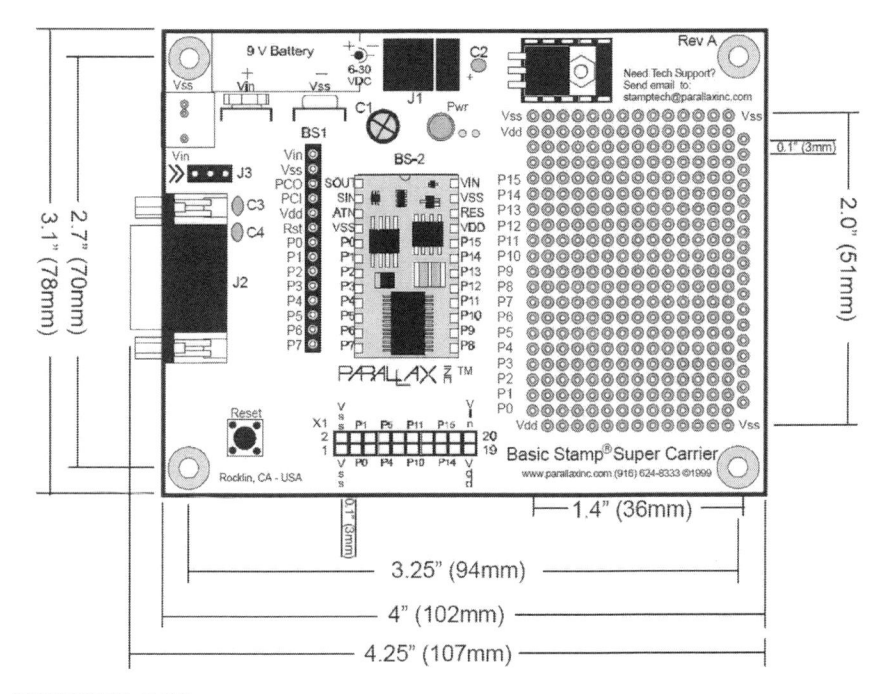

BASIC Stamp Microcontrollers 75

FIGURE 2.11
BASIC Stamp Super Carrier (Rev. A) (Copyright 2006 Parallax Inc.)

2.2.2 BASIC Stamp Super Carrier (Rev. A)

The BASIC Stamp Super Carrier board, shown in Figure 2.11, is designed to accommodate the BS1-IC, BS2-IC, BS2e-IC, and BS2sx-IC modules. This board provides ample prototyping space for simple or moderate circuits. Care should be taken not to power up the board with a BS1-IC and a BS2-IC, BS2e-IC, or BS2sx-IC inserted at the same time.

2.2.3 Board of Education (Rev. B)

The Board of Education is designed to accommodate the BS2-IC, BS2e-IC, and BS2sx-IC modules. This board, shown in Figure 2.12, provides a small breadboard for quickly prototyping simple or moderate circuits.

2.2.4 BASIC Stamp Activity Board (Rev. C)

This BASIC Stamp Activity Board (BSAC), shown in Figure 2.13, is designed to accommodate the BS1-IC, BS2-IC, BS2e-IC, BS2sx-IC, and BS2p24-IC modules. The BASIC Stamp Activity Board is excellent for projects requiring buttons, LEDs, a speaker, etc. All the components are prewired, and labels

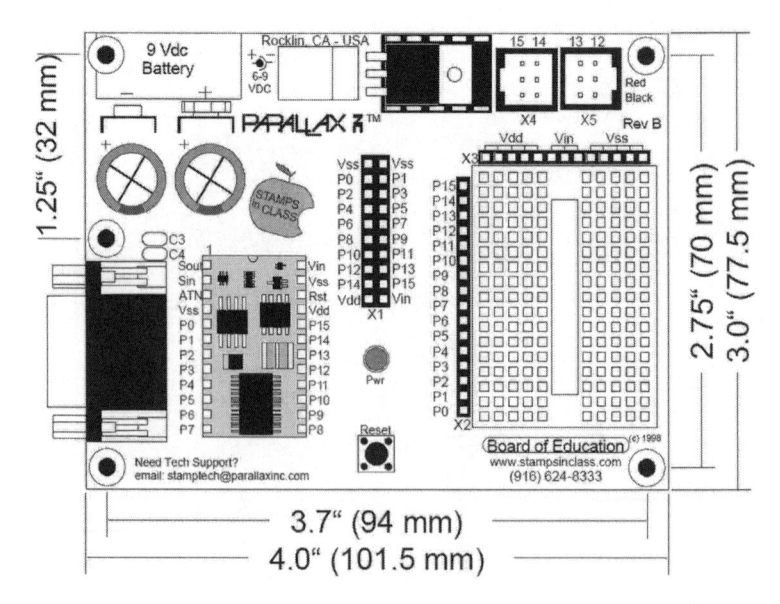

FIGURE 2.12
Board of education (Rev. B) shown with BS2-IC inserted (Copyright 2006 Parallax Inc.)

next to them indicate the I/O pin they are connected to.

2.3 BASIC Stamp Editor

In this section, we will explore the BASIC Stamp Editor. We'll cover connecting BS2e to a PC, installing the editor, and software interfaces for WINDOWS and DOS.

2.3.1 Connecting BS2e to the PC

As shown in Figure 2.14, the female side of a 9-pin serial cable is connected to one of the COM ports of a PC, and the male side is connected to the DB-9 connector on the carrier board of the BS2e.

2.3.2 Installing the BASIC Stamp Editor

The BASIC Stamp Editor software is available for Windows, Macintosh, and DOS operating systems. Parallax does not support the BASIC Stamp Editor

BASIC Stamp Microcontrollers

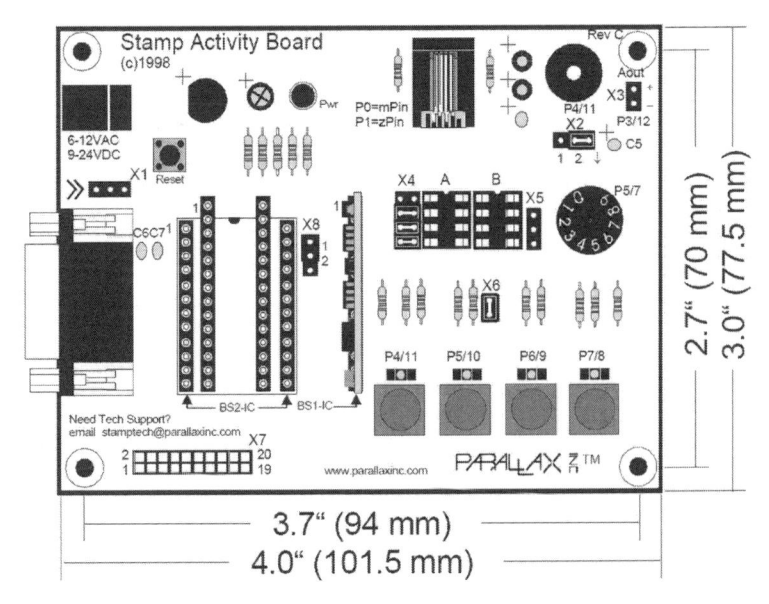

FIGURE 2.13
BASIC Stamp Activity Board (Rev.C) (Copyright 2006 Parallax Inc.)

FIGURE 2.14
Male side of the DB-9 connector connected to the carrier board of the BS2e

software for the Macintosh operating system, but there is a PBASIC tokenizer available. Application software for the Mac OS was written by a customer of Parallax, Inc. Mac users can download this software from the Parallax Web site. The following system requirements are a minimum for using the BASIC Stamp Editor.

System Requirements for Windows and DOS Operating System

- 80486 (80286 for DOS) (or higher) IBM or compatible PC

- Windows 95/98/NT/2000/XP operating system (DOS 5.0 or higher for DOS versions)

- 32 Mb of RAM (1 Mb for DOS)

- 1 MByte of available hard drive space

- CD-ROM drive

- One available serial port

System Requirements for Macintosh Operating System

- G3 or higher PC

- Mac OS X 10.2.8 or newer

- 64 Mb of RAM (1 Mb for DOS)

- 1 MByte of available hard drive space

- CD-ROM drive

- One available serial port

To install the BASIC Stamp Editor on Windows or DOS:

- Insert the Parallax CD into the CD-ROM drive. The CD should autostart. If using DOS, explore it with the CD (change directory) and DIR (directory list) commands.

- Select the Software BASIC Stamp section.

- Select the DOS or Windows version you wish to use, and click the Install button. If exploring the CD through DOS, use the COPY command to copy it to a desired directory on the hard drive.

- Close the CD and run the BASIC Stamp Editor program from the directory it was copied to.

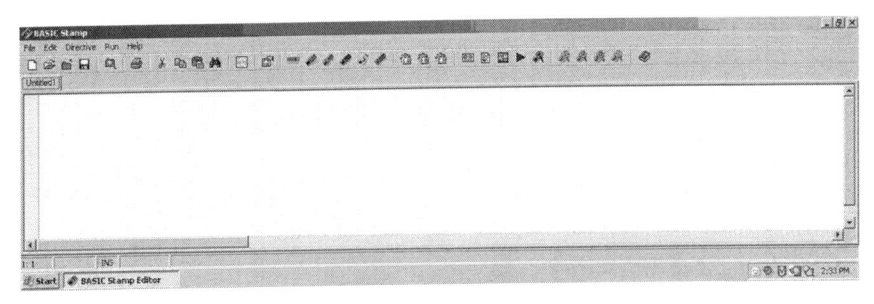

BASIC Stamp Microcontrollers 79

FIGURE 2.15
          The BASIC Stamp Editor

FIGURE 2.16
          A zoomed-in view of the BASIC Stamp Editor toolbar

2.3.3 Software Interface for Windows

The editor consists of one main editor window, shown in Figure 2.15 and 2.16, which you can use to view and modify up to 16 source-code files at once.

A user can switch between source-code files by simply pointing and clicking on a file's tab. After entering the desired source code in the editor window, select the Run menu and click the Run button or press Ctrl+R to tokenize and download the code to the BASIC Stamp. However, check the syntax of the code before hitting Run-select the Check Syntax option in the Run menu or simply press Ctrl+T. If the program is free of syntactic errors and all the connections to the BS2e are proper, a tab at the bottom right of the editor goes green and displays the message "Tokenize successful," and the code gets downloaded into the BS2e.

The Windows editor supports more than one model of the BASIC Stamp; hence it is necessary to tell the editor which model is being programmed. The editor uses three methods to determine the model of the BASIC Stamp being programmed:

- The STAMP directive. For a BS2e, the directive, '$STAMP BS2E, should be entered as the first line of the code.

- The extension on the file name of the source code. For a BS2e it is .bse.

- The default Stamp mode (as set by preferences).

Whenever a file is loaded, tokenized, downloaded, or viewed in the memory map, shown in Figure 2.17, the BASIC Stamp looks for the STAMP directive first. The STAMP memory map has three input-output registers (red) and 12

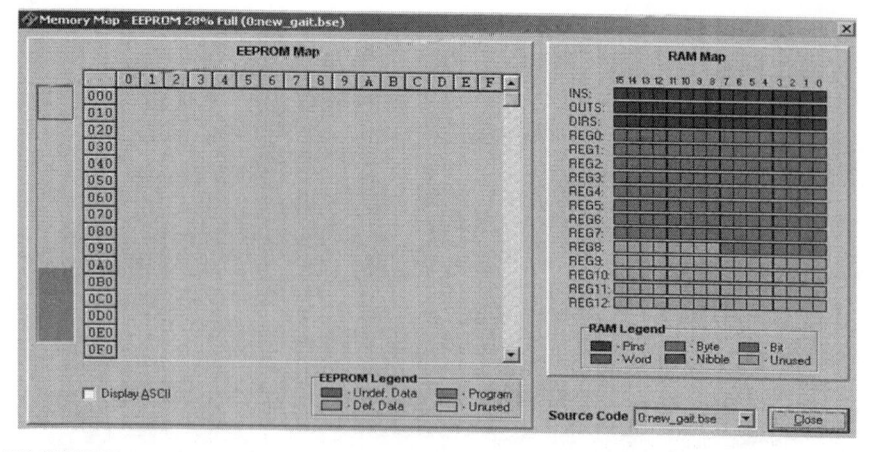

FIGURE 2.17

Memory map

regular registers (green). If it cannot find the STAMP directive in the source code, it looks at the extension on the file name (for a .bs2, .bse, .bsx, or .bsp). If it does not understand the extension, then it uses the default Stamp mode, as defined by preferences.

Each editor page can be a separate project or part of a single project. A project is a set of up to eight files that should all be downloaded to the BASIC Stamp for a single application. Each of the files within the project is downloaded into a separate "program slot." A program slot is a specific region in the memory for the specified components (file) of the project. For BASIC Stamp projects (consisting of multiple programs), the STAMP directive has an option to specify additional file names, as demonstrated in the following syntax:

' $STAMP BS2e, file2, file3, , file8

This form of the STAMP directive is used if a project consisting of multiple files is desired. This directive must be entered into the first program (it gets downloaded into program slot 0), and not into any of the other files in the project. The file2, file3, etc. items should be the actual names (and optionally the path) of the other files in the project. file2 refers to the program that should be downloaded into program slot 1, file3 is the program that should be downloaded into program slot 2, etc. If no path is given, the path of program 0 (the program in which the STAMP directive is entered) is used.

The editor has the ability to treat projects as one logical unit, and can download each of the associated source-code files to the BS2e, BS2sx, or BS2p at once. The BASIC Stamp Windows Editor also features a memory map that displays the layout of the current PBASIC program: DATA usage and RAM register usage. Pressing Ctrl+M can activate this window. When the memory map is activated, the editor will check the program for syntax errors and, if the

TABLE 2.2
Keyboard shortcuts for File functions

Shortcut	Description
Ctrl+O	Open a source-code file into the Editor window.
Ctrl+S	Save the current source-code file to disk.
Ctrl+P	Print the current source code.

TABLE 2.3
Keyboard shortcuts for Edit functions

Shortcut	Description
Ctrl+Z	Undo the last action.
Ctrl+X	Cut selected text to the clipboard.
Ctrl+C	Copy selected text to the clipboard.
Ctrl+V	Paste text from the clipboard to the selected area.
Ctrl+A	Select all text in the current source code.
Ctrl+F	Find or replace text.
F3	Find text again.
F5	Open the Preferences window.

program's syntax is OK, a color-coded map of the read-only memory (ROM) and electrically erasable programmable read-only memory (EEPROM) will be displayed.

Note: The map indicates only how the program will be downloaded to the BASIC Stamp; it does not "read" the BASIC Stamp's memory. Also, fixed variables and any aliases do not show up on the memory map as memory used. The fixed variables are variables specific to the STAMP controller, such as port names and register names. The editor ignores fixed variables when it arranges automatically allocated variables in memory.

Tables 2.2, 2.3, and 2.4 list the keyboard shortcuts used in the BASIC Stamp Editor for File functions, Edit functions, and Coding functions, respectively.

2.3.4 Software Interface for DOS

The DOS versions support only one BASIC Stamp module; a separate DOS editor is available for each model of the BASIC Stamp. For BS2e the version of the BASIC Stamp editor for DOS is Stamp2e.exe. The BASIC Stamp DOS Editor can load and edit only one source-code file at a time. Source code can be loaded into the editor by pressing Alt+L and selecting a file from the menu.

Note: The Browse menu shows only files in the current directory, the directory that the BASIC Stamp DOS Editor is run from.

TABLE 2.4
Keyboard shortcuts for Coding functions

Shortcut	Description
F6 or Ctrl+I	Identify BASIC Stamp firmware.
F7 or Ctrl+T	Perform a syntax check on the code and display any error messages.
F8 or Ctrl+M	Open the Memory Map window.
F9 or Ctrl+R	Tokenize code, download to BASIC Stamp, and open the Debug window if needed.
F11 or Ctrl+D	Open a new Debug window.
F12	Switch to the next window (Editor, Debug #1, Debug #2, Debug #3, or Debug #4)
Ctrl+1 ... Ctrl+9	Switch to Debug Terminal #1 ... Debug Terminal #9 if that terminal window is open.
Ctrl+	Switch to the Editor window.
ESC	Close the current window.

The BS2e, BS2sx, and BS2p models support up to eight programs to be downloaded into separate program slots. From here on, any application for these models of the BASIC Stamp will be called a project. Each of the files within the project must be downloaded into a separate program slot. For BASIC Stamp projects (consisting of multiple programs), the BASIC Stamp DOS Editor must be used to individually load and download each of the files into the appropriate slot. The DOS editor can load only one source-code file at a time. The sequence of keystrokes to load and download two programs into two separate program slots would be as follows:

1. Alt+L loads a program into the editor.

2. Alt+0 sets the editor to program ID 0.

3. Alt+R downloads this program into program slot 0 of the BASIC Stamp's EEPROM. The shortcut Alt+R downloads only one program at a time.

4. Alt+L loads another program into the editor.

5. Alt+1 sets the editor to program ID 1.

6. Alt+R downloads this program into program slot 1 of the BASIC Stamp's EEPROM.

Note: Each program must be loaded separately.

The BASIC Stamp DOS Editors for the BS2, BS2e, BS2sx, and BS2p also feature a memory map that displays the layout of the current PBASIC program, data usage, and RAM register usage. Typing Alt+M activates this

TABLE 2.5
Keyboard shortcuts for File functions

Shortcut	Description
Alt+L	Open a source-code file into the Editor window.
Alt+S	Save the current source-code file to disk.
Alt+Q	Close the editor.

TABLE 2.6
Keyboard shortcuts for Edit functions

Shortcut	Description
Alt+X	Cut selected text to the clipboard.
Alt+C	Copy selected text to the clipboard.
Alt+V	Paste text from the clipboard to the selected area.
Alt+F	Find or replace text.
Alt+N	Find text again.

window. When the memory map is activated, the editor will check the program for syntax errors and, if the program's syntax is OK, a color-coded map of the RAM will appear.

Note: As with the Windows Stamp Editor media map, the map indicates only how the program will be downloaded to the BASIC Stamp; it does not "read" the BASIC Stamp's memory. Also, fixed variables and any aliases do not show up on the memory map as memory used.

Tables 2.5, 2.6, and 2.7 list the keyboard shortcuts in the BASIC Stamp DOS Editor for DOS for File functions, Edit functions, and Coding functions, respectively.

TABLE 2.7
Keyboard shortcuts for Coding functions

Shortcut	Description
Alt+0	through Alt+7 Set program slot to download to.
Alt+I	Identify the BASIC Stamp firmware.
Alt+M	Open the Memory Map window.
Alt+R	Tokenize code, download it to the BASIC Stamp, and open the Debug window if necessary.
Alt+P	Open the potentiometer calibration window.

2.4 PBASIC Programming Fundamentals

In this section, we will discuss PBASIC programming fundamentals such as how to declare variables, define arrays, create aliases and modifiers, declare constants and expressions. We'll also cover BASIC Stamp math and some important PBASIC commands to interface with the BS2e development board.

2.4.1 Declaring Variables

Before a variable can be used in a PBASIC program, it must be declared. The following is the syntax for declaring variables.

```
Name VAR Size
```

In the declaration, Name is the name of the variable, VAR is the BASIC Stamp directive to set up the variable, and Size is the number of bits of storage for the variable. The rules for building variable names are as follows:

1. Variable names must begin with a letter and can be a mixture of letters, numbers, and underscores.

2. Variable names must not be the same as PBASIC keywords.

3. Variable names can be up to 32 characters in length.

You have the following four choices for the size argument:

1. BIT: Value can be 0 or 1.

2. NIB (i.e., Nibble): Value can be any number between 0 and 15.

3. BYTE: Value can be any number between 0 and 255.

4. WORD: Value can be any number between 0 and 65535.

Here are two examples:

```
myvar    VAR  BIT
new_name VAR  NIB
```

Here myvar is a variable of size 1 bit, and can have a value of either 0 or 1. Similarly, new_name is a variable of size 1 nibble and can have any value between 0 and 15.

Note: If a value exceeds the size of its variable, the excess bits will be lost. In the previous example, suppose new_name were assigned a value of 260 (binary = 100000100). new_name will hold only the lowest 8 bits of 260 (i.e., 0100), which means that actually new_name = 4.

2.4.2 Defining Arrays

An array is a group of variables of the same size and sharing a single name, but broken up into numbered cells called elements. The syntax is as follows:

```
Name    VAR    Size(n)
```

Name is the name of the variable, and Size is the type of storage for the variable. The letter n is the number of elements of the array. In this example,

```
my_arr VAR BYTE(5)
```

The variable my_arr is an array of five elements and each element has a size of 1 byte.

The elements of an array can be accessed by index numbers. Numbering starts from 0 and ends at n-1. In the preceding example, my_arr(0) would give the value of the first element of the array, and my_arr(4) would give the value of the fifth (and last, in this instance) element of the array. The first element is referred to as the 0th element. Another way to access the 0th element of the array is to simply give the name of the array without the index number. This means that my_arr will give the value of the 0th element of the array.

Note: If the index value exceeds the maximum value for the size of the array, PBASIC will not respond with an error message. Instead, it will access the next RAM location past the end of the array, which can cause all sorts of bugs in the program. There is no error message because the BASIC Stamp does not always have a display device connected to it for displaying error messages. This means that for the 5 byte array my_arr described earlier, the allowable index numbers are 0 through 4.

2.4.3 Alias

An alias is an alternative name for an existing variable. The following is an example:

```
Car      VAR    BYTE
Eclipse  VAR    Car
```

In this example, Eclipse is an alias to the variable Car. Anything stored in Car shows up in Eclipse and vice versa. Both names refer to the same physical piece of RAM. This kind of alias can be useful when a temporary variable is to be used in different places in the program, and when its function has to be reflected in each place.

TABLE 2.8
List of modifiers

Symbol	Definition
LOWBYTE	Low byte of a word
HIGHBYTE	High byte of a word
BYE0	Byte 0 (low byte) of a word
BYTE1	Byte 1 (high byte) of a word
LOWNIB	Low nibble of a word or byte
HIGHNIB	High nibble of a word or byte
NIB0	Nib 0 of a word or byte
NIB1	Nib 1 of a word or byte
NIB2	Nib 2 of a word
NIB3	Nib 3 of a word
LOWBIT	Low bit of a word, byte, or nibble
HIGHBIT	High bit of a word, byte, or nibble
BIT0	Bit 0 of a word, byte, or nibble
BIT1	Bit 1 of a word, byte, or nibble
BIT2	Bit 2 of a word, byte, or nibble
BIT3	Bit 3 of a word, byte, or nibble
BIT4..BIT7	Bits 4 though 7 of a word or byte
BIT8..BIT15	Bits 8 through 15 of a word

2.4.4 Modifiers

An alias can also allow access into a portion of another variable using modifiers. Table 2.8 lists all the available modifiers.

In this example,

```
Dog     VAR     BYTE
Head    VAR     Dog.HIGHNIB
Tail    VAR     Dog.LOWNIB
```

If the variable Dog is given the value 10110000, then Head will contain 1011 and Tail will contain 0000.

Here is another example:

```
Room    VAR     WORD
Door    VAR     Room.HIGHBYTE.LOWNIB.BIT3
```

Here if the variable Room is given the value 1001110101100001, then Door will contain the third bit of the lower nibble of the higher byte of the Room value. This means that the Door value is 1. Lower byte or nibble means the leftmost byte or nibble of the value. So the higher byte of the value of Room is 10011101. Similarly, the lower nibble of this byte is 1101. Finally, the third bit of this nibble is 1 since the 0th bit is the rightmost bit of the binary value.

The following example shows how a variable's nibbles can be accessed and manipulated.

```
my_arr      VAR     BYTE(10)
my_first    VAR     my_arr.LOWNIB(0)
my_second   VAR     my_arr.LOWNIB(1)
```

Here my_arr is an array of 10 bytes. Suppose the 0th element of my_arr is given a value 10001100 (i.e., my_arr = my_arr(0) = 10001100). my_first and my_second are aliases used to access the lower and higher nibbles of my_arr(0). Modifier LOWNIB(0) is used to access the lower nibble of my_arr(0) so that my_first = 1100 and modifier LOWNIB(1) is used to access the higher nibble of my_arr(0) so that my_second = 1000. Note that instead of HIGHNIB, LOWNIB(1) is used to access the higher nibble. This is because though my_arr() is a 10-byte array and has 10 byte-sized cells numbered 0 through 9, when it is addressed as a nibble array using my_arr.LOWNIB(), it has 20 nibble-sized cells numbered 0 through 19. Hence, my_arr.LOWNIB(1) will correspond to the higher nibble of the 0th element of my_arr.

Now consider this example:

```
my_arr      VAR     BYTE(10)
my_first    VAR     my_arr.HIGHNIB(0)
my_second   VAR     my_arr.HIGHNIB(1)
```

Here again my_arr is an array of 10 bytes. Suppose the 0th element of my_arr is given a value 10101100 (i.e., my_arr = my_arr(0) = 10001100) and the first element is given by my_arr(1) = 11110001. my_first and my_second are aliases used to access the lower and higher nibbles of my_arr(0). Modifier HIGHNIB() works a bit differently from modifier LOWNIB(). my_arr(0).HIGHNIB(0) will access the higher nibble of the 0th element of my_arr so that my_first = 1010 and modifier HIGHNIB(1) will access the lower nibble of my_arr(1) so that my_second = 0001.

2.4.5 Constants and Expressions

The following is the syntax for declaring a constant.

```
Name    CON     ConstantValue
```

Name is the name of the constant, CON is the BASIC Stamp directive to set up a constant, and ConstantValue is the value given to the constant.

The following are examples of constant declarations.

```
my_const    CON     9
new_val     CON     my_const*3-5
```

Here my_const and new_val are constants. my_const has been given a value 9. The value of new_val = 9 * 3 - 5 = 22. This means that expressions can also be used while declaring constants.

The BASIC Stamp editor software solves the expressions used to define constants from left to right. The operators that are allowed in constant expressions are shown in Table 2.9.

TABLE 2.9
The operators that are allowed in constant expressions

Operator	Description
+	Add
-	Subtract
*	Multiply
/	Divide
<<	Shift left
>>	Shift right
&	Logical AND
|	Logical OR
^	Logical XOR

2.4.6 BASIC Stamp Math

This section explores the fundamentals of BASIC Stamp math. Numbering systems, arithmetic operations, and arithmetic operators are discussed and explained with examples.

2.4.6.1 Numbering Systems

Decimal, hexadecimal, and binary are the three numbering systems that can be used with the BASIC Stamp. Since some of the symbols used in decimal, hex, and binary numbers overlap (1 and 0 are used by all; 0 through 9 apply to both decimal and hex) the BASIC Stamp Editor software needs prefixes to tell the numbering systems apart, as shown below:

```
99      Decimal (no prefix)
$1A6    Hex
1101    Binary
```

2.4.6.2 Arithmetic Operations and Operators

The BASIC Stamp solves math problems in the order they are written; from left to right. The result of each operation is fed into the next operation. So to compute 12+3*2/4, the BASIC Stamp goes through the following sequence:

12 + 3 = 15
15 * 2 = 30
30 / 4 = 7

The result is given as 7 in the last line since the Stamp does not do decimal math. The BASIC Stamp performs all math operations by the rules of positive integer math. That is, it handles only whole numbers, and drops any fractional portions from the results of computations. The BASIC Stamp handles negative numbers using two's complement rules. Two's complement

TABLE 2.10
Binary operators available in BS2e

Operator	Description
+	Add.
-	Subtract.
*	Multiply.
**	The Multiply High operator.
*/	The Multiply Middle operator.
/	Divide.
//	Modulus operator; returns the remainder left after dividing one value by another.
MIN	The Minimum operator; limits a value to a specified 16-bit positive minimum. The syntax of MIN is Value MIN Limit.
MAX	The Maximum operator; limits a value to a specified 16-bit positive maximum. The syntax of MAX is Value MAX Limit.
DIG	The Digit operator.
>>	The Shift Right operator.
<<	The Shift Left operator.
REV	The Reverse operator.
&	Bitwise AND.
|	Bitwise OR.
^	Bitwise XOR.

of a binary number is obtained by adding logic 1 to the logic NOT of the number. For example, if the binary number is 0101, the two's complement of the number is 1011 (1010 + 0001).

The BS2e allows parentheses to change the order of evaluation. Enclosing a math operation in parentheses gives it priority over other operations. So to compute 12+((3*2)/4) the BSe2 goes through the following sequence.

3 * 2 = 6
6 / 4 = 1
12 + 1 = 13

BASIC Stamp arithmetic and logical operators are of two kinds: Binary and Unary. Operators that take one argument are called unary operators and those that take two are called binary operators. Table 2.10 lists the binary operators available in BS2e.

The Multiply High operator multiplies variables and/or constants, returning the high 16 bits of the result. When two 16-bit values are multiplied, the result can be as large as 32 bits. Since the largest variable supported by PBASIC is 16 bits, the highest 16 bits of a 32-bit multiplication result are usually lost. When the numbers are small, the * may give correct result since the multiplication result may be stored in 16 bits. The ** instruction gives these upper 16 bits of the multiplication result. Thus, the data loss is minimized because the most significant 16 bits of the result are stored.

The Multiply Middle operator multiplies variables and/or constants, returning the middle 16 bits of the 32-bit result. This has the effect of multiplying a value by a whole number and a fraction. The whole number is the upper byte of the multiplier (0 to 255 whole units), and the fraction is the lower byte of the multiplier (0 to 255 units of 1/256 each).If a value is to be multiplied by 1.5, the whole number, and therefore the upper byte of the multiplier, would be 1; and the lower byte (fractional part) would be 128, since 128 / 256 = 0.5.

The Digit operator returns the specified decimal digit of a 16-bit positive value. Digits are numbered from 0 (the rightmost digit) to 4 (the leftmost digit of a 16-bit number; 0 to 65535). Therefore, DIG 2 of 5467 is 4.

The Shift Right operator shifts the bits of a variable to the right a specified number of places. Bits shifted off the right end of a number are lost; bits shifted into the left end of the number are 0s. Shifting the bits of a value right n number of times has the same effect as dividing that number by 2 to the nth power. For example, 100 >> 3 shift the bits of the number 100 right three places, making it equivalent to 100 / 2^3.

The Shift Left operator shifts the bits of a value to the left a specified number of places. Bits shifted off the left end of a number are lost; bits shifted into the right end of the number are 0s. Shifting the bits of a value left n number of times has the same effect as multiplying that number by 2 to the nth power. For example, 100 << 3 shift the bits of the number 100 left three places, making it equivalent to 100 * 2^3.

The Reverse operator returns a reversed (mirrored) copy of a specified number of bits of a value, starting with the rightmost bit. For instance, %10101101 REV 4 would return %1011, which is a mirror image of the first four bits of the value.

The unary operators available in BS2e are listed in Table 2.11.

2.4.7 Important PBASIC Commands Used while Interfacing the BS2e with the Lynxmotion 12 Servo Hexapod

This section presents some of the basic commands used while interfacing the BS2e development board with the Hexapod robot, discussed in Chapter 8. Debugging, programming, and serial communication commands are explored with example programs.

TABLE 2.11
Unary operators available in BS2e

Operator	Description
ABS	Converts a signed (two's complement) 16-bit number to its absolute value. The absolute value of a number is a positive number representing the difference between that number and 0. For instance, the absolute value of -99 is 99. The absolute value of 99 is also 99.
COS	Returns cosine in two's complement binary radians.
DCD	The Decoder operator; a 2n-power decoder of a four-bit value. DCD accepts a value from 0 to 15 and returns a 16-bit number with the bit, described by value, set to 1. For instance, DCD of 12 = 0001000000000000.
	Complements (inverts) the bits of a number. Each bit that contains a 1 is changed to 0, and each bit containing a 0 is changed to 1. This process is also known as a bitwise NOT and one's complement. An example: (11001100) = 00110011.
-	The Negative operator; negates a 16-bit number (converts to its two's complement).
NCD	The Encoder operator; is a "priority" encoder of a 16-bit value. NCD takes a 16-bit value, finds the highest bit containing a 1, and returns the bit position plus one (1 through 16). If no bit is set (the input value is 0) NCD returns 0. (Therefore, NCD of 1101 is 4.)
SIN	Returns cosine in two's complement binary radians.
SQR	Computes the integer square root of an unsigned 16-bit number. Most square roots have a fractional part that the BASIC Stamp discards when doing its integer-only math. For instance, SQR computes the square root of 99 as 9 and not 9.9498.

2.4.8 DEBUG

This command displays information on the PC screen within the BASIC Stamp editor program. The syntax is as follows:

```
DEBUG   OutputData   {,OutputData}}
```

Here, OutputData is a variable/constant/expression that specifies the information to output. Valid data can be ASCII characters (text strings and control characters), decimal numbers (0-65535), hexadecimal numbers (0000-FFFF), or binary numbers (up to 1111111111111111). This command can be used to display text or numbers in various formats on the PC screen.

An example is

```
DEBUG   "Hello World"
```

After this one-line program is downloaded, the BASIC Stamp Editor will open a debug terminal on your PC screen and wait for a response from the BASIC Stamp. A moment later, the phrase "Hello World" will appear. If the debug terminal is closed, the program keeps executing but the DEBUG data cannot be seen anymore. The sample line of code can also be written as DEBUG "Hello", "World". Multiple pieces of data can be sent with just one DEBUG command by separating them with commas.

Consider another example:

```
x       VAR     BYTE
x = 65
DEBUG x
```

This program will display A on the PC screen. This is because even though x is specified as 65, what gets displayed is the ASCII code of 65, which is the letter A. However, you'll achieve a different result by simply adding "DEC" to the final line:

```
x       VAR     BYTE
x = 65
DEBUG DEC    x
```

Now the PC screen will display the number 65. This is because the decimal formatter DEC has been used before x in the DEBUG command.

Expressions are allowed with the DEBUG command. In the following example, the addition of 65 and 100 (i.e., 165) is displayed on-screen:

```
x       VAR     BYTE
x = 65
DEBUG DEC    x+100
```

And here the hexadecimal value of 65 (i.e., 41) is displayed on-screen:

```
x       VAR     BYTE
x = 65
DEBUG HEX    x
```

2.4.9 FOR...NEXT

The syntax for this command is as follows:

```
FOR Counter = StartValue TO EndValue {STEP StepValue}...NEXT}
```

Create a repeating loop that executes the program lines between FOR and NEXT, incrementing or decrementing Counter according to StepValue until the value of the Counter variable passes the EndValue.

Counter is a variable (usually a byte or a word) used as a counter. StartValue is a variable/constant/expression that can have any value between 0 and 65535, and that specifies the initial value of the variable (Counter).

EndValue is a variable/constant/expression that can have any value between 0 and 65535 that specifies the end value of the variable (Counter). When the value of Counter is outside of the range StartValue to EndValue, the FOR...NEXT loop stops executing and the program goes on to the instruction after NEXT.

StepValue is an optional variable/constant/expression that can have any value between 0 and 65535, by which the Counter increases or decreases with each iteration through the FORNEXT loop. If StartValue is larger than EndValue, PBASIC understands StepValue to be negative even though no minus sign is used.

FOR...NEXT loops let the program execute a series of instructions for a specified number of repetitions (called iterations). By default, each time through the loop, the counter variable is incremented by 1 unless set by the StepValue argument. It will continue to loop until the result of the counter is outside of the range set by StartValue and EndValue.

Here is an example:

```
i       VAR     BYTE
FOR     i = 1 TO 10
DEBUG   "Hello World", CR
NEXT
```

This program will display "Hello World" on the PC screen 10 times. Each time the display will be on a new line due to the CR formatter used. By default the loop is incremented by 1 even though no Step value is specified. After the value of i exceeds 10, the control automatically exits the loop and the program ends.

The following example shows how to create a counter using a FOR loop:

```
x       VAR     byte
FOR     x = 1 TO 10 STEP 3
DEBUG DEC ? x
NEXT
```

The question mark in the DEBUG command causes "x = " to be displayed on the PC screen. In this example the counter x is incremented in steps of 3. The program will output the following:

x = 1
x = 4
x = 7
x = 10

2.4.10 END

This command ends the program, placing the BASIC Stamp into low-power mode indefinitely. The syntax is as follows:

```
END
```

This command is optional, because once the BASIC Stamp reaches the end of the program it enters low-power mode indefinitely. However, END can prevent the program from looping continuously, as illustrated in the GOSUB example.

2.4.11 RETURN

The syntax for the RETURN command is as follows:

```
RETURN
```

RETURN sends the program back to the address (instruction) immediately following the most recent GOSUB. If RETURN is executed without a prior GOSUB, the BASIC Stamp will return to the first executable line of the program.

2.4.12 PAUSE

The syntax for this command is as follows:

```
PAUSE   Period
```

Period is a variable/constant/expression that can be any number between 0 and 65535, which specifies the duration of the pause. The unit of time for Period is 1 millisecond. PAUSE delays the execution of the next program instruction for the specified number of milliseconds.

2.4.13 GOSUB

The syntax for this command is as follows:

```
GOSUB    Address
```

Address is the label that specifies where to go. When a PBASIC program reaches a GOSUB, the program executes the code beginning at the specified address label. GOSUB also stores the address of the instruction immediately following itself. When the program encounters a RETURN command, it interprets it to mean, "Go to the instruction that follows the most recent GOSUB." GOSUB is used mainly to execute the same piece of code from multiple locations. This technique is used to save program space. A maximum of 255 GOSUBs are allowed per program, four of which may be nested GOSUBs.

Here is an example of a GOSUB in action:

```
GOSUB    beauty
DEBUG    " is a joy forever"
END
beauty:
DEBUG    "A thing of beauty"
RETURN
```

In this example when the control reaches the command GOSUB, the control will branch to the section of code beginning with the label "beauty." The DEBUG command is then used in this subroutine to print "A thing of beauty" on the screen. On reaching the RETURN command, the control will go back to the line just after GOSUB (so the control is now at the first DEBUG command, which prints "is a joy forever" on the screen). The END command signifies that the program has reached its end.

If the END command is removed, once "is a joy forever" has been printed on-screen the control will fall into the "beauty" subroutine by itself. Hence, "A thing of beauty" gets printed once more; when the program reaches the RETURN command, the control goes back to the line below GOSUB, again prints "is a joy forever," and then (due to the absence of END) again falls into the "beauty" subroutine and the same procedure repeats infinitely. It is of vital importance to ensure that the program does not fall into a subroutine.

2.4.14 SEROUT

The function of this command is to transmit asynchronous serial data (e.g., RS-232 data). The syntax is as follows:

```
SEROUT Tpin {\Fpin},Baudmode,{Pace,}{Timeout,Tlabel,}[OutputData]
```

Tpin is a variable/constant/expression that can have any value between 0 and 16. It specifies the I/O pin through which the serial data will be transmitted. This pin will be set to output mode. If Tpin is set to 16, the BASIC Stamp uses the dedicated serial-output pin (SOUT, physical pin 1).

Fpin is an optional variable/constant/expression that can have any value between 0 and 15. It specifies the I/O pin to monitor for flow-control status. This pin will be set to input mode. If Fpin is used, the optional Timeout and Tlabel arguments in the SEROUT command have to be used.

Baudmode is a variable/constant/expression that can have any value between 0 and 65535. It specifies serial timing and configuration.

Pace is an optional variable/constant/expression that can have any value between 0 and 65535, and determines the length of the pause between transmitted bytes. Pace cannot be used simultaneously with Timeout. The unit of Pace is the millisecond (ms).

Timeout is an optional variable/constant/expression that can have any value between 0 and 65535. It tells SEROUT how long to wait for Fpin permission to send data. If permission does not arrive in time, the program will jump to the address specified by Tlabel. The unit of Timeout is the millisecond (ms).

Tlabel is an optional label that must be provided along with Timeout. Tlabel indicates where the program should go if permission to send data is not granted within the period specified by Timeout.

OutputData is a list of variables, constants, expressions, and formatters that tells SEROUT how to format outgoing data. SEROUT can transmit individual or repeating bytes, and convert values into decimal, hex, or binary text representations. These actions can be combined in any order in the OutputData list.

The SEROUT command is used to send asynchronous serial data. The term asynchronous means "no clock." More specifically, asynchronous serial communication means data is transmitted and received without the use of a separate "clock" wire. Data can be sent using only two wires: one for data and one for the ground. In contrast, synchronous serial communication uses at least three wires: one for the clock, one for data, and one for the ground. The PC's serial ports (also called COM ports or RS-232 ports) use asynchronous serial communication. RS-232 is the electrical specification for the signals that PC serial ports use. Unlike normal logic, where 5 volts is logic 1 and 0 volts is logic 0, RS-232 uses -12 volts for logic 1 and +12 volts for logic 0. This specification allows communication over longer wire lengths without amplification.

Asynchronous serial communication relies on precise timing. Both the sender and the receiver must be set for identical timing, usually expressed in bits per second (bps), called baud. SEROUT requires a value called Baudmode that tells it the important characteristics of the incoming serial data: the bit period, the number of data and parity bits, and the polarity. The Baudmode argument for SEROUT accepts a 16-bit value that determines its characteristics: 1-stop bit, 8-data bits/no-parity or 7-data bits/even parity. The steps for calculating the Baudmode are presented here:

TABLE 2.12
Common baud rates and baudmodes

Baud Rate	8-Bit No-Parity Inverted	8-Bit No-Parity True	7-Bit Even-Parity Inverted	7-Bit Even-Parity True
300	19697	3313	27889	11505
600	18030	1646	26222	9838
1200	17197	813	25389	9005
2400	16780	396	34972	8588
4800	16572	188	24764	8380
9600	16468	84	24660	8276

- Determine the bit period; i.e., calculate INT(1000000 / baud rate) - 20.

- Set the data bits and parity. For 8 bits and no parity, the value chosen is 0. For 7 bits and even parity the value chosen is 8192.

- Select the polarity. For True (i.e., noninverted) the value chosen is 0, and for inverted the value chosen is 16384.

- Select driven or open output. For driven output the value chosen is 0, and for open output the value chosen is 32768.

- Add the results of steps1 through 4 to find the Baudmode.

Note: INT means "convert to integer;" and drop the numbers to the right of the decimal point.

Consider a baud rate of 300. The Baudmode for 7-bit even parity with inverted polarity and driven output will be as calculated in the following steps.

1. INT(1000000/300) - 20 = INT(3333.333)-20 = 3333-20 = 3313. (INT gets rid of the digits after the decimal point).

2. For 7 bits with even parity, the value chosen = 8192.

3. For inverted polarity, the value chosen = 16384.

4. For driven output, the value chosen = 0.

Hence Baudmode = 3313 + 8192 + 16384 + 0 = 27889. Table 2.12 lists common baud rates and the corresponding Baudmodes.

When you are using existing software or hardware for communication, its speed(s) and mode(s) will determine the choice of baud rate and mode. In general, 7-bit/even-parity (7E) mode is used for text, and 8-bit/no-parity (8N) mode is used for byte-oriented data. The most common mode is 8-bit/no-parity mode, even when the data transmitted is just text. Most devices that use a 7-bit data mode do so to take advantage of the parity feature. Parity can detect some communication errors, but to use it you lose one data bit. This means that incoming data bytes transferred in 7E mode can represent only values from 0 to 127, rather than the 0 to 255 of 8N mode. Consider three examples.

```
SEROUT   1, 16780, [65]
```

Here SEROUT will transmit a byte equal to 65 through pin 1 using the Baudmode 8-bit, no parity with inverted polarity (i.e., 16780) at a baud rate of 2400. If the BASIC Stamp were connected to a PC set to the same baud rate, the character A, which is the ASCII code for 65, would appear on-screen.

In the following code, DEC is a decimal formatter that causes 65 to appear as 65 on the PC screen instead of appearing as A.

```
SEROUT   1, 16780, [DEC = 65]
```

The following code will display "Slowly" on one line and "but surely" on the next line since the formatter CR (carriage return) has been used.

```
SEROUT   1, 16468, 1000, ["Slowly",CR]
SEROUT   1, 16468, ["but surely"]
```

The Baudmode chosen is 8-bit, no parity with inverted polarity at a baud rate of 9600. Here, BS2e transmits the word "Slowly" with a one-second delay between each character since the value for the Pace argument is 1000 (ms).

Table 2.13 lists the available conversion formatters. All of the conversion formatters work similar to the decimal formatter.

TABLE 2.13
Available conversion formatters

Conversion Formatter	Number Type
DEC1..5	Decimal; can be fixed to 1-5 digits
SDEC1..5	Signed decimal; can be fixed to 1-5 digits
HEX1..4	Hexadecimal; can be fixed to 1-4 digits
SHEX1..4	Signed hexadecimal; can be fixed to 1-4 digits
IHEX1..4	Indicated hexadecimal; can be fixed to 1-4 digits ($ prefix)
ISHEX1..4	Signed, indicated hexadecimal; can be fixed to 1-4 digits ($ prefix)
BIN1..16	Binary; can be fixed to 1-16 digits
SBIN1..16	Signed binary; can be fixed to 1-16 digits
IBIN1..16	Indicated binary; can be fixed to 1-16 digits (% prefix)
ISBIN1..16	Signed, indicated binary; can be fixed to 1-16 digits (% prefix)

Note: Fixed-digit formatters like DEC4 will pad the number with leading 0s if necessary; for example, DEC4 65 sends 0065. If a number is larger than the specified number of digits, the leading digits will be dropped; so DEC4 56422 sends 6422.

Note: Signed modifiers work under two's complement rules. This means the actual value of a negative number is found by calculating its two's complement. You can find two's complement of a binary number by adding 1 to the inverted version of the number. For example, two's complement of a negative number 1101 is 0010 + 1 = 0011. Thus, the number is -3.

Table 2.14 lists special formatters.

TABLE 2.14
Special formatters for BS2e

Special Formatter	Action
?	Displays "symbol = x" and a carriage return, where x is a number. Default format is decimal, but may be combined with conversion formatters (for instance, enter BIN ? x to display "x = binary_number").
ASC ?	Displays "symbol = x" and a carriage return, where x is an ASCII character.
STR ByteArray {\L}	Sends a character string from an array. The optional \L argument can be used to limit the output to L characters. Otherwise, characters will be sent up to the first byte equal to 0 or until the end of RAM space is reached.
REP Byte\L	Sends a string consisting of Byte repeated L times (for example, REP "X"\10 sends "XXXXXXXXXX").

Parity is a simple error-checking feature. When the SEROUT command's Baudmode is set for even parity, it counts the number of 1s in the outgoing byte and uses the parity bit to make that number even. For instance, if the 7-bit value = 0011010, the parity bit is set to 1 to make an even number of 1s (four).

The Fpin argument is used to monitor flow control. This means that it is used while sending and receiving data. The following example shows how Fpin and Tpin are used to monitor the flow control.

```
serdata   VAR   BYTE
SEROUT    1\0,  84, [serdata]
```

TABLE 2.15
The relationship between serial polarity and Fpin states

Type Ready to Receive	Not Ready to Receive
Inverted Fpin is high (1)	Fpin is low (0)
Noninverted Fpin is low (0)	Fpin is high (1)

Here serdata is a variable of type BYTE. The Tpin argument is 1. This means that I/O pin 1 is set as output. The Fpin argument is 0. This means that I/O pin 0 is set as an input. The Baudmode is 8-bit, no parity with noninverted polarity, see Table 2.12. When the receiver is ready to receive the data from the BASIC Stamp, pin 0 (Fpin) will go low; this means that BS2e can now transmit the data. Table 2.15 shows the relationship between serial polarity and Fpin states

If the polarity is inverted, when the receiver is ready to receive data, Fpin will go high and vice versa. In the preceding example if the Fpin permission never occurs, the program is stuck. This means that if the receiver is never ready to receive, the program will wait forever for the ready signal. To get around this bug, the Timeout and Tlabel arguments are used:

```
serdata    VAR    BYTE
SEROUT     1\0, 84, 2000, Noreception, [serdata]
Noreception:
DEBUG      "timed out"
```

In the sample program, if no permission is received in two seconds (Timeout = 2000ms), the program aborts SEROUT and continues to the label Noreception (specified by Tlabel).

2.4.15 SERIN

The function of this command is to receive asynchronous serial data (e.g., RS-232 data). The syntax is as follows:

SERIN Rpin {\Fpin},Baudmode,{Plabel,}{Timeout,Tlabel,}[InputData]

The BASIC Stamp halts when the SERIN command is used so that decimal text can be entered. The processor waits until the data are entered. Rpin is a variable/constant/expression (0-16) that specifies the I/O pin through which the serial data will be received. Using the SERIN command, this pin will be set to input mode. On all BASIC Stamp BS2 models, if Rpin is set to 16, the BASIC Stamp uses the dedicated serial-input pin (SIN, physical pin 2), which is normally used by the Stamp Editor during the download process. Fpin, Baudmode, Timeout, and Tlabel are explained in the "SEROUT" section earlier in this chapter. They have the same functionality for the SERIN command. Plabel is an optional label that indicates where the program should go in the event of a parity error. This argument should be provided only if Baudmode indicates 7 bits and even parity. The following examples illustrate different ways of inputting data from the serial port.

```
serStr      VAR    Byte(10)              ' make 10-byte array
serStr      (9) = 0                      ' put 0 in last byte of array
SERIN       1, 16468, [STR serStr\9]     ' get nine bytes
DEBUG       STR    serStr                ' display
```

In this example, the BASIC Stamp receives nine bytes through I/O pin 1 at 9600 bps, 8-bit no parity inverted, and stores them in an array of 10 bytes. We store only 9 bytes since we would like to reserve space for the 0 byte that many BASIC Stamp string-handling routines regard as an end-of-string marker. This is especially important when dealing with variable-length arrays. Here is an example of a variable-length array:

```
serStr  VAR    Byte(10)                    ' make 10-byte array
serStr(9) = 0                              ' put 0 in last byte of array
SERIN   1, 16468, [STR serStr\9\"*"]       ' stop at "*" or nine bytes
DEBUG   STR    serStr                      ' display
```

If the serial input is "Hi!*" the DEBUG subroutine would display "Hi!" because it gets the bytes up to (but not including) the end character (*). It fills the unused bytes with 0s. The SERIN command can compare incoming data with a predefined sequence of bytes using the WAIT formatter. The simplest form waits for a sequence of up to 6 bytes given as part of the InputData list. The following example shows how this property can be used to create a password check in a system.

```
SERIN   1, 16468, [WAIT("PASS1970")]
DEBUG   "Password accepted."
```

SERIN waits for the password, and the program halts until it is received. The password is case-sensitive since WAIT looks for an exact match for a sequence of bytes. Thus, if "Pass1970" is entered, it will be ignored. Using this functionality, the program could have several passwords (depending on the operation of the main program). In this chapter, we have explored several BASIC Stamp controllers often used in robotic kits. More-detailed explanations of the complete list of commands can be seen in the BASIC Stamp manuals available online. Also, some robotic kits suppliers, such as Lynxmotion, Inc., provide BASIC Stamp manuals for their robotic kits or Stamp evaluation boards.

3

PC Interfacing

In order to be able to program a robot for repetitive tasks or to integrate with sensors like cameras, we need to be able to connect the robot to a controller. We will use a PC as the robot controller for some robots in this book. Therefore, we need to interface the robot with a PC. There are many ways the robot can be connected to a PC. We can control the robot using relays by developing a sensor board that connects to some computer port, such as the parallel port or a USB port or a serial port. Serial ports are getting obsolete, so we will not discuss those in detail. In order to develop interface, we need to know how the parallel port works and how we can use it to connect to the robot. This is discussed below.

3.1 Parallel Port Interface

The parallel port on a PC has a DB25 female pin connector. The pins are related to hardware in the PC. The connector pin relationship to the hardware registers is shown in Figure 3.1.

The parallel port can be controlled using three registers. We can read and write to these registers using software to access and control what happens to the pins of the parallel port. Table 3.1 describes the parallel port pins.

There are three registers associated with each parallel port. These are

- Data register

- Status register

- Control register

In order to use the parallel port for control, we need to know the address of the port registers so that we can use them to read and write to and from the port. During the Power On Self Test (POST) the BIOS is programmed to check for the printer ports and store their addresses at memory locations 0040:0008 to 0040:000FH. Each address takes two bytes following the little endian convention of the 8086 processors which means low byte is followed by high byte. These memory locations can be read to see which LPT (line

103

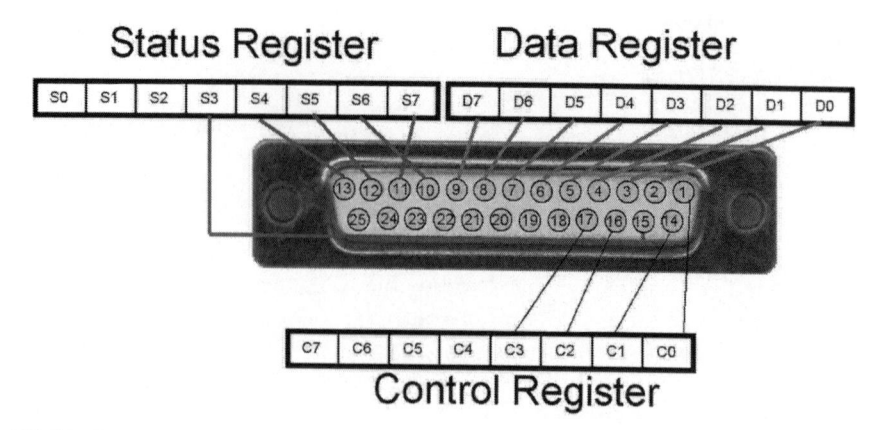

FIGURE 3.1

Parallel port DB25F

TABLE 3.1
Parallel port pin description

Pin No.	Signal name	Direction	Register bit	Inverted
1	nStrobe	In/Out	Control-0	Yes
2	Data0	In/Out	Data-0	No
3	Data1	In/Out	Data-1	No
4	Data2	In/Out	Data-2	No
5	Data3	In/Out	Data-3	No
6	Data4	In/Out	Data-4	No
7	Data5	In/Out	Data-5	No
8	Data6	In/Out	Data-6	No
9	Data7	In/Out	Data-7	No
10	nAck	In	Status-6	No
11	Busy	In	Status-7	Yes
12	Paper-Out	In	Status-5	No
13	Select	In	Status-4	No
14	Linefeed	In/Out	Control-1	Yes
15	nError	In	Status-3	No
16	nInitialize	In/Out	Control-2	No
17	nSelect-Printer	In/Out	Control-3	Yes
18-25	Ground	-	-	-

PC Interfacing

printer) ports are available. Zeros (logic 0) are found in memory locations if that corresponding port is missing. When we find the address of any available printer port, that address is the one for the data register of that printer. The next memory location is that of the status register of the same printer port, followed by the control register of the same printer port. For example, let us say that we write a program to read the memory address 0040:0008 and read two bytes and those two bytes are 0378H. Now this will be the address for the data register for LPT1. The address for its status register will be 0379H and its control register will be 037AH. These, in fact, are the most popular addresses for parallel port on a PC. Table 3.2 shows the BIOS addresses for LPT.

TABLE 3.2
BIOS Address for LPT

I/O Base Address	LPT
0040:0008-0040:0009	LPT1
0040:000A-0040:000B	LPT2
0040:000C-0040:000D	LPT3

Note: We have used the hexadecimal system here. In hexadecimal system (unlike the decimal system where 10 digits are used, namely 0, 1, 2, 3, 4, 5, 6, 7, 8 and 9) sixteen symbols are used. These are: 0, 1, 2, 3, 4, 5, 6, 7, 8, 9, A, B, C, D, E, and F. These are used together to express all numbers. For example 27 in decimal system means. In the hexadecimal system 2F is, which is 47 in decimal. We use an H to indicate a hexadecimal system. For example 0378H means 0378 is a hexadecimal system.

Example addresses for LPT ports are shown in Table 3.3.

TABLE 3.3
Printer port address

Line Printer	Data Register (R/W)	Status Register (Read Only)	Control Register (R/W)
LPT1	03BCH	03BDH	03BEH
LPT2	0378H	0379H	037AH
LPT3	0278H	0279H	027AH

Block diagram view of the three ports associated with a parallel port are shown in Figure 3.2.

We will now study how to write a program to write to and read from the

Data Port

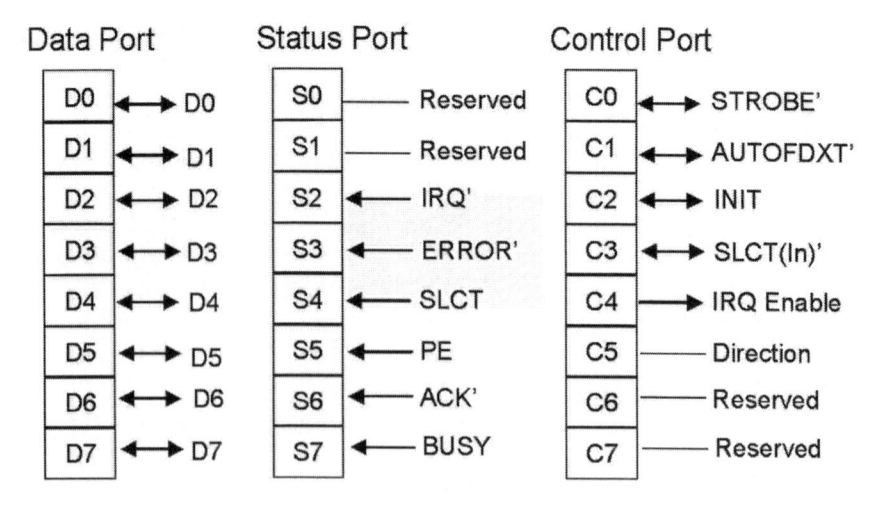

D0	←→	D0
D1	←→	D1
D2	←→	D2
D3	←→	D3
D4	←→	D4
D5	←→	D5
D6	←→	D6
D7	←→	D7

Status Port

S0	—	Reserved
S1	—	Reserved
S2	←	IRQ'
S3	←	ERROR'
S4	←	SLCT
S5	←	PE
S6	←	ACK'
S7	←	BUSY

Control Port

C0	←→	STROBE'
C1	←→	AUTOFDXT'
C2	←→	INIT
C3	←→	SLCT(In)'
C4	→	IRQ Enable
C5	—	Direction
C6	—	Reserved
C7	—	Reserved

FIGURE 3.2

Parallel port breakdown

parallel port. The first thing to do is to find the port address of the parallel port on your computer. Each operating system allows its own way for doing that. For example for Windows XP, we can do it by following the steps shown below. First right click over "My Computer" and then click on "Properties" (see Figure 3.3).

After that click on the hardware tab in the system property window as shown in Figure 3.4. This will bring the next window. Click the "device manager" button on this window. Then open up the "ports" and then double click on the printer port. There, click on the "resources" tab. The resulting window (Figure 3.5) will show the resources used by the printer port. In the example shown below we see that the printer port is LPT1 and starts at memory 0378H. Now we can write code that accesses this memory location. However Windows XP does not allow programs direct access to the ports. In order for our program to access the printer port we can use some third party drivers that will allow our program to access the port. There are some free libraries available that can do this job.

3.1.1 Port Access Library for Windows XP

One free library is called WinIo and is available for free from Internals web site [37]. You can download winio.zip from that website and then unzip it to a directory. The help file in the directory shows how to use the library in different programming languages and platforms. For example, to use the library in Visual C++ you must do the following:

1. Place winio.dll, winio.vxd, and winio.sys in the directory where your

PC Interfacing

FIGURE 3.3

Finding my computer properties

FIGURE 3.4

Clicking on hardware tab

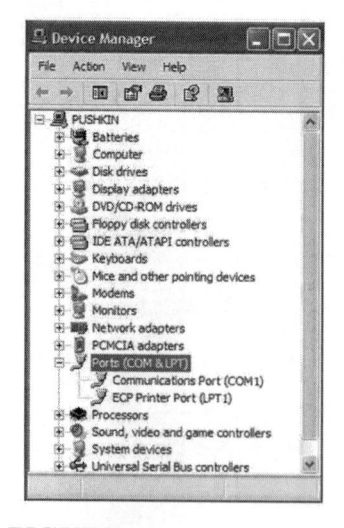

FIGURE 3.5

Resources for LPT1

application's executable file resides.

2. Add winio.lib to your project file by right clicking on the project name in the Visual C++ workview pane and selecting "Add Files to Project...".

3. Add the #include "winio.h" statement to your source file.

4. Call InitializeWinIo.

5. Call the library's functions to access I/O ports and physical memory.

6. Call ShutdownWinIo.

The following is a sample program to show how the parallel port is accessed.

```
#include <windows.h>
#include <stdio.h>
#include "winio.h"

void main()
{
  DWORD dwPortVal;
  bool bResult;

  // Call InitializeWinIo to initialize the WinIo library.

  bResult = InitializeWinIo();
```

```
  if (bResult)
  {
    // Under Windows NT/2000/XP, after calling InitializeWinIo,
    // you can call _inp/_outp instead of using GetPortVal/SetPortVal

    GetPortVal(0x378, &dwPortVal, 4);

    SetPortVal(0x378, 10, 4);
    // When you're done using WinIo, call ShutdownWinIo

    ShutdownWinIo();
  }
  else
  {
    printf("Error during initialization of WinIo.\n");
    exit(1);
  }
}
```

The following is the description of the GetPortVal function in the help file of the library. This function reads a BYTE/WORD/DWORD value from an I/O port:

```
bool _stdcall GetPortVal(
   WORD wPortAddr,
   PDWORD pdwPortVal,
   BYTE bSize
);
```

In the description, wPortAddr [in] is I/O port address; pdwPortVal [out] is a pointer to a DWORD variable that receives the value obtained from the port; bSize [in] is the number of bytes to read (Can be 1 (BYTE), 2 (WORD) or 4 (DWORD)).

If the function succeeds, the return value is true. Otherwise, the function returns false. The **GetPortVal** function reads a byte, a word, or a double word from the specified I/O port.

Note: Under Windows 98/ME, an application must use the **GetPortVal** function to read values from an I/O port. Under Windows NT/2000/XP, it is possible to use the _inp/_inpw/_inpd functions instead of using **GetPortVal**, provided that the InitializeWinIo function has been called beforehand.

3.1.2 Hardware Signals

The parallel port of a PC uses TTL signals. That means that the port has been designed following TTL logic and the corresponding hardware. Now, let

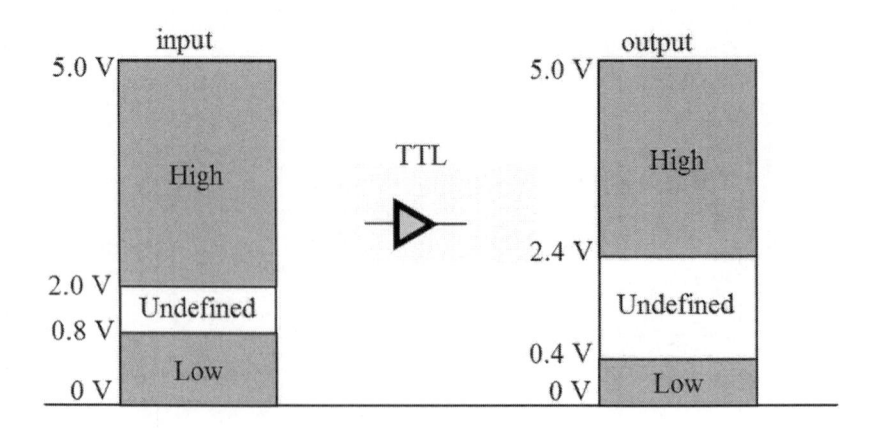

FIGURE 3.6

TTL voltage levels for input and output

FIGURE 3.7

TTL inverter circuit

us try to understand what this means in detail. We will understand what the voltage levels are for logic high and logic low, what currents can the TTL devices handle, as well as the construction and operation principle of TTL logic.

3.1.2.1 TTL Voltage Signals

A TTL logic chip accepts voltage levels from 2.0 V and higher to be considered a logic "high" signal (or logic 1), and any voltage below 0.8V to be a logic "low" (or logic 0). However, a TTL chip guarantees that when it signals a logic 1 on its output, the voltage at the output pin will be greater than or equal to 2.4 V and it guarantees that when it signals a logic 0 on its output, the voltage at the output pin will be less than or equal to 0.4 V. The range of voltages for TTL logic is from 0V to 5V. These values are shown in Figure 3.6.

Let us try to understand the significance of these different voltages for handling noise in the system. Let us say that we are connecting one TTL inverter to another. The circuit for this is shown in Figure 3.7.

We are using 7404 inverter chip. You can get the details about this chip from Texas Instruments' web site [38]. In the circuit, if we input a logic 0

PC Interfacing

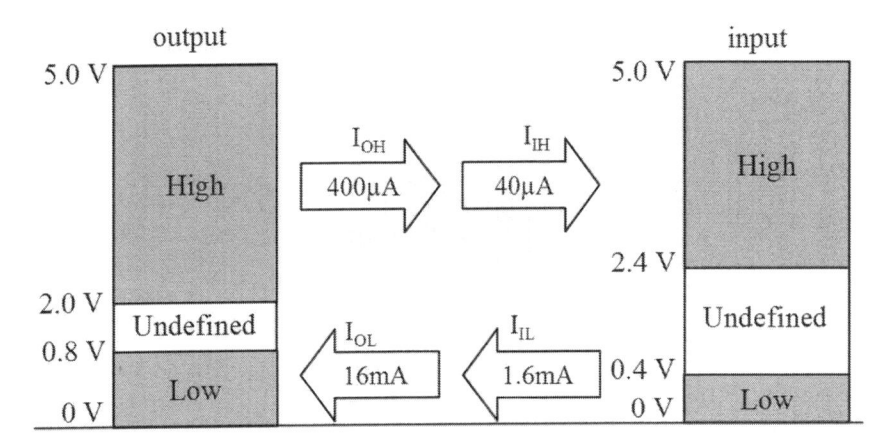

FIGURE 3.8
TTL current capabilities

signal at the input of the first chip, the output of that chip will be at logic 1. This means that the voltage will be 2.4 volts or higher. Now this signal goes as input to the next inverter. For this inverter to consider a signal as high, the signal must be 2.0 volts or higher. That means that a noise of -0.4 volts can be handled that could change the voltage between one output to the next input. This is called **noise margin**. We can also find the same when the output from the first inverter is a logic 0 signal.

3.1.2.2 TTL Current Capabilities

We also need to know the current capabilities of TTL hardware. We must make sure that when we interface with a parallel port that we do not connect components that will force more current into or out of the port than TTL logic can handle. When the current comes out of the port or the pin then, we say that the port is **sourcing current**, and if the current goes into the port or the pin, we say it is **sinking current**. The current capabilities of TTL are shown in Figure 3.8.

This figure shows that when the pin outputs a high it can source up to 400 μA of current and when it outputs a low, then it can sink up to 16 mA of current. When the pin takes an input that is a high signal then it can sink 40 μA of current and when it takes an input which is a low signal, then it can source 1.6 mA of current. This shows that the output of one TTL chip can be connected to 10 other TTL chips. This number is called **fanout** of the chip. We need to control the motors from the signals of the parallel port. We can accomplish that by using output pins. We can also use some pins for input. Therefore, we need to understand how to interface with these pins.

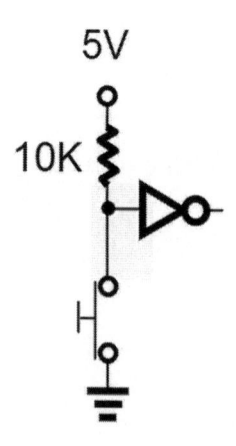

FIGURE 3.9

Active-Low switch with pull-up resistor

3.1.2.3 TTL Interfacing

Figure 3.9 and Figure 3.10 shows examples of how to interface a switch that goes into the input of a TTL chip.

In order to input a logic 1 into the port we can directly connect 5V DC signal to the pin, and if we want input a logic 0 into the port we can directly connect the ground signal to the pin. However, with a switch without using the resistors shown in the figure above, there would be a direct path without resistance from the DC power signal to ground causing high current on that path. Therefore, we need resistors, as shown, to limit the current. For using the port for output, we have to be careful about how much current would be involved. Therefore, we have to use resistors to limit the current, otherwise we can damage the parallel port. Figure 3.11 shows direct interface for driving LEDs from the port.

We have seen how to directly drive LEDs from TTL output pins. However, it is always safer to use transistors to drive LEDs rather than doing a direct drive. That way we can provide more current to the LEDs without having that current interact with the port directly. We can either use an NPN transistor or a PNP transistor for driving. In the case of NPN, shown in Figure 3.12, a high output on the pin turns the transistor on and the LED lights up, and when the output is low, the LED is off. For a PNP transistor, shown in Figure 3.13, the situation is the opposite. When the pin output is high, the PNP transistor is off and consequently the LED is also off. When the pin output is low, the PNP turns on and that turns the LED also on.

We can also use a transistor to drive a relay that can be used to switch a motor or a solenoid on and off. We can not drive a motor or a solenoid directly from a digital output, because the current capability of these devices can not be met by the digital outputs. The relay should have a diode in parallel as

PC Interfacing

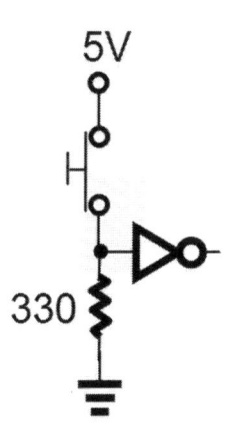

FIGURE 3.10

Active-High switch interfacing

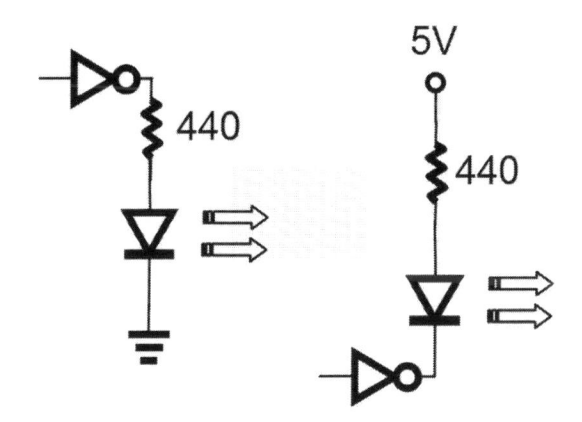

FIGURE 3.11

TTL direct LED driving

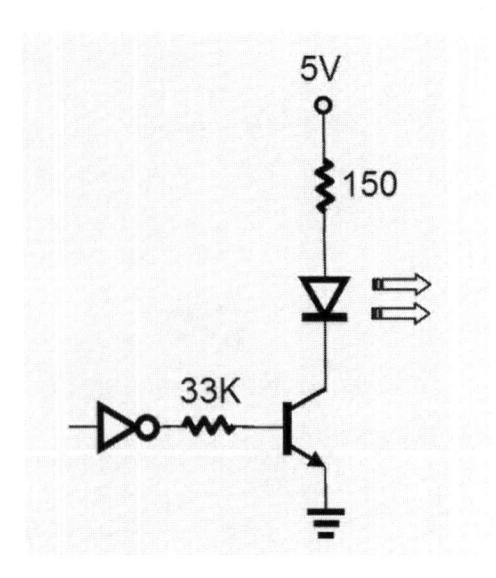

FIGURE 3.12
    TTL transistor LED driving with an NPN transistor

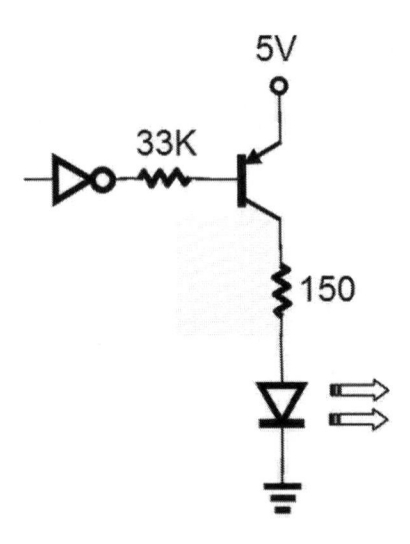

FIGURE 3.13
    TTL transistor LED driving with a PNP transistor

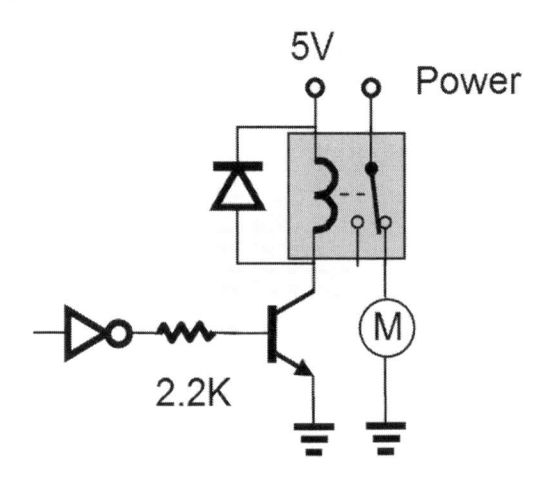

PC Interfacing 115

FIGURE 3.14
TTL transistor relay driving

shown Figure 3.14, so that when the relay is turned off the current still flowing through the relay coil has a path to circulate and dissipate through. Without the diode there would be a high voltage produced due to the instantaneous switching of an inductor coil.

3.1.3 PC Interfacing Board

In this section we describe a board that uses a safe input/output interface for a PC parallel port. We will interface the parallel port for 8-bit input and 8-bit output. For input and output, we will use the output pins as shown in Figure 3.15.

We need to protect the parallel port pins, and therefore, we will use buffers to accomplish that. We can buffer the output pins using 74LS244 chips. However, in order to be able to drive higher current outputs, we can use ULN2803 chips. Let us study these two chips to understand how to interface with them.

3.1.3.1 74LS244

The pinout of the chip is shown in Figure 3.16.

The logic diagram of the chip is shown in Figure 3.17.

The schematic equivalent of the output of the chip is shown in Figure 3.18. The nominal value of the resistance R is 50 Ohms. Schematic for Parallel Port Output Interface with 74LS244 is shown in Figure 3.19.

The letter G indicates that the corresponding pins should be connected to ground. LEDs series with 440 Ohm resistors are attached to the output so that the parallel port can be observed. The code for controlling the LEDs

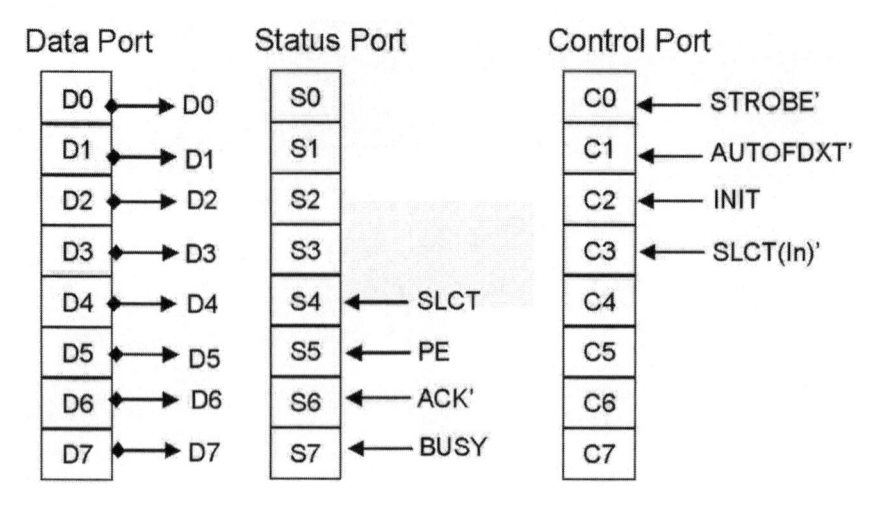

FIGURE 3.15

Input output pins for the board

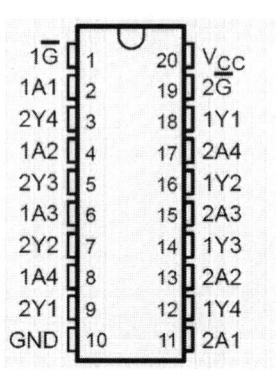

FIGURE 3.16

Pinout for 74LS244 (Courtesy of Texas Instruments Incorporated)

FIGURE 3.17
Logic diagram of 74LS244 with input, output, and control lines of Schmidt triggers (Courtesy of Texas Instruments Incorporated)

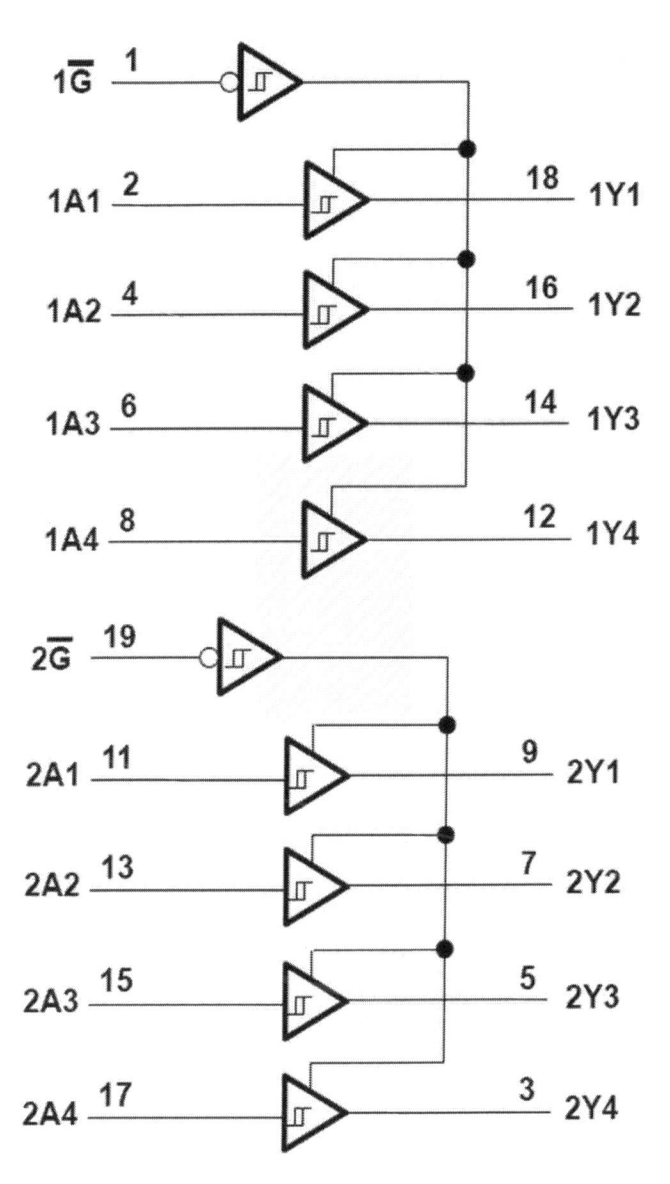

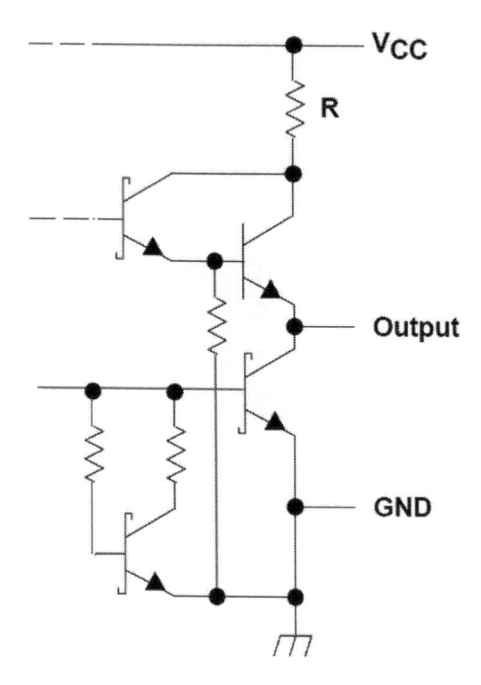

FIGURE 3.18
Output for 74LS244 (Courtesy of Texas Instruments Incorporated)

PC Interfacing

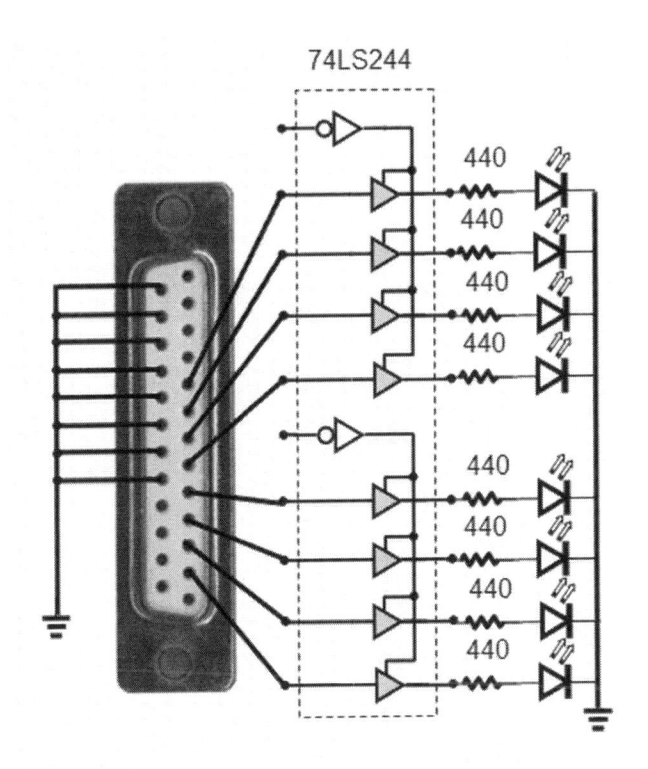

FIGURE 3.19
Schematic for parallel port output interface with 74LS244 (Courtesy of Texas Instruments Incorporated)

is given in the below code. The same setup procedures can be followed as explained early in the chapter.

```
#include <windows.h>
#include <stdio.h>
#include "winio.h"

void main()
{
  BYTE bPortVal;
  bool bResult;

  bResult = InitializeWinIo();

  if (bResult)
  {
    SetPortVal(0x378, 0, 1);
    ShutdownWinIo();
  }
  else
  {
    printf("Error during initialization of WinIo.\n");
    exit(1);
  }
}
```

We can write different numbers using SetPortVal to control what LEDs should light up. For example, using SetPortVal(0x378, 0xFF,1) we can turn on all the LEDs. The number 0xFF is the hexadecimal for the binary number 11111111 and writes 1 onto each output pins of the data port at the address 0x378.

3.1.3.2 ULN2803

The pinout of the chip is shown in Figure 3.20.

The logic diagram of the chip is shown in Figure 3.21.

The transistor level schematic for the chip is shown in Figure 3.22.

This chip can handle up to 50V of output and 500 mA current. To see the use of this chip driving LED outputs from a parallel port, we present the following circuit (see Figure 3.23).

The zener diode used in this design is for protecting the chip against voltage spikes while driving inductive loads like relays or motors. The code for controlling LEDs with ULN2803 uses inverted logic, i.e., to turn on an LED, we must write logic 0 on that pin.

PC Interfacing

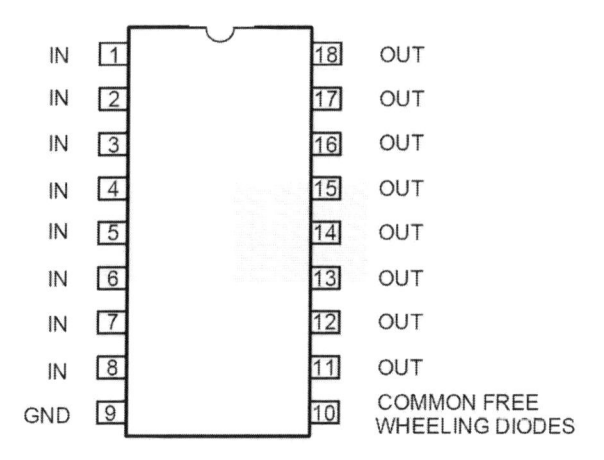

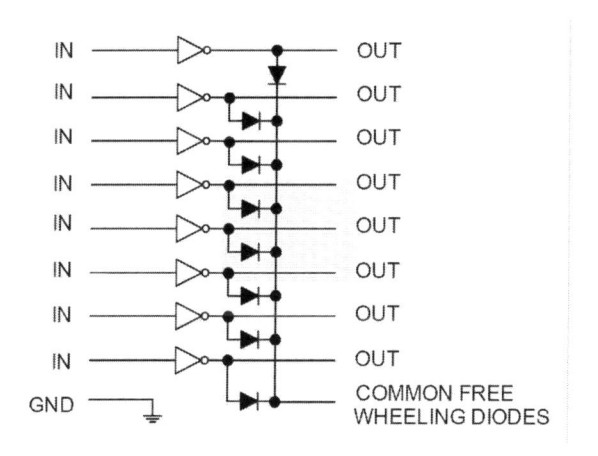

FIGURE 3.20
Pinout for ULN2803 (Courtesy of Texas Instruments Incorporated)

FIGURE 3.21
Logic diagram for ULN2803: logic inverters and protective diodes (Courtesy of Texas Instruments Incorporated)

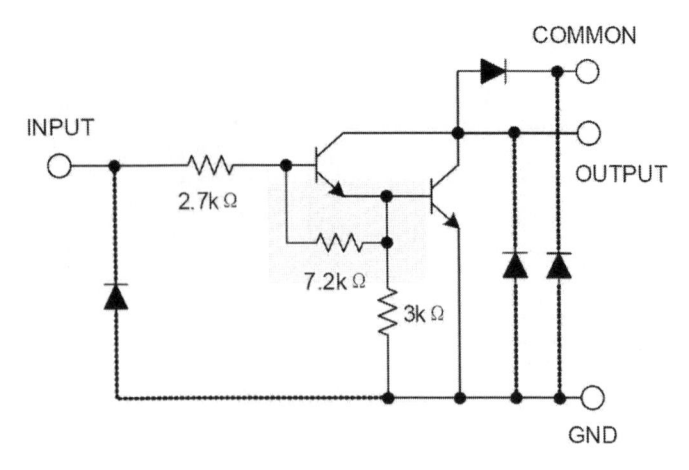

FIGURE 3.22
Schematic for ULN2803 (Courtesy of Texas Instruments Incorporated)

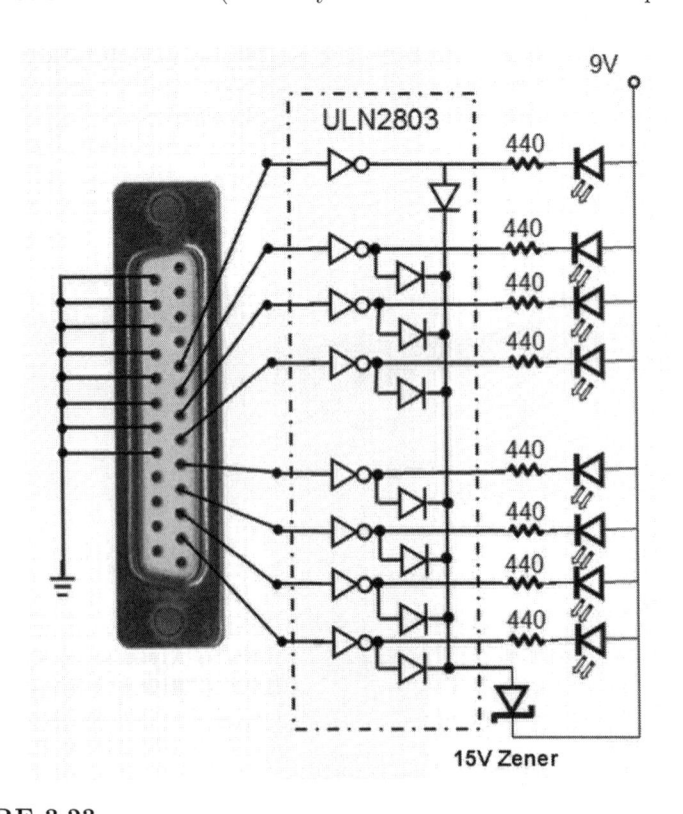

FIGURE 3.23
Schematic for parallel port output interface with ULN2803 (Courtesy of Texas Instruments Incorporated)

PC Interfacing   123

FIGURE 3.24
Schematic for parallel port input output interface (Courtesy of Texas Instruments Incorporated)

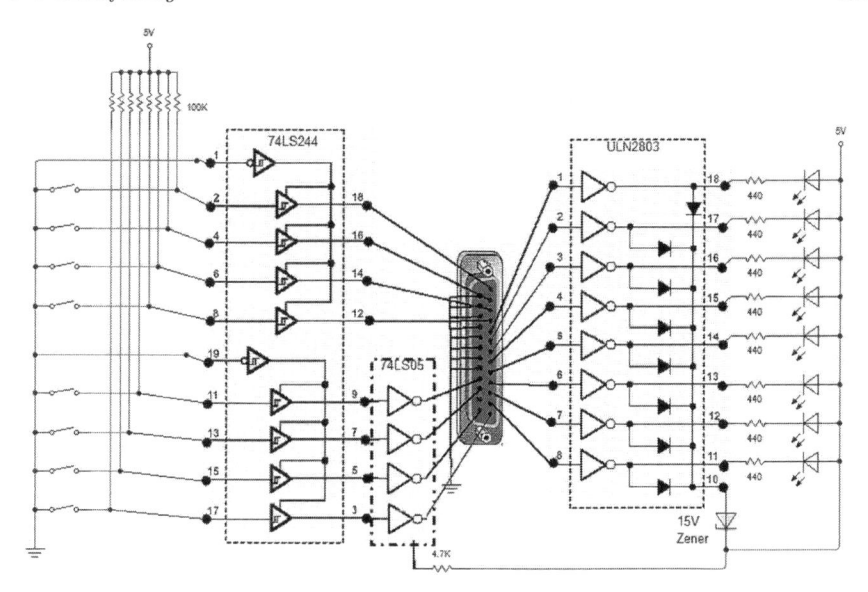

3.1.3.3 Input-Output Interfacing

We can use 74LS244 as a buffer for input to the parallel port. As shown in Figure 3.15, we can use 4 status and 4 control pins to get 8 bits of input. The status pins can be read through the output of the 74LS244 buffer as shown below. The control pins have open collector outputs. Therefore, to use those pins for input require us to use an open collector buffer that has pins pulled to high with 4.7 K-Ohm resistor. Also, during initialization of the software, you should write xxxx0100 to the control port before we can start reading from it. This makes the port turned on, and after that if the external device turns it low, we can read that, and if it does not, then we read a high. We use a 74LS05 as the open collector inverter as the input buffer for these pins. Schematic for Parallel Port Input Output Interface is given in Figure 3.24.

The open collector interfacing is shown in Figure 3.25.

We can drive relays instead of LEDs using the ULN2803 outputs. This is shown in Figure 3.26.

3.1.4 Visual Basic Access to Parallel Port

There is a library that can be used to access parallel port in windows. It is called IO.dll and available at Geek Hideout web site [40]. That site also provides details on how to use it. It also provides a sample visual basic code

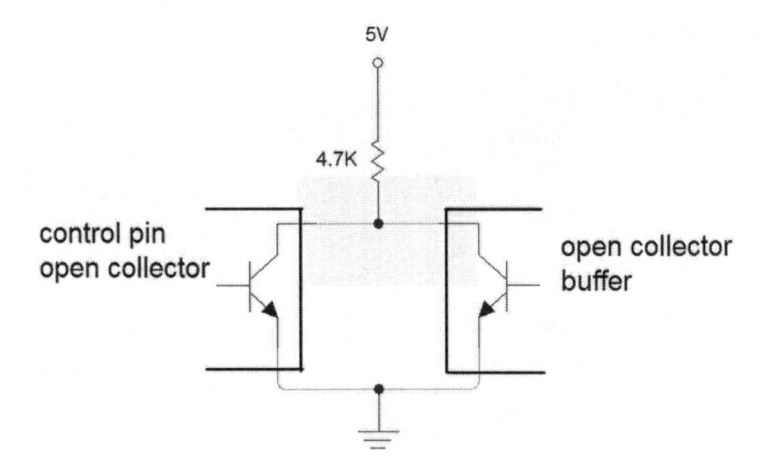

FIGURE 3.25

Open collector interface

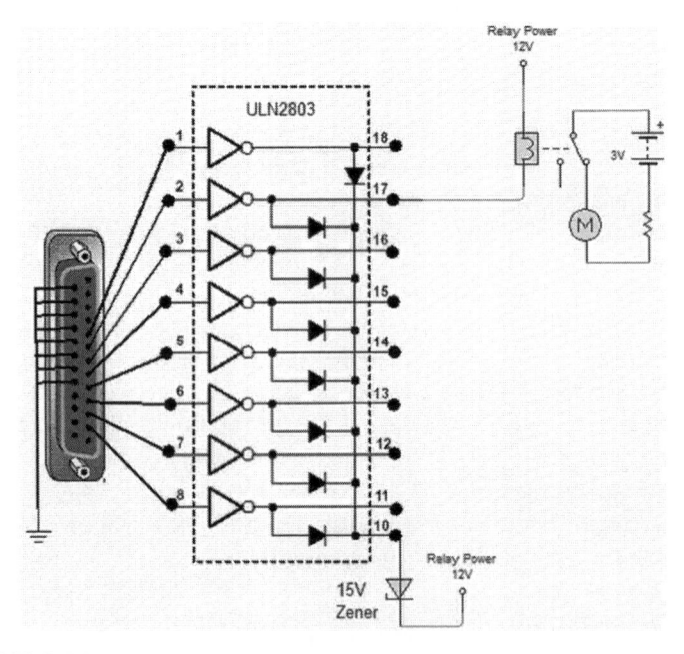

FIGURE 3.26
Relay interface (Courtesy of Texas Instruments Incorporated)

PC Interfacing 125

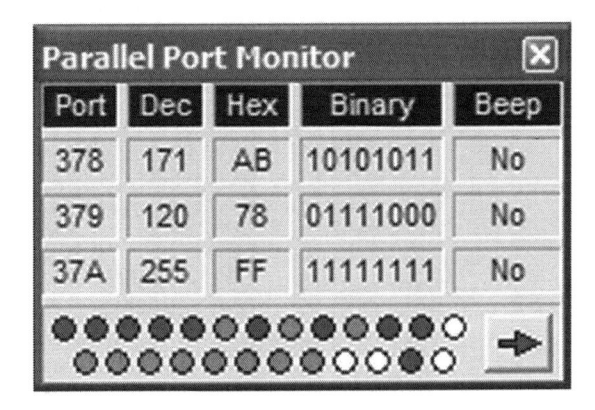

FIGURE 3.27
Parallel port monitor

to monitor the parallel port.

The screen shot of the parallel port monitor is shown in Figure 3.27.

One other library that is freely available is inpout32.dll. One of the sites it is available at is Programmers Heaven web site [41]. Next section describes a breadboarded circuit used for output using this library.

3.1.5 Breadboarded Output Circuit

The following circuit, shown in Figure 3.28, was built on a solderless breadboard.

The breadboard circuit is shown in Figure 3.29. Notice that we are using a 5V power for the circuit.

The bottom breadboard has the circuit for parallel port output interfacing with the LEDs and the top breadboard has the 5V power generation from a 6V AC adapter. The circuit for that is shown in Figure 3.30.

The AC adapter used is shown in Figure 3.31.

To get the wires from the parallel port cable, we cut the cable to get the wires out and then stripped the leads out. Then we soldered a 22-gauge wire with each wire and put heat shrink tubing on it. All the ground wires were soldered together with one 22-gauge wire, so that all the required ground connections would be accomplished by a single wire on the breadboard. The other side of the cable is a standard male DB25 that connects to the DB25F on the computer. The cable wiring is shown in Figure 3.32.

The Visual Basic .NET code we use here uses inpout32.dll. You can obtain information about this dll file from Logix4u web site [42]. You can also download the code from that site. The programming development is as follows. We design the user form as shown in Figure 3.33.

The code for the "in" button of the user interface form is given below.

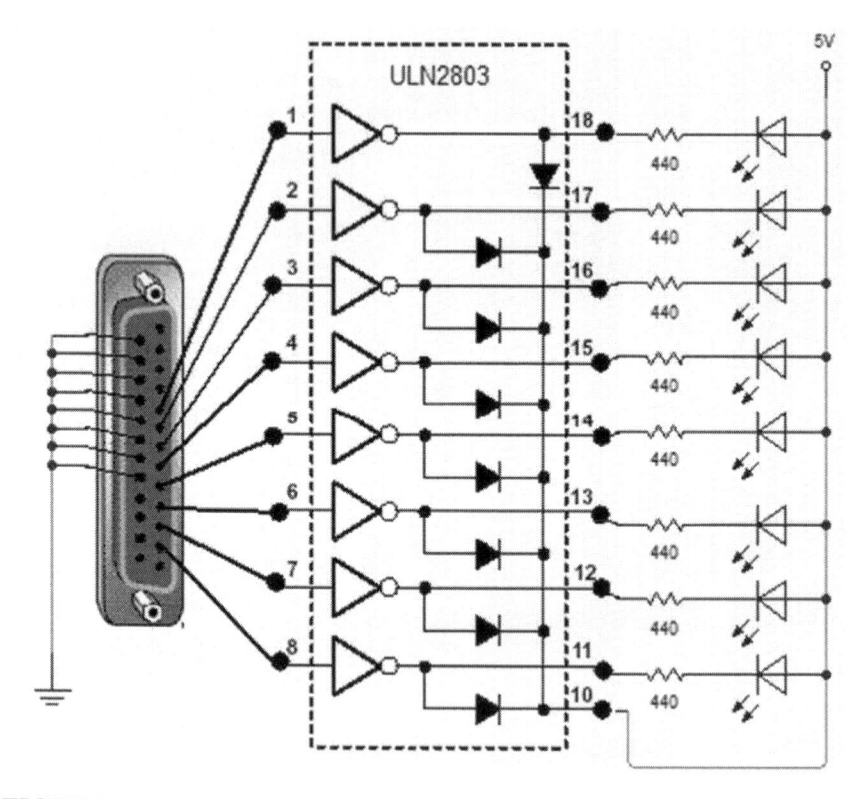

FIGURE 3.28
Schematic for parallel port output interface with ULN2803 (Courtesy of Texas Instruments Incorporated)

PC Interfacing

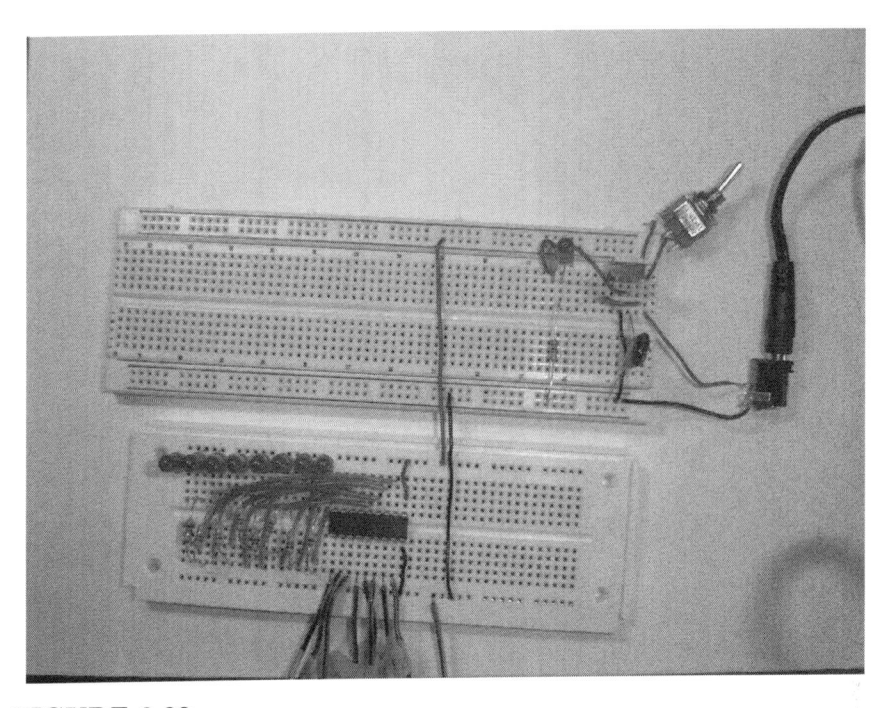

FIGURE 3.29
                Breadboarded circuit

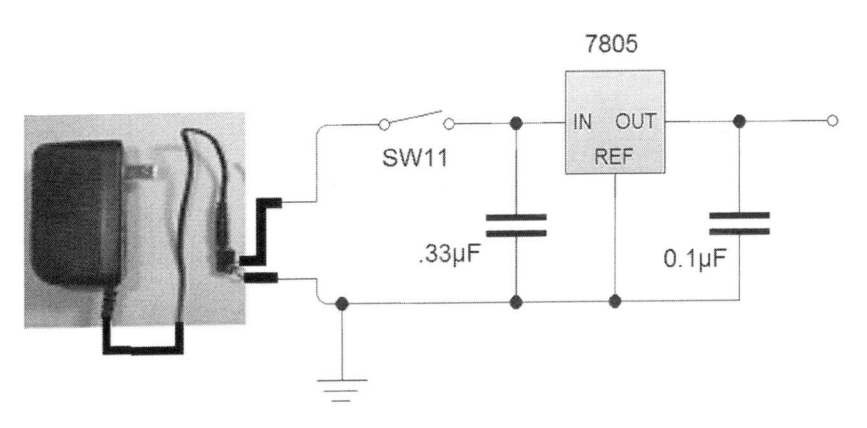

FIGURE 3.30
                5V Power supply circuit

FIGURE 3.31

AC power adapter

```
Private Sub Command1_Click()
Text2.Text = Str(Inp(Val("&H" + Text1.Text)))
End Sub
```

The following is the code for the "out" button of the user interface form.

```
Private Sub Command2_Click()
Out Val("&H" + Text1.Text), Val(Text2.Text)
End Sub
```

The following code is the main program and will be in the project file code whereas previous program segments are in the form which created the form in Figure 3.34. It sets the necessary library, inpout.dll, as well as defining aliases and variables.

```
Public Declare Function Inp Lib "inpout32.dll" _
Alias "Inp32" (ByVal PortAddress As Integer) As Integer
Public Declare Sub Out Lib "inpout32.dll" _
Alias "Out32" (ByVal PortAddress As Integer, ByVal Value As Integer)
```

The actual output of the program is shown in Figure 3.34. The user can type in the address of the parallel port and then type in the value to be output followed by a click of the "out" button. The user can also click on the "in" button to read the input at the port.

PC Interfacing

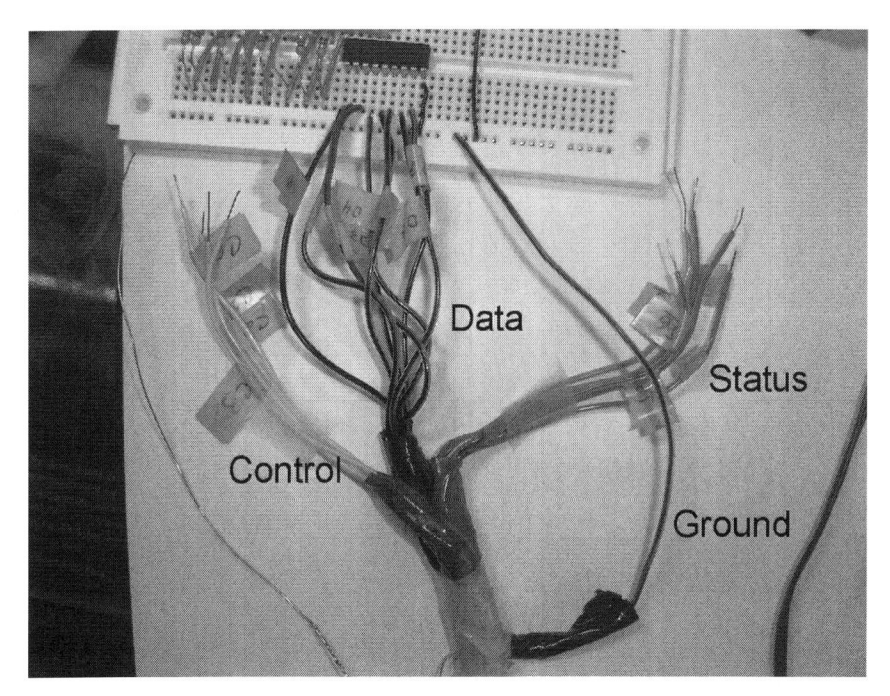

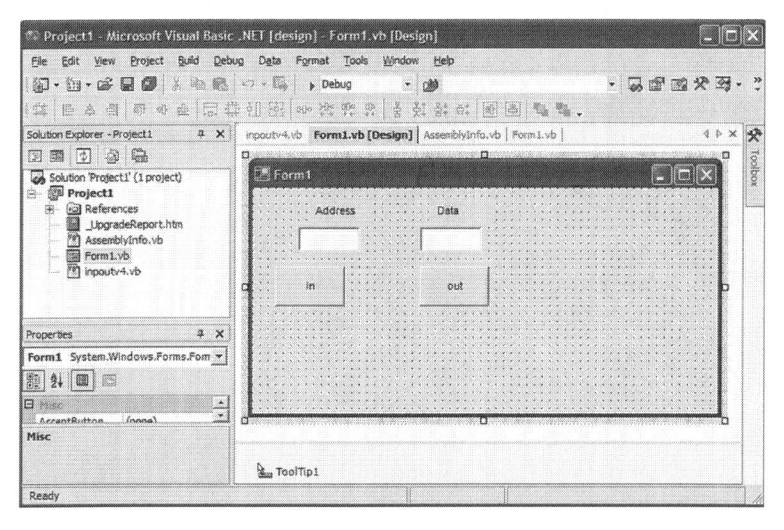

FIGURE 3.32
    Parallel port cable wiring for breadboarding

FIGURE 3.33
    Visual Basic form

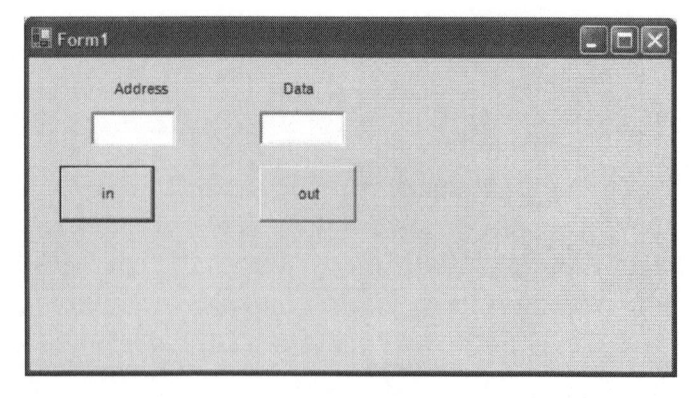

FIGURE 3.34

Window of Visual Basic form

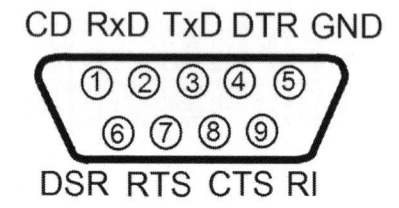

CD RxD TxD DTR GND

① ② ③ ④ ⑤
⑥ ⑦ ⑧ ⑨

DSR RTS CTS RI

FIGURE 3.35

Serial port DB9 pins

3.2 Serial Port Interfacing

Serial port is used for transferring data serially as compared to the parallel port that is used for parallel transfer of data. The DB9 pins of a serial port are shown in Figure 3.35 and described in Table 3.4.

3.2.1 PC to PC Communication

This section describes cabling (Null Modem) and necessary software components for PC to PC serial communication.

3.2.1.1 Cabling (Null Modem)

Let us study how to have serial communication between two PCs. We can have a null modem connecting the serial ports of the two computers. A null modem connects serial ports of two computers as shown in Figure 3.36. This null modem is a null modem with full handshaking. We can also have null modems with fewer connections. The simplest and the cheapest one is a

TABLE 3.4
Serial port pin description

Pin No.	Signal	Signal Name
1	CD	Carrier Detect
2	RxD	Received Data
3	TxD	Transmitted Data
4	DTR	Data Terminal Ready
5	GND	Ground
6	DSR	Data Set Ready
7	RTS	Request to Send
8	CTS	Clear to Send
9	RI	Ring Indicator

three wire connection based null modem. There can be other null modem connections. The choice of which one to use depends upon what software will be used on both sides.

The three wire minimal null modem is shown in Figure 3.37.

One can also make a PC communicate with itself for example by using a terminal program and connecting pins as shown in Figure 3.38.

3.2.1.2 Software

The .NET framework version 2 (beta) consists of access for providing serial port communication. We can write a C# application for that. In order to program the serial port using C#, you can download the Visual C# Express edition. Details of an example for a serial port communication between two PCs is provided at the Code Project web site [46].

Some details of that example are provided here. In order to use serial port features, we use the serial port namespace as:

```
using System.IO.Ports;
```

To create a serial port object we write the following line of code.

```
// Create a Serial Port Object
SerialPort sp = New SerialPort();
```

The most important methods of the serial port class are:

1. Open( ) : Opens a new serial port connection.

2. Close( ) : Closes a serial port connection.

3. ReadLine( ) : Reads upto the NewLine in the input buffer. If there is a timeout, then it returns Null.

4. WriteLine(string) : Appends the string with a NewLine and writes it to the output buffer.

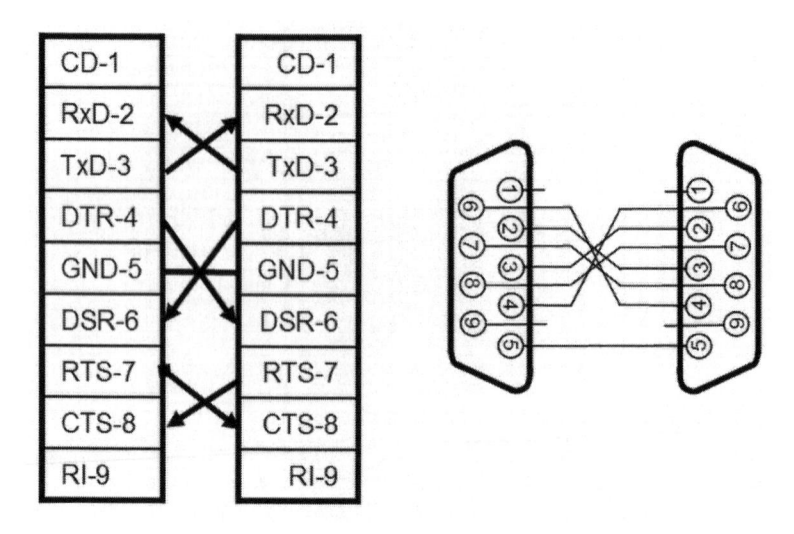

FIGURE 3.36

Null modem with full handshaking

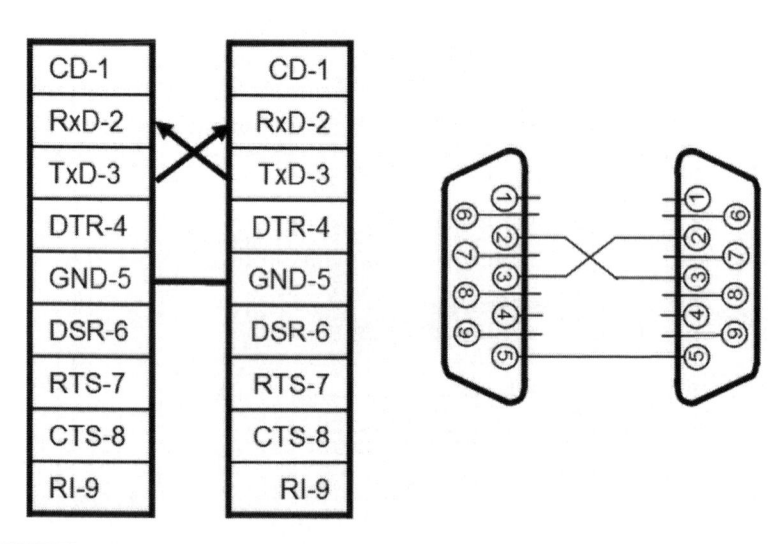

FIGURE 3.37

Null modem without full handshaking

PC Interfacing 133

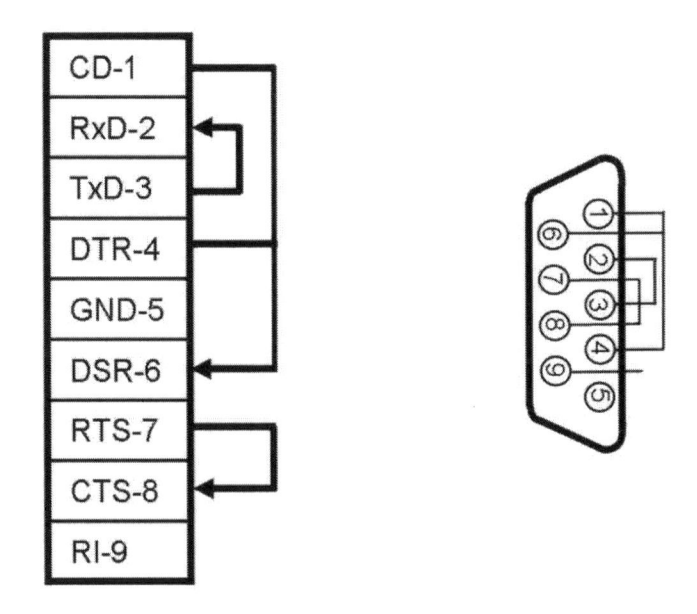

FIGURE 3.38

Self loop connection

Important public properties of the serial port class are:

1. Baudrate : Gets or sets the serial baud rate.

2. Stopbits : Gets or sets the stop bits.

3. ReadTimeout : Specifies in milliseconds the timeout for a read operation.

3.2.2 PC to Microcontroller Serial Communication

Let us study how to have serial communication between a PC and a microcontroller. In Figure 3.39, we see a PIC16F84A microcontroller in a minimal circuit to perform serial communication. It uses pin RA1 to receive and RA2 to transmit serial data to a PC. The PC side can be programmed as above using C#, or can be programmed in many other languages such as Visual Basic, or C++. We can also use a terminal program for the serial communication. On the microcontroller side, we can program the serial communication using many different languages too, such as PICBasic, C, assembly, etc. Assembly is the most cumbersome. Basic and C languages are easier for serial programming.

In PICBasic, we can use **SERIN** and **SEROUT** commands to perform serial communication. The descriptions of these two are given below and are taken

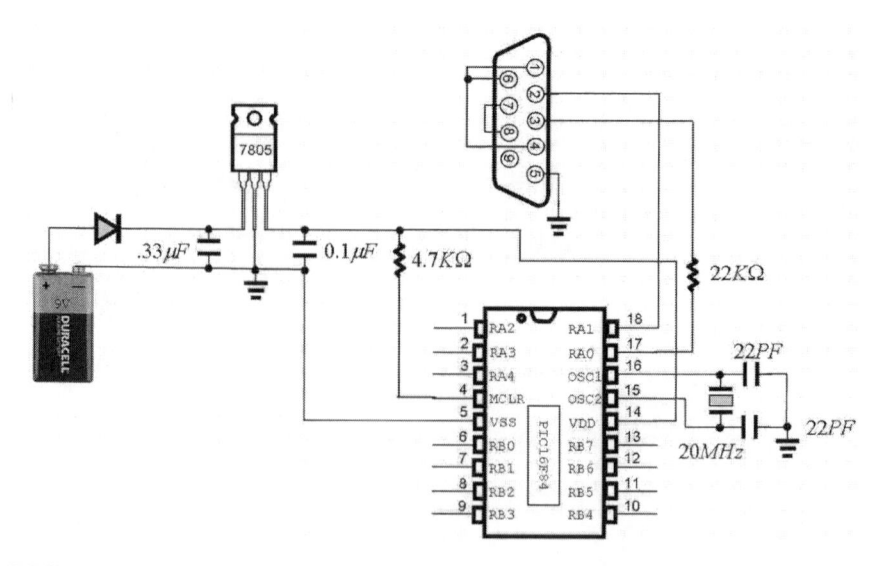

FIGURE 3.39
Serial communication between a PC and a PIC16F84A microcontroller

directly from the PICBasic manual.

3.2.2.1 SERIN in PICBasic

The following is the format for the SERIN command in PICBasic to set up the serial port for input.

SERIN Pin, Mode,{(Qual{,Qual}),} Item{,Item}

The PC receives one or more items on Pin in standard asynchronous format using 8 data bits, no parity, and one stop bit. Mode is one of the following shown in Table 3.5.

Note: 9600 baud is an addition to the PICBasic Compiler and is not available on the BASIC Stamp I.

The list of data items to be received may be preceded by one or more qualifiers enclosed within parentheses. SERIN must receive these bytes in exact order before receiving the data items. If any byte received does not match the next byte in the qualifier sequence, the qualification process resets (i.e., the next received byte is compared to the first item in the qualifier list). A Qualifier can be a constant, variable, or a string constant. Each character of a string is treated as an individual qualifier.

Once the qualifiers are satisfied, SERIN begins storing data in the variables associated with each Item. If the variable name is used alone, the value of the received ASCII character is stored in the variable. If variable is preceded by a pound sign (#), then SERIN converts a decimal value in ASCII and

PC Interfacing

TABLE 3.5
Serial communication modes for SERIN command

Mode Symbol	Value	Baud Rate	Communication Type
T2400	0	2400	TTL True
T1200	1	1200	TTL True
T9600	2	9600	TTL True
T300	3	300	TTL True
N2400	4	2400	TTL Inverted
N1200	5	1200	TTL Inverted
N9600	6	9600	TTL Inverted
N300	7	300	TTL Inverted

FIGURE 3.40
Current limiting resistor for inverted serial inputs with their pin numbers in DB9 and DB25 connectors

stores the result in that variable. All nondigits received prior to the first digit of the decimal value are ignored and discarded. The nondigit character which terminates the decimal value is also discarded. The following is an example of SERIN command.

SERIN 1, N2400, ("A"), B0

This command lets the PC wait until the character "A" is received serially on Pin1 and put the next character into B0. The Baud rate is set for 2400. While single-chip RS-232 level converters are common and inexpensive, the excellent I/O specifications of the PICmicro MCU allow most applications to run without level converters. Rather, inverted input (N9600..N300) can be used in conjunction with a current limiting resistor as shown in Figure 3.40.

3.2.2.2 SEROUT in PICBasic

The following is the format for the SEROUT command in PICBasic to set up the serial port for output.

SEROUT Pin, Mode, (Item{,Item})

This command sends one or more items to Pin in standard asynchronous format using 8 data bits, no parity, and one stop. Mode is one of the following shown in Table 3.6.

TABLE 3.6

Serial communication modes for SEROUT command

Mode Symbol	Value	Baud Rate	Communication Type
T2400	0	2400	TTL True
T1200	1	1200	TTL True
T9600	2	9600	TTL True
T300	3	300	TTL True
N2400	4	2400	TTL Inverted
N1200	5	1200	TTL Inverted
N9600	6	9600	TTL Inverted
N300	7	300	TTL Inverted
OT2400	8	2400	Open Drain
OT1200	9	1200	Open Drain
OT9600	10	9600	Open Drain
OT300	11	300	Open Drain
ON2400	12	2400	Open Source
ON1200	13	1200	Open Source
ON9600	14	9600	Open Source
ON300	15	300	Open Source

SEROUT supports three different data types which may be mixed and matched freely within a single SEROUT statement.

1. A string constant is output as a literal string of characters.

2. A numeric value (either a variable or a constant) will send the corresponding ASCII character. Most notably, 13 is carriage return and 10 is line feed.

3. A numeric value preceded by a pound sign (#) will send the ASCII representation of its decimal value. For example, if W0 = 123, then #W0 (or #123) will send '1', '2', '3'.

The following is an example of a SEROUT command.

```
Serout 0,N2400,(#B0,10)
```

This command sends the ASCII value of B0 followed by a linefeed out to Pin0 serially. The Baud rate is set to 2400 with the TTL True mode. While single-chip RS-232 level converters are common and inexpensive, thanks to current RS-232 implementation and the excellent I/O specifications of the PICmicro MCU, most applications do not require level converters. Rather, inverted TTL (N300..N9600) can be used. A current limiting resistor is suggested (RS-232 is suppose to be short-tolerant) as shown in Figure 3.41.

PC Interfacing

```
          1K                         DB9    DB25
Pin ─────/\/\/\─────── RS-232 RX     Pin 2  Pin 3

                ┌───── RS-232 GND    Pin 5  Pin 7
                │
                ═
```

FIGURE 3.41
Current limiting resistor for inverted serial outputs

3.3 USB Interfacing

If a PC does not have a serial port, and only has USB ports, then we can use a USB to serial converter to perform communication between the host PC and a microcontroller. On the PC side we can use a converter such as the one by Parallax, shown in Figure 3.42.

FIGURE 3.42
USB to serial converter (Copyright 2006 Parallax Inc.)

This product is based on USB interfacing chips by FTDI and come with software drivers so that one can program the communication on the PC. One could also provide a USB interface to a microcontroller using this device. Details about this device and FTDI chips can be obtained from the Parallax Inc. web site [26]. Another very easy solution to providing a USB interface is to use the following Pololu USB to RS232 serial interface, shown in Figure 3.43. Information about this interface can be obtained from Pololu Inc. web site [45].

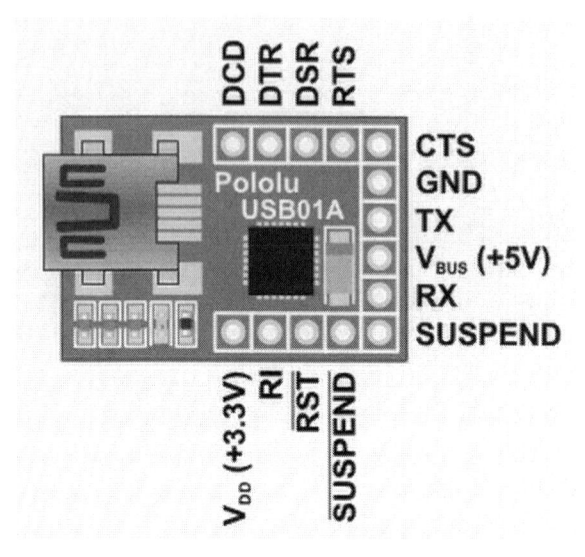

FIGURE 3.43
Pololu USB to serial board (Copyright 2006 Pololu Corporation)

This device comes with a software driver, which allows the PC software to use standard serial code (as the one using C#). The output pins have the standard RS232 interface.

4

Robotic Arm

In this chapter we will study a robotic arm that is built using DC motors and is controlled by switches. The robotic arm we will study is the OWI-007 robotic arm trainer. The robot is shown in Figure 4.1 below.

From the figure we can see that the robot has five degrees of freedom. That means we can have it move in five independent ways. The five different movements are created in five different joints as described below.

1. **Base Joint**: This joint allows movement of $350°$ rotational motion.

2. **Shoulder Joint**: This joint allows movement of $120°$ rotational motion.

3. **Elbow Joint**: This joint allows movement of $135°$ rotational motion.

4. **Wrist Joint**: This joint allows movement of $340°$ rotational motion.

5. **Gripper**: This joint allows movement of $2''$ linear motion (open and close actions).

The arm is made from lightweight plastic. Most of the stress-bearing parts are also made of plastic. The DC motors used in the robotic arm are small, high rpm, low torque motors. To increase the motor's torque, each motor is connected to a gearbox. The motor gearbox assemblies are used inside the construction of the robotic arm. While the gearboxes increase the motor's

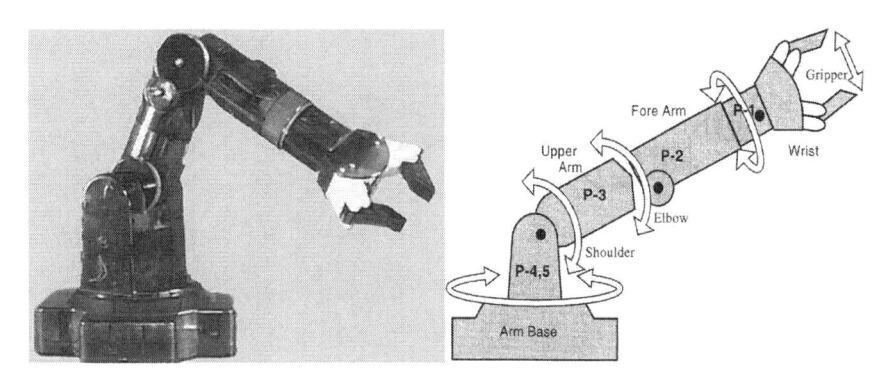

FIGURE 4.1
OWI-007 robotic arm trainer (Copyright 2007 OWI ROBOTS)

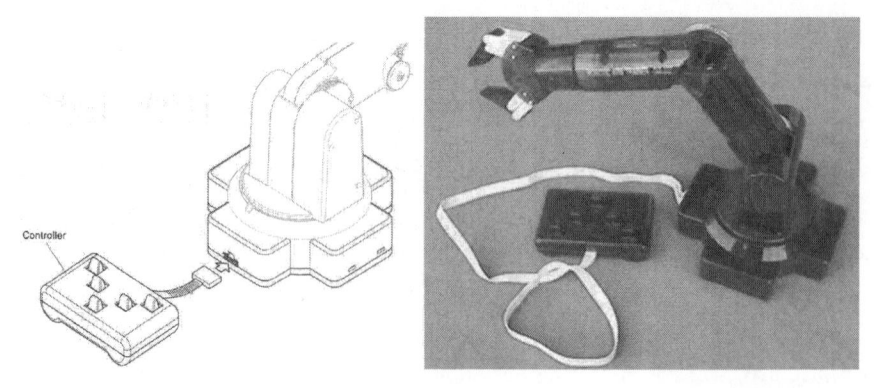

FIGURE 4.2
OWI-007 control panel (Copyright 2007 OWI ROBOTS)

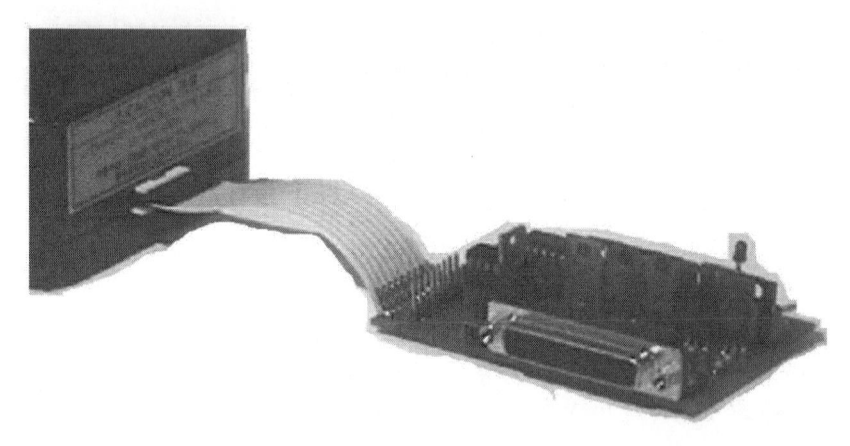

FIGURE 4.3
OWI-007 PC interface kit (Copyright 2007 OWI ROBOTS)

torque, the robotic arm is not capable of lifting or moving a great amount of weight. The maximum recommended lifting capacity is 4.6 ounces (130 grams).

The robotic arm can be controlled by hand using a control panel. The control panel has five switches to control the five DC motors that control the various joints of the robot. The control panel is shown in Figure 4.2.

We can interface the robot with a PC using its parallel port and an interface kit. The robotic arm interface kit connects OWI's 007 Robotic Arm Trainer (tm) to a personal computer (IBM PC or compatible). The interface connects to the PC's parallel port. The interface kit is shown in Figure 4.3.

Robotic Arm

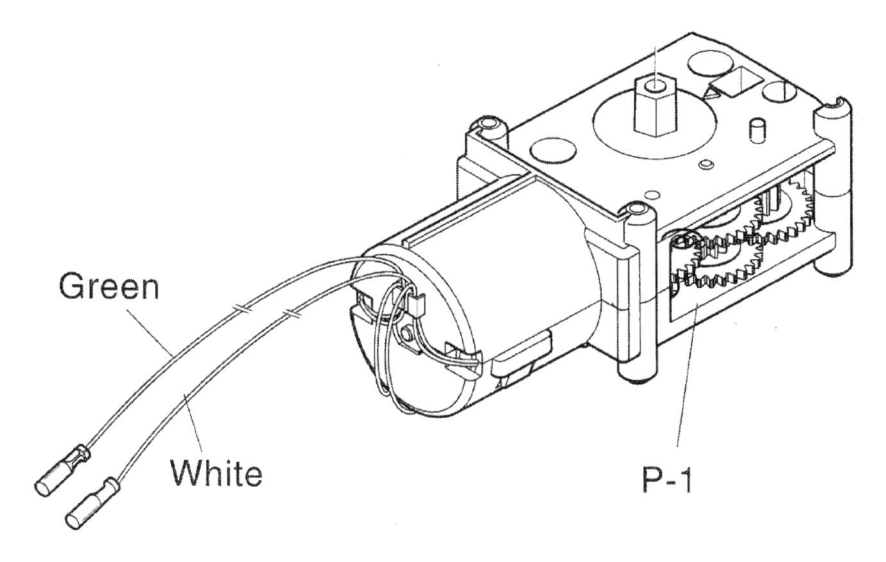

FIGURE 4.4
OWI-007 DC motor and gear box (Copyright 2007 OWI ROBOTS)

4.1 Construction and Mechanics

We will understand the mechanical construction of the robotic arm by first understanding the DC motor and the gearbox. After that we will study each section between consecutive joints in more details.

4.1.1 DC Motor and Gear Box

The DC motor and the gearbox that are used in the robot are shown in Figure 4.4.

The motors used in the robot are low torque but high speed motors. Therefore, we use a gear box that converts the output of the motor to lower speed and higher torque. The gears used in the gear box are shown in Figure 4.5.

The first gear that is connected to the motor shaft changes the motion from the motor axis to an axis that is perpendicular to the motor shaft. This is more clearly shown in Figure 4.6. The motor output shaft is connected to a worm gear that moves and makes the spur gear rotate. It is worthwhile to note that only the worm gear motion can make the other gear move and not the other way around.

The next four gears connect to each other in the same plane. Figure 4.7 shows the two gears connected together.

Output axis A gives the rotation motion in the same plane as these gears.

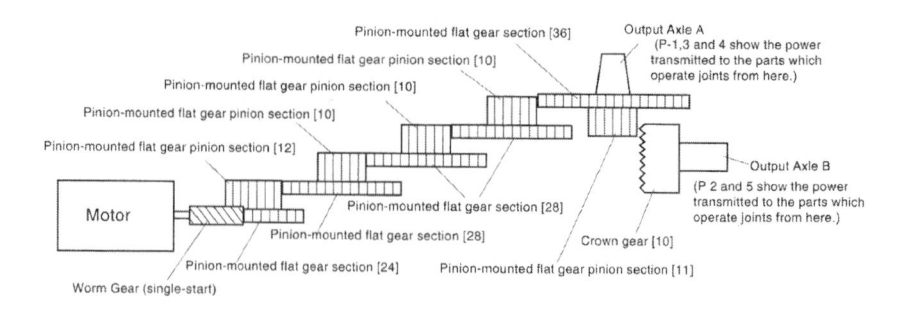

The figures in [ ] indicate the number of teeth on the gears.

FIGURE 4.5
OWI-007 DC motor and gear box details (Copyright 2007 OWI ROBOTS)

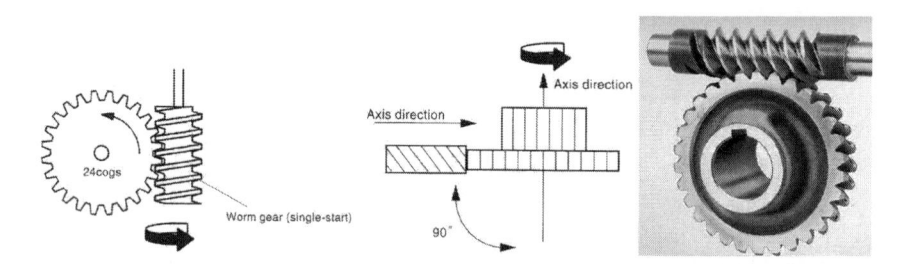

FIGURE 4.6
OWI-007 motor and worm gear (Copyright 2007 OWI ROBOTS and Emerson Power Transmission Corporation)

Robotic Arm 143

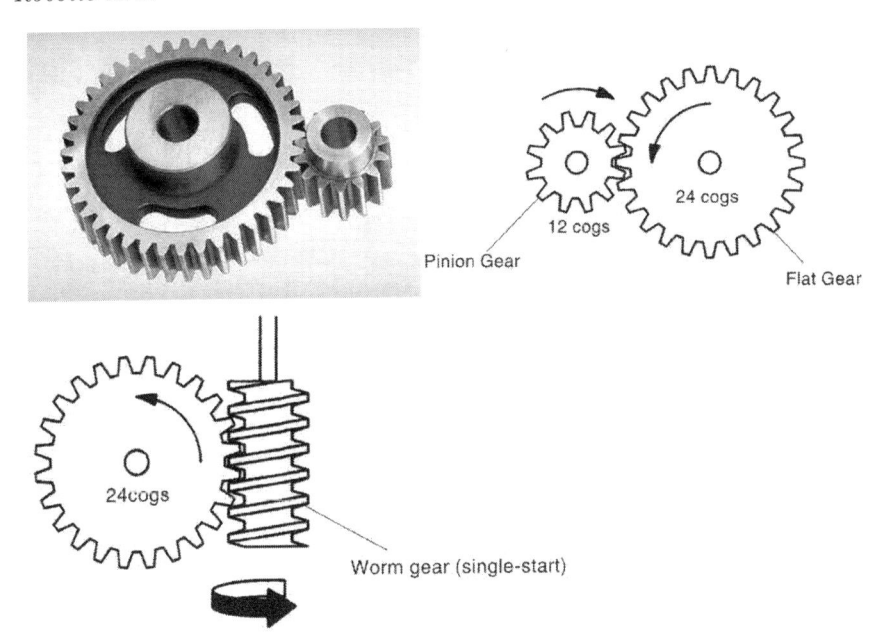

FIGURE 4.7
OWI-007 spur gears (Copyright Emerson Power Transmission Corporation)

We can also obtain output from axis B by changing the rotation motion again by ninety degrees by using bevel gears as shown in Figure 4.8.

4.1.2 Gear Torques and Speed

Now, let us understand how a gear works in terms of transferring torque and speed. To study this, we will refer to Figure 4.9.

Gear-1 (pinion) has radius R_1, exerts a force F_2 on gear-2 (driven gear) at the point of contact of the two gears and the speed of the point of contact is v_1. Gear-2 has radius R_2, exerts a reaction force F_1 on gear-1 at the point of contact of the two gears and the speed of the point of contact is v_2. For gear action, we have the following two balance equations. The first states that the speed of the two gears at the point of contact is the same.

$$v_1 = v_2 \tag{4.1}$$

The other, due to Newton's third law, states that the action reaction forces are equal.

$$F_1 = F_2 \tag{4.2}$$

Let ω_1 be the angular speed of gear-1 and ω_2 be the angular speed of gear-2. Since we know that linear speed is equal to the product of angular speed and

FIGURE 4.8
OWI-007 bevel gears (Copyright Emerson Power Transmission Corporation)

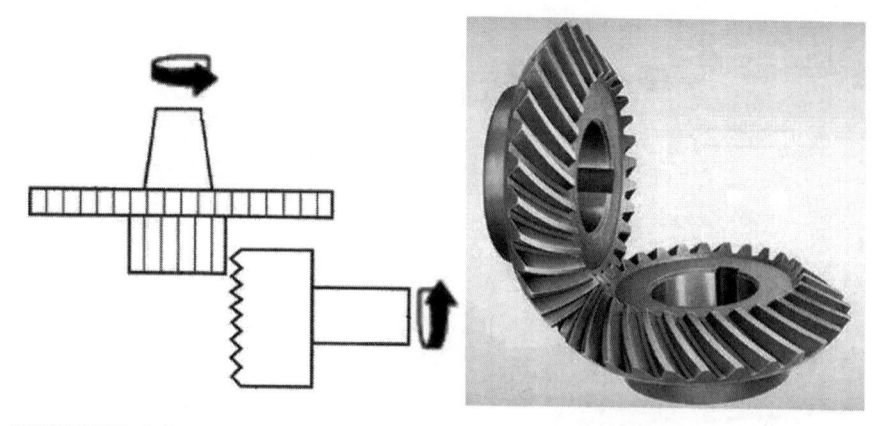

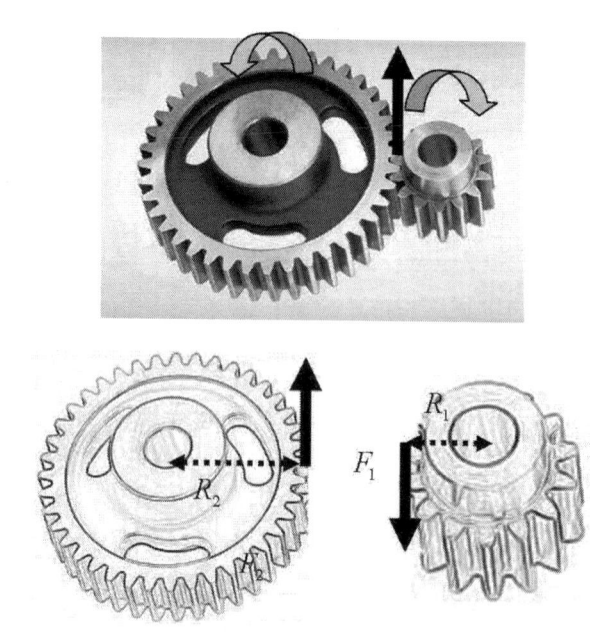

FIGURE 4.9
OWI-007 spur gears (Copyright Emerson Power Transmission Corporation)

Robotic Arm

radius, we can re-write equation 4.1 as:

$$R_1\omega_1 = R_2\omega_2 \qquad (4.3)$$

We can use this equation as:

$$\frac{\omega_1}{\omega_2} = \frac{R_1}{R_2} \qquad (4.4)$$

The ratio $\frac{R_2}{R_1}$ is called the gear ratio. Torque (τ) is equal to the product of force and radius, or alternately, force is the quotient of torque when it is divided by radius. Using this in condition 4.2 gives:

$$\frac{\tau_1}{R_1} = \frac{\tau_2}{R_2} \qquad (4.5)$$

We can also write this as:

$$\frac{\tau_1}{\tau_2} = \frac{R_1}{R_2} \qquad (4.6)$$

Combining equation 4.4 and equation 4.6 and rearranging terms yields:

$$\tau_1\omega_1 = \tau_2\omega_2 \qquad (4.7)$$

This shows that in an ideal gear, the input power is equal to the output power. Therefore, if we reduce the speed, the torque increases and vice versa.

This is also true for the entire gear train. We can find out the overall gear ratio for the entire gear train and then find out the overall velocity reduction (torque increase) for the gear in our robot. By having a gear train we can achieve very low overall gear ratio, since the overall gear ratio is just the product of all the gear ratios in the gear train, and very low gear ratio produces a high torque while reducing the output angular speed.

4.1.3 Gripper Mechanism

To understand the finger mechanism, please refer to Figure 4.10.

The pinion gear is attached to the motor shaft. As that rotates, the motion is translated into a linear motion of the rack gear. As the rack gear moves in a straight line, it pushes (or pulls depending on the rotation direction) on the finger base. The finger base rotates about the fixed point shown in Figure 4.10. The fixed center of the rotation of the finger is attached to the outside shell of the robot hand, so that it stays fixed. As we can see, the finger tip moves in an arc since the finger base and the finger are rigidly connected. We would like the finger surface to be perpendicular to the object we want the gripper to hold. To see the difference, look at Figure 4.11 and Figure 4.12.

In Figure 4.11, we see that the forces that the fingers exert on the object push the object away from the fingers if there is not enough friction. On the

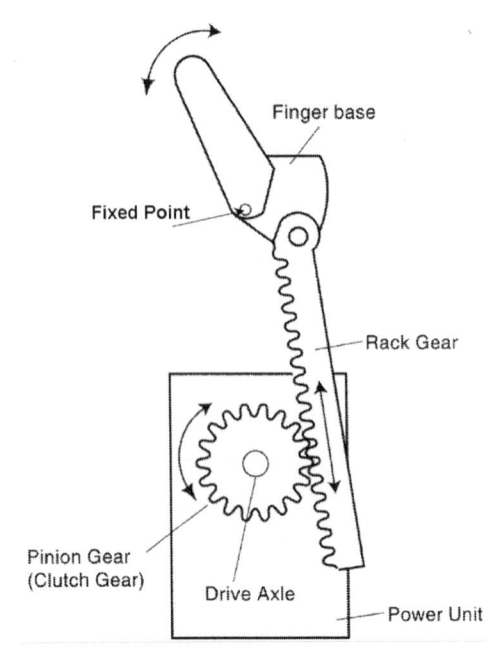

FIGURE 4.10
OWI-007 rack and pinion gear for finger (Copyright 2007 OWI ROBOTS)

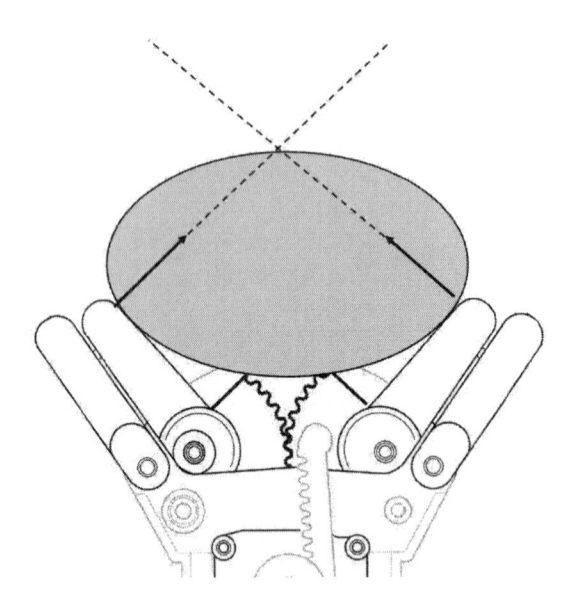

FIGURE 4.11
Nonperpendicular fingers (Copyright 2007 OWI ROBOTS)

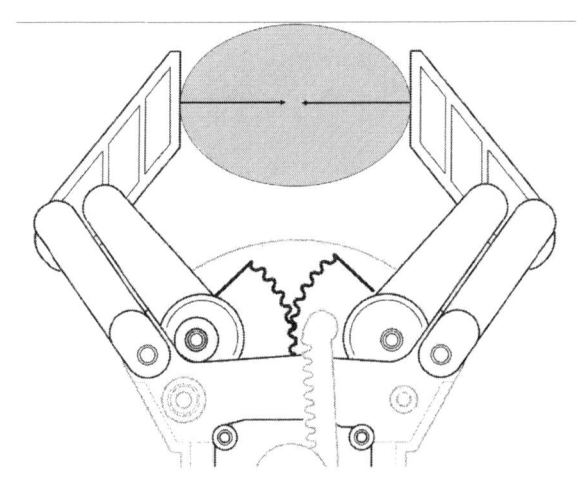

FIGURE 4.12
Perpendicular fingers (Copyright 2007 OWI ROBOTS)

other hand, if we use the mechanism shown in Figure 4.12, the normal forces do not push the object away.

To obtain this mechanism for the gripper, we need to use a four-link mechanism as shown in Figure 4.13.

Using the four-link mechanism provides the parallel movement we need as compared to the circular one we were getting before. Four-link mechanism is ideal as compared to other ones, as a three link mechanism provides no movement and five and more have extra flexibility that is not desired at all.

When the gripper is closed or open all the way, and we still apply power to the motor, then we could cause damage if we do not limit the amount of torque (or current in the motor). We can do this by using a clutch gear. We can control the amount of torque limit by how tightly we screw this gear. When the gripper has reached its limit and we still apply more power to the motor, then the motor shaft will rotate without the gear moving (the gear slips), since the gear works with contact friction with the clutch disk (and hence has the clutch action). This produces "idling" rotation without moving the gears as shown in Figure 4.14.

The gripper has the rack and pinion gear mechanism for one finger. The other finger works as a mirror image of the other finger because of the gear mechanism that makes the second side of the gripper follow the motion of the first side as shown in Figure 4.15.

4.1.4 Wrist Mechanism

The wrist also uses a DC motor with the gearbox. The gear on the motor shaft connects to the wrist with the help of a clutch plate. The wrist has a

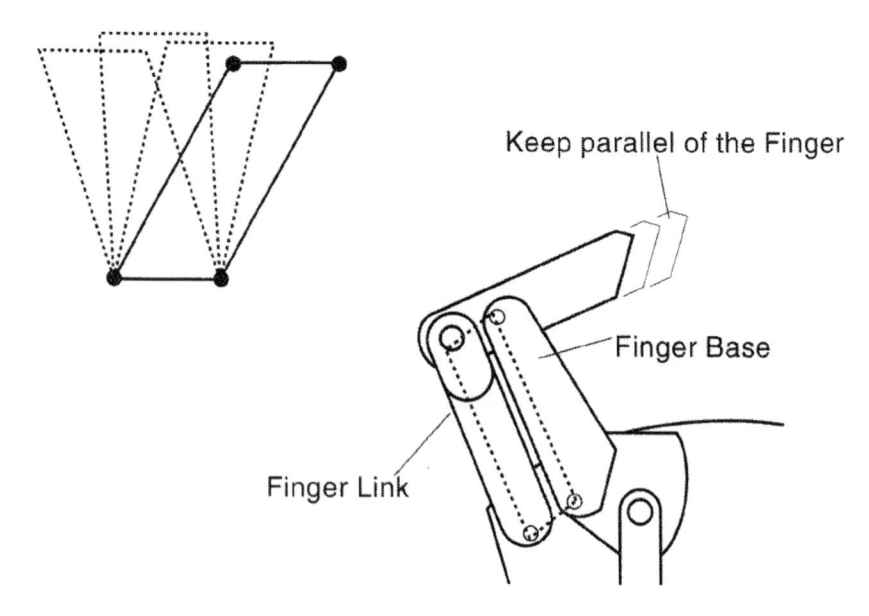

FIGURE 4.13
Four-link mechanism (Copyright 2007 OWI ROBOTS)

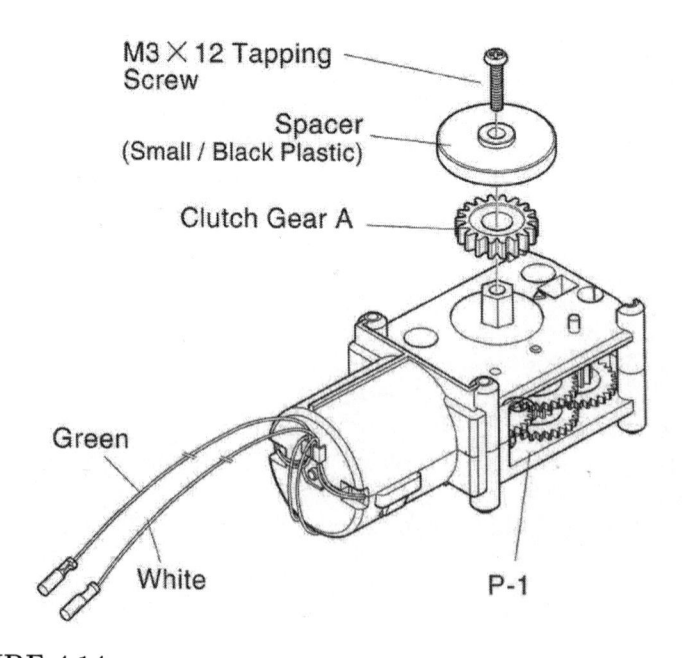

FIGURE 4.14
Torque limiting by clutch gear (Copyright 2007 OWI ROBOTS)

Robotic Arm 149

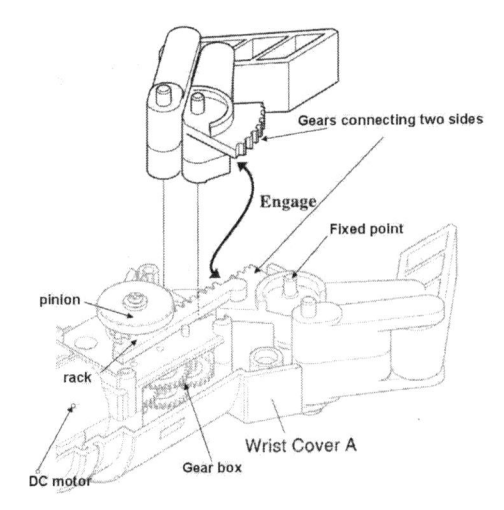

FIGURE 4.15
Mechanism for both sides (Copyright 2007 OWI ROBOTS)

clutch and rotor stopper to limit the rotation of the wrist. The clutch plate limits the torque on the wrist when the wrist has rotated all the way till it has hit the stopper. The clutch plate gets pushed forward when that happens and disengages from the driving motor gear. This mechanism can be seen in Figure 4.16.

4.1.5 Elbow and Shoulder Mechanism

The elbow and shoulder joints also use the same type of DC motor with gearbox. However, since the elbow and shoulder have more load to handle, they need more torque capability. In order to do that, they need to have further reduction in speed (and corresponding increase in torque) and that can be accomplished by additional gears as shown in Figure 4.17. These joints also have clutch based torque limiting, as well as rotation stoppers, so that the joints do not keep moving and entangle the cables. Term gear cog is used interchangeably with gear "teeth".

4.1.5.1 Elbow Details

The gear assembly for the elbow joint is shown in Figure 4.18.

The DC motor is on the other side of the upper arm. Its output shaft's gear is attached to the mediation gear on the other side. Figure 4.19 shows the motor inside the upper arm.

The gear mechanism and the DC motor are on the opposite sides of the upper arm. This is shown clearly in the Figure 4.20.

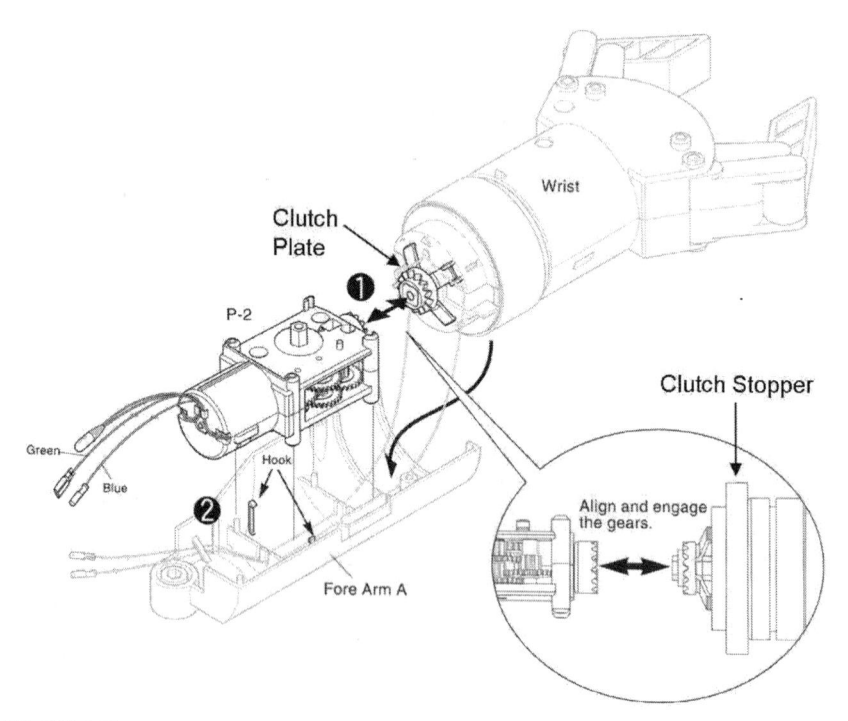

FIGURE 4.16
Wrist mechanism (Copyright 2007 OWI ROBOTS)

Robotic Arm

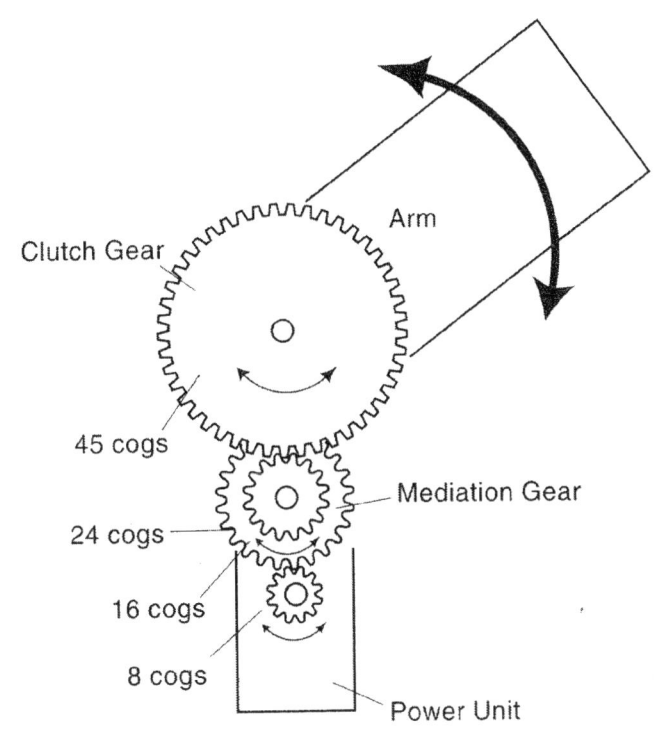

FIGURE 4.17
Elbow and shoulder mechanism (Copyright 2007 OWI ROBOTS)

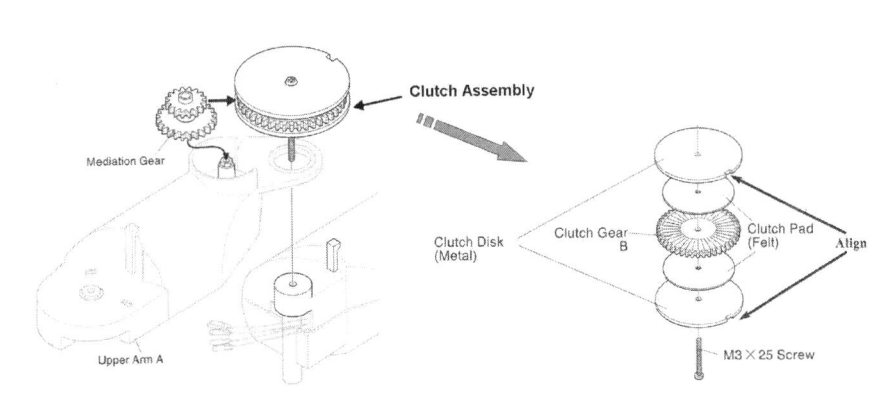

FIGURE 4.18
Elbow gear assembly (Copyright 2007 OWI ROBOTS)

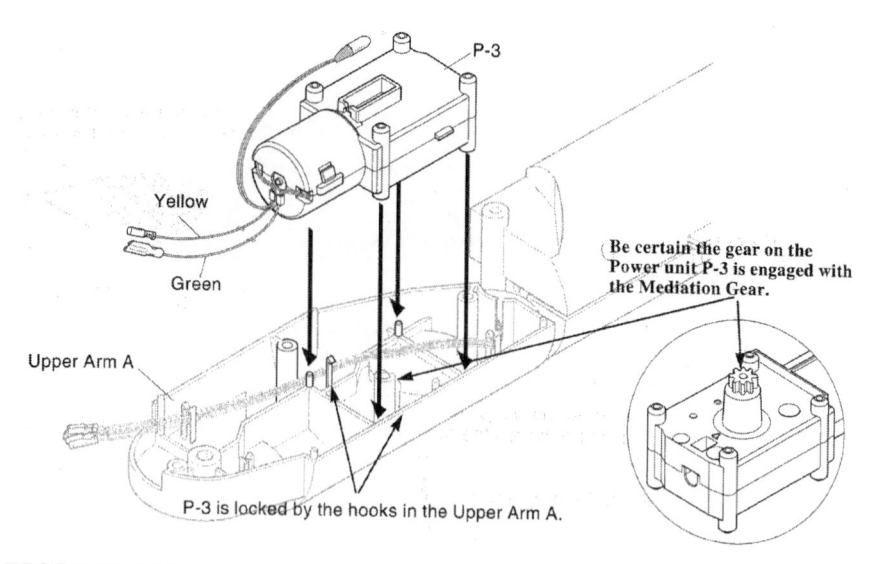

FIGURE 4.19
Elbow DC motor placement (Copyright 2007 OWI ROBOTS)

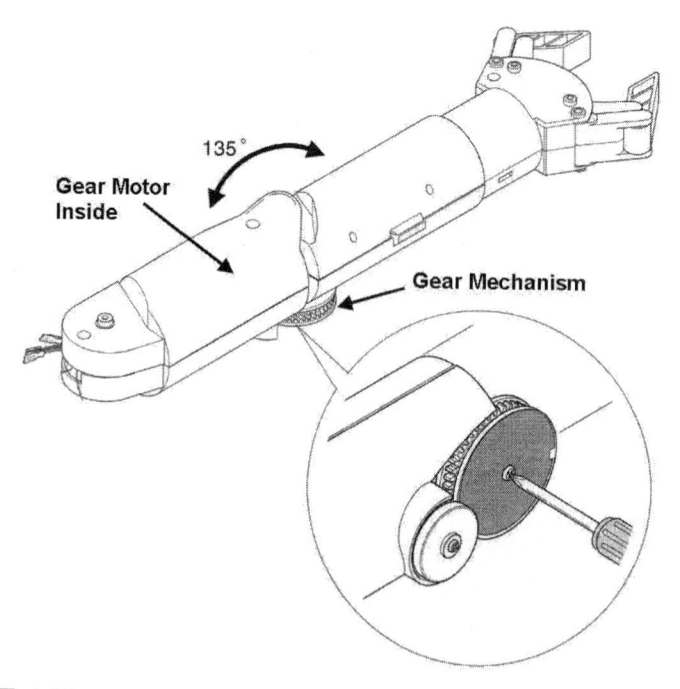

FIGURE 4.20
Upper arm details (Copyright 2007 OWI ROBOTS)

Robotic Arm

FIGURE 4.21
Shoulder DC motor and gear mechanism placement (Copyright 2007 OWI ROBOTS)

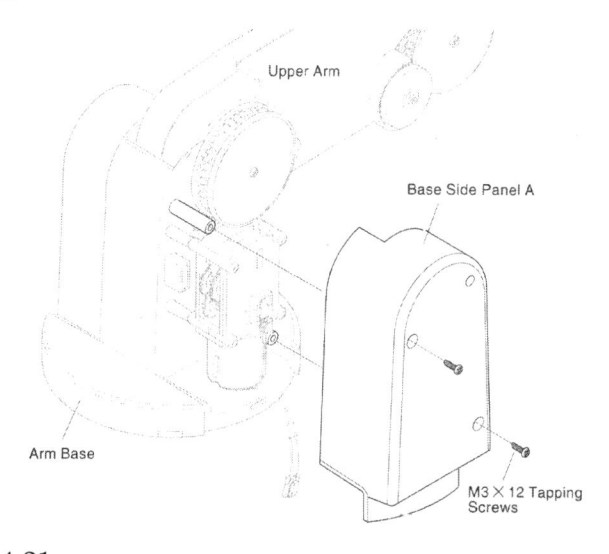

4.1.5.2 Shoulder Details

The gear assembly and the DC motor placement for the shoulder joint are shown in Figure 4.21.

4.1.6 Robot Base Mechanism

The robot base is a static base. The shoulder has another DC motor on the opposite side of where the shoulder motor is placed. That motor is for rotating the entire robot around on the base. This motor is placed as shown in Figure 4.22.

The motor is connected to an internal gear on the robot base that travels around a big static circular gear. The base remains fixed, and as the base motor is powered, its rotation rotates the shaft and that makes the robot arm rotate around. The gear mechanism is illustrated in Figure 4.23.

4.2 Electrical Control

The electrical control of the robot motors with the hand help control panel is simple. The overall schematic is shown in Figure 4.24.

There is a light bulb connected in parallel to each motor. Therefore when-

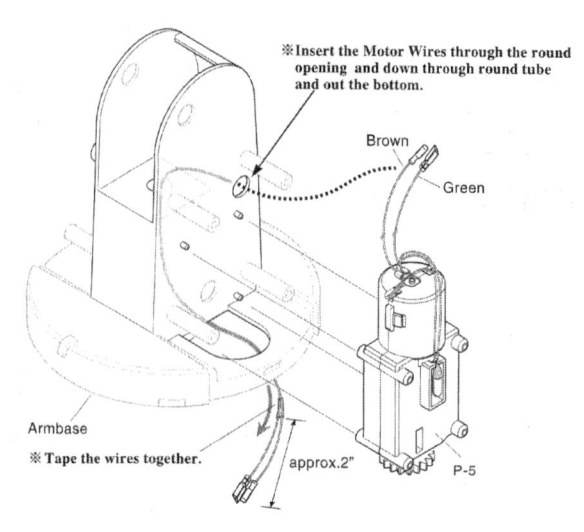

FIGURE 4.22
Robot base DC motor placement (Copyright 2007 OWI ROBOTS)

FIGURE 4.23
Robot base gear mechanism (Copyright 2007 OWI ROBOTS)

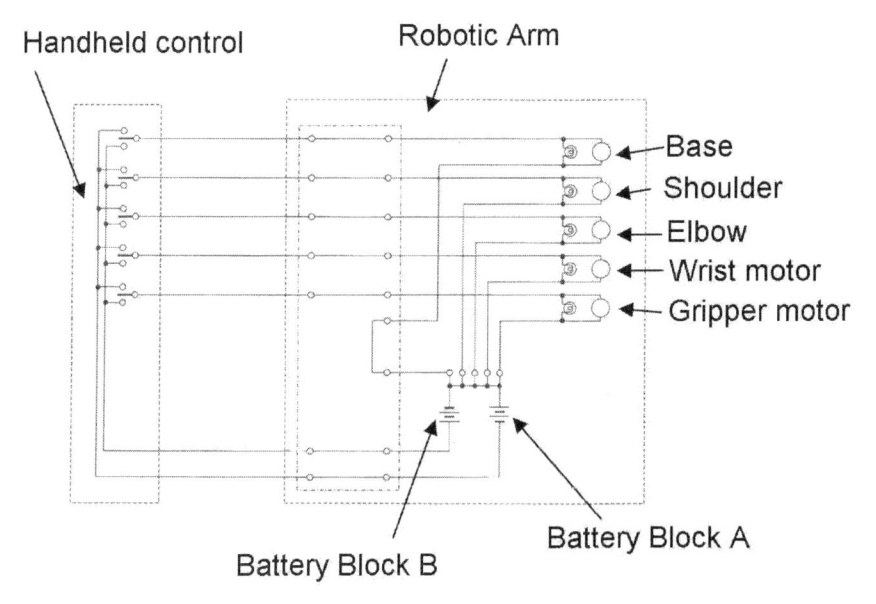

Robotic Arm

FIGURE 4.24
Electrical system schematic (Copyright 2007 OWI ROBOTS)

ever any motor is powered, the corresponding bulb lights up so that the user can see what motor/joint is moving. The hand help control panel has five switches for five motors. Each switch has three positions. The middle position is the neutral position, where the motor circuit is not completed and therefore the corresponding motor does not move. In the up position of the switch the motor is connected to the battery block A and it starts moving in one direction, and in the down position the motor is connected to the battery block B which has the reversed polarity causing the motor to rotate in the opposite direction. Therefore the three switch positions give no motion, and forward and reverse motion for each motor. These switch positions and control are shown in the Figure 4.25.

4.2.1 Robot Programming

As we have seen we can control the robot by hand using the hand held control panel. However, we would like to design a computer control for the robot. There are many reasons for that. When we have a PC or microprocessor control of a robot, then it can be used in automation as well as for doing more intelligent tasks. These are described in more detail below:

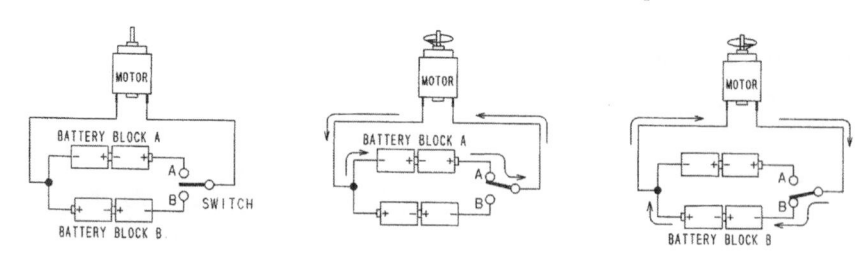

FIGURE 4.25
Motor direction control (Copyright 2007 OWI ROBOTS)

4.2.1.1 Programmed/Repetitive Tasks

In some industrial applications, such as welding on an assembly line, the robots need to do the same task over and over again. For instance, on an assembly line we might have to pick an item from one place and put it in a big bin, shown in Figure 4.26. In order to do that, we need to program the robot to learn the path the robot needs to take for that task. In order to do that we can manually move the robot to various points on the path and then click a button to make the robot learn those points. Then the robot can create its own trajectory to follow from those points. After this teaching task is done, and these points have been stored in the robot memory, we can start the repetitive task by clicking some other button. The hand held controller for a robot is called a **teach pendant** because we use it to teach the robot its tasks. Also the process the robot microprocessor takes to generate trajectories using the taught points is called **trajectory planning**.

4.2.1.1.1 Sensor (Vision) based Real-time Control
In some industrial applications, there are tasks that are not predictable. For instance in the pick and place task, the location where the object to be picked arrives might not be predictable. In those cases, we have to take the use of sensors that tell the robot where the object is and then the robot has to go get that object and then take it to the placement spot. We could use a camera that is connected to a computer as shown in Figure 4.27. The computer performs image processing to find where the object is and then the robot is commanded to perform the task accordingly.

There is one major difference in the repetitive task of the previous section and the vision based control of this section. The difference is about coordinates. When we teach the robot different points, we move the robot joints and then once it has the desired configuration, we click a button so that the microprocessor can learn that point. Now that point essentially records all the joint variable values and stores them. For example, in the OWI-007 robot we could have sensors on each joint to measure what the angles on each motor are, and when we click the button on the teach pendant for the microprocessor

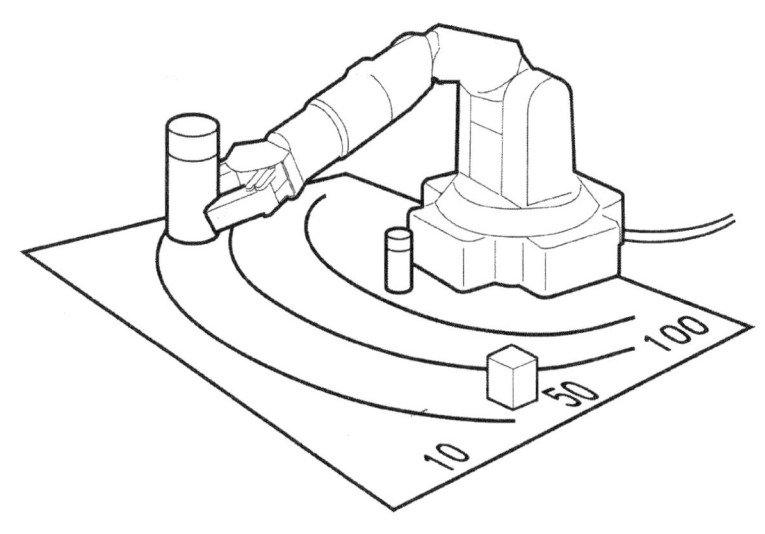

Robotic Arm

FIGURE 4.26
Programmed repetitive tasks (Copyright 2007 OWI ROBOTS)

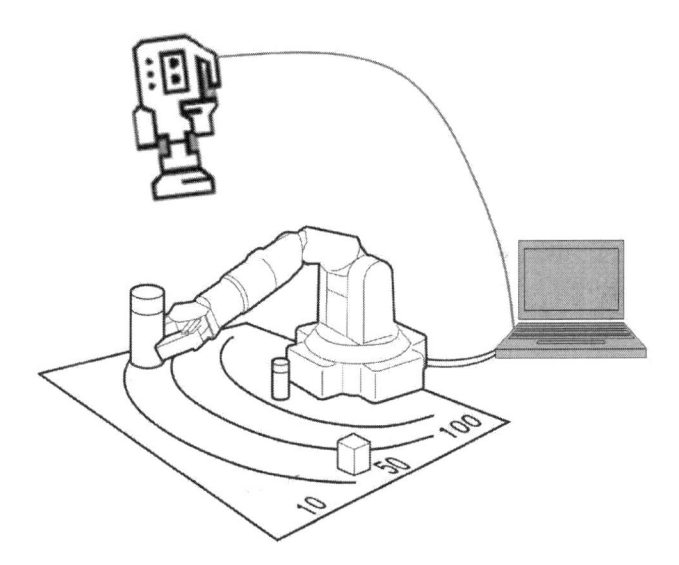

FIGURE 4.27
Vision based robotic task (Copyright 2007 OWI ROBOTS)

to learn that joint, it will read the values from all the sensors and store all those angle values.

Now, let us consider the case when we have the camera and that the object to be picked has been identified by the camera. However, the camera knows the location of the object in terms of some coordinates that we use for the camera. For instance we can identify some fixed point and three perpendicular directions to be the x-y-z axis for the camera. Now, the robot needs to know what angles to move all its joints based on what camera has measured. That means we need a way to convert what the camera sees in terms of its coordinates into robot joint angles. If we are given the robot angles, we can calculate where the gripper of the robot will be by using some mathematical functions. That function (transformation) is called **forward kinematics**. If we are given the camera coordinates of the object and we need to calculate the robot angles, we use some mathematical transformation called **inverse kinematics**. The microprocessor/computer connected to the camera needs to calculate the joint coordinates using the inverse kinematics, so that the robot can be commanded to move its joints to those values.

4.3 Parallel Port Interface Circuit using Relays

There are many ways we can create computer interface for the robotic arm. One way is to replace the switches in the hand held controller with computer-controlled relays. We could control the relays from a parallel port of a computer as shown in Figures 4.28 and 4.29. The figures show how to do bidirectional control of one of the motors on this robotic arm. When the relay is on, the motor rotates in one direction, and when it is off, it rotates in the reverse direction. Therefore, in this scheme, each motor either rotates clockwise or anti-clockwise, and the motor can not be turned off by the computer. In Figure 4.28, it is understood that each cell represents two cells in series. In other words, each cell in the figures is 6V. Therefore, in the robotic arm there are four cells that are used and the center tap point is taken between the second and the third cell.

We see in Figure 4.28 and 4.29 that when the output from the ULN2803 driver is high, the current is flowing in one direction, and when the output is low, the relay turns on, which makes the current flow in the opposite direction through the motor, causing it to rotate in the opposite direction.

We can use two relays for the motor so that we can make it rotate clockwise, anti-clockwise, or turn it off. This scheme is shown in Figures 4.30, 4.31, and 4.32.

Now, if we want to control five motors of the robotic arm, we will need ten servos to have independent control of each motor. However, if we want to

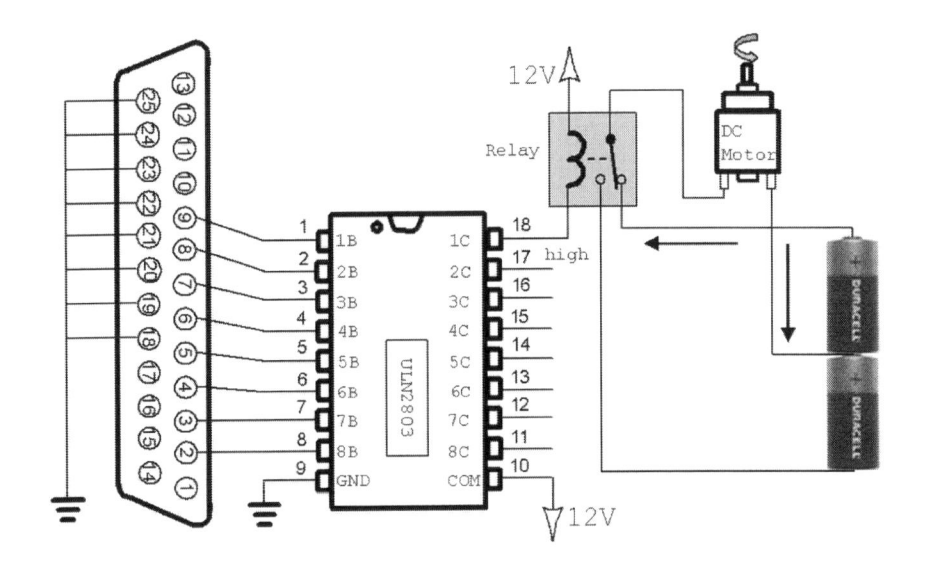

FIGURE 4.28
Forward reverse control of a motor for the robotic arm (Forward) (Courtesy of Texas Instruments)

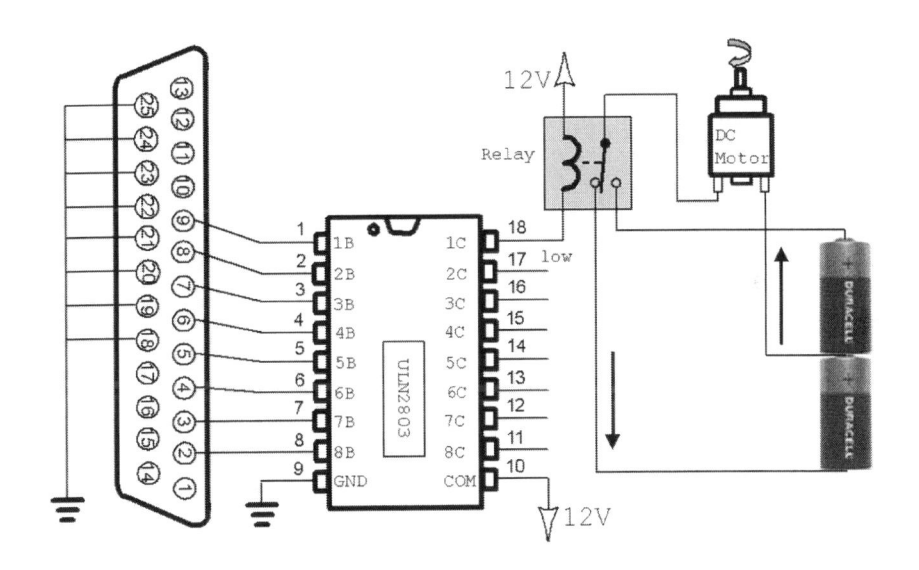

FIGURE 4.29
Forward reverse control of a motor for the robotic arm (Reverse)(Courtesy of Texas Instruments)

FIGURE 4.30
Forward reverse and off control of a motor for the robotic arm (Forward)
(Courtesy of Texas Instruments)

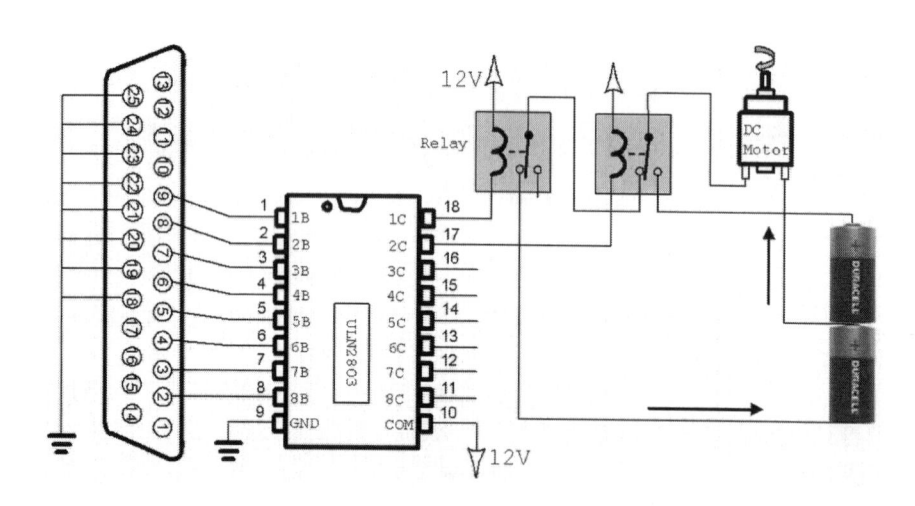

FIGURE 4.31
Forward reverse and off control of a motor for the robotic arm (Reverse)
(Courtesy of Texas Instruments)

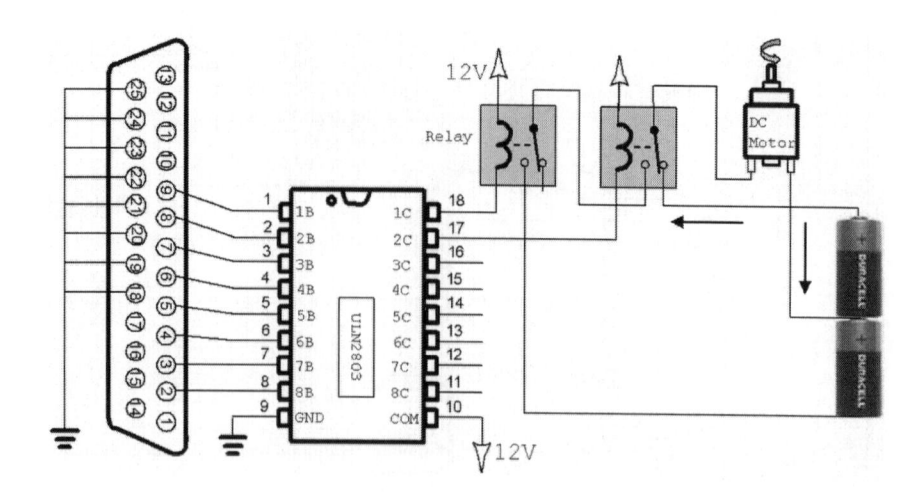

Robotic Arm 161

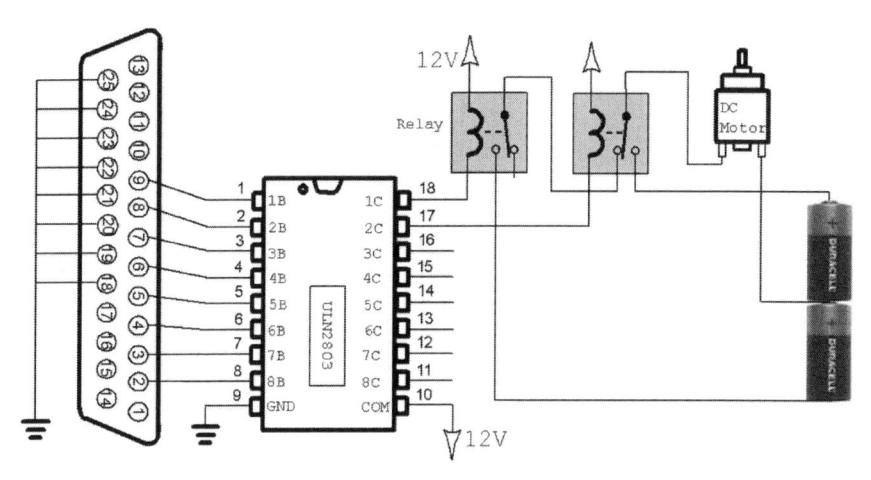

FIGURE 4.32
Forward reverse and off control of a motor for the robotic arm (Off)
(Courtesy of Texas Instruments)

control only one motor at a time, we can use the circuit shown in Figure 4.33 to accomplish that. In this circuit, there is one servo for overall forward and reverse, and then, each motor has independent on-off control.

The following software code is written in C for the Borland compiler version 5.5, available at Borland's web site [43]. On the site, click the link for Borland 5.5 compiler. You need to register to the site and download the free compiler. After the download, you can copy and paste the code shown in 4.3.1 into a C++ file and compile it.

4.3.1 Code for PC Robotic Arm Control

```
#include <stdio.h>/* standard I/O Library */
#define DATA 0x0378     /* Value to access the data pins of
   the parallel port   */

void main()
{
/* Declare all variables                                  */
int GrpCl,GrpOp,ElbDn,ElbUp,BasCW,BasCCW;
int ShdUp,ShdDn,WstCCW,WstCW,Stop,finish;
char comd;

/* Initialize the variables */
finish =0;  /* Boolean value to end the program          */
comd ='n';  /* Initialize the command variable           */
```

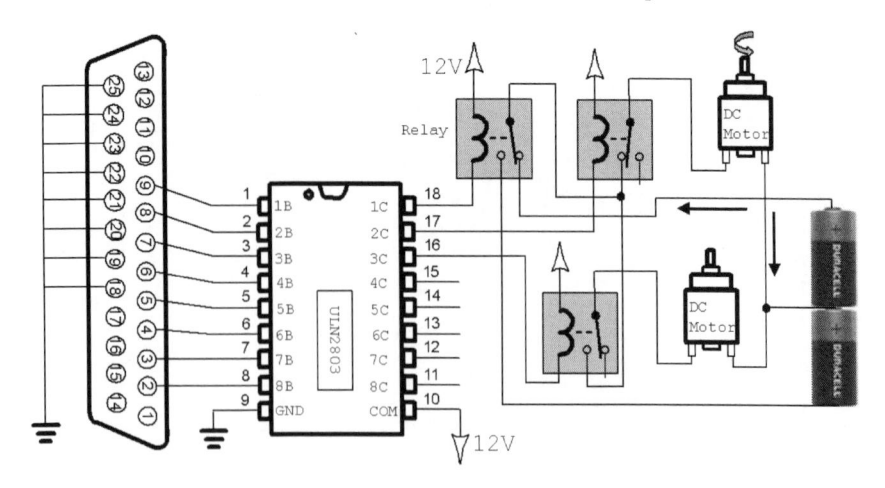

FIGURE 4.33

Forward reverse and off control for multiple motors for the robotic arm
(Courtesy of Texas Instruments)

```
/* The following are the variables used to store the
   appropriate bit values to control the motors
   NOTE: Depending on which relay is connected to which
   output pin, the following codes for various commands
   will change*/
GrpCl  =0x02;  /* send 0000 0010; to close gripper     */
GrpOp  =0x03;  /* send 0000 0011; to open gripper      */
ElbDn  =0x04;  /* send 0000 0100; move elbow down      */
ElbUp  =0x05;  /* send 0001 0101; move elbow up        */
BasCCW =0x08;  /* send 0000 1000; rotate base CCW      */
BasCW  =0x09;  /* send 0010 1001; rotate base CW       */
ShdUp  =0x10;  /* send 0001 0000; move shoulder up     */
ShdDn  =0x11;  /* send 0001 0001; move shoulder down   */
WstCCW =0x20;  /* send 0010 0000; rotate wrist CCW     */
WstCW  =0x21;  /* send 0010 0001; rotate wrist CW      */
Stop   =0x00;  /* stop all motors                      */

/* Output message to user */
clrscr(); /* Clear the Screen */
printf("Please enter a command for the robot to execute");
printf("\n:"); /* Linefeed and colon */

/* Start the loop */
while(!finish)
```

```c
{
comd = getch(); /* Get command from user          */

putch(comd); /* Echo the command to the screen   */

if (comd == 'r')   /* Close Gripper               */
outportb(DATA,GrpCl); /* output data to
parallel port   */
else if (comd == 'f') /* Open Gripper             */
outportb(DATA,GrpOp);
else if (comd == 'a') /* Elbow Down               */
outportb(DATA,ElbDn);
else if (comd == 'q') /* Elbow Up                 */
outportb(DATA,ElbUp);
else if (comd == 'd') /* Base CCW                 */
outportb(DATA,BasCCW);
else if (comd == 's') /* Base CW                  */
outportb(DATA,BasCW);
else if (comd == 't') /* Shoulder Up              */
outportb(DATA,ShdUp);
else if (comd == 'g') /* Shoulder Down            */
outportb(DATA,ShdDn);
else if (comd == 'e') /* Wrist CCW                */
outportb(DATA,WstCCW);
else if (comd == 'w') /* Wrist CW                 */
outportb(DATA,WstCW);
else if (comd == 'n') /* Stop all motors          */
outportb(DATA,Stop);
else if (comd == 'p') /* Exit the program         */
{
finish = 1; /* Set value as true         */
outportb(DATA,Stop); /* Stop all motors */
}
}
}
```

4.4 USB Interface using Relays

We can design our own USB based controller for the robotic arm. Our own scheme is based on the Pololu USB to serial chip discussed in the previous chapter. The serial receive and transmit pins are connected to two pins of

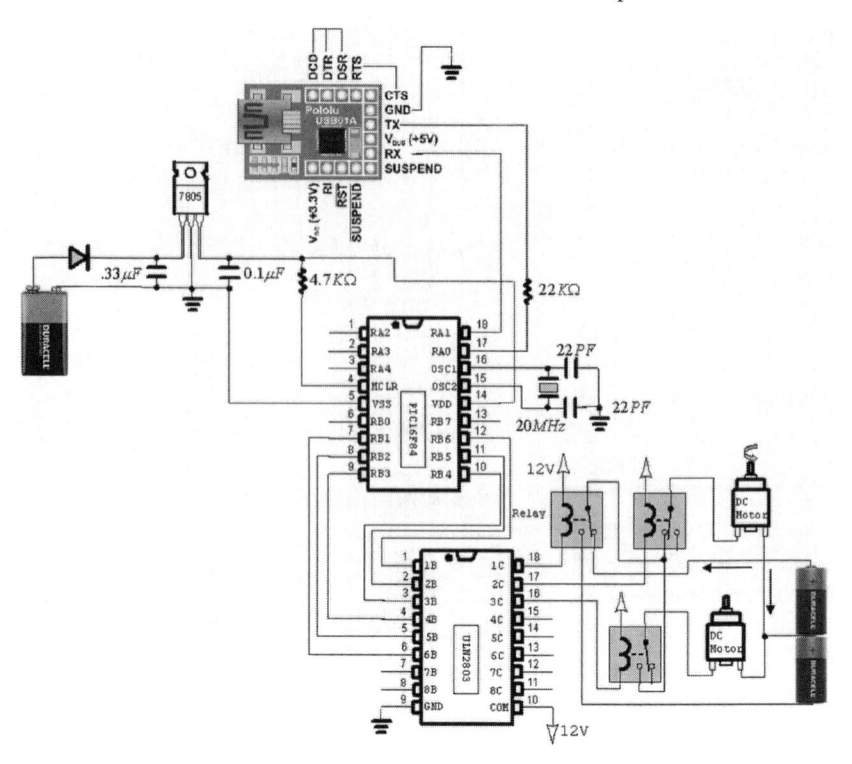

FIGURE 4.34

USB based robotic arm control (Courtesy of Texas Instruments and Pololu Inc.)

a PIC16F84. The PIC uses port B six pins to control the six relays for controlling the five degrees of freedom for the robotic arm. The schematic for this circuit is shown in Figure 4.34. The schematic only shows three of the six relays. The PC software for this would use the driver for the Pololu that comes with that board, and then we would program the PIC16F84 to read the commands from the serial communication and then control the relays using its portB output pins. The 7805 chip is a regulator chip to have 5 V supply for the circuitry.

Instead of designing our own USB based controller, we can use an off the shelf system to control the relays. One such system is the ADU208 USB based relay controller, shown in Figure 4.35. It comes with its own software development kit.

Robotic Arm 165

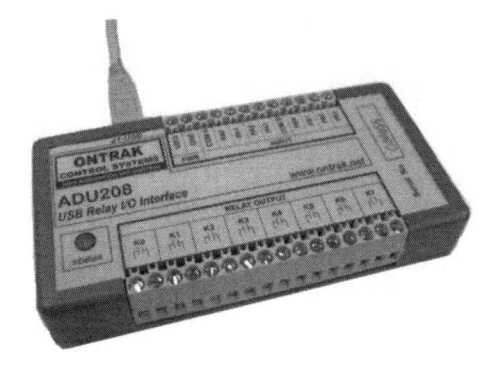

FIGURE 4.35
USB based off the shelf robotic arm control (Copyright 2003 Ontrak Control Systems)

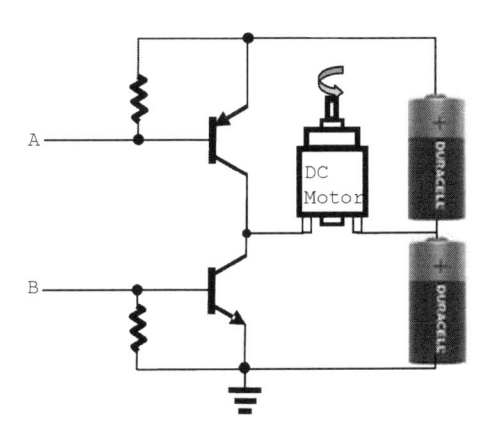

FIGURE 4.36
Motor transistor control

4.5 Parallel Port Interface Circuit using Transistors

We can turn a motor on and off using relays and control the direction as shown above. However, if we want to also control the speed, it is better to use transistors. That way we can use PWM (Pulse Width Modulation) to turn transistors on and off to control the motor speed by choosing appropriate duty cycle for the desired speed. Duty cycle is the percentage of time in a single cycle the transistor is on. The following circuit, shown in Figure 4.36, shows the basic on-off motor control using transistors.

We can see from the figure that when we turn the top transistor (PNP) on

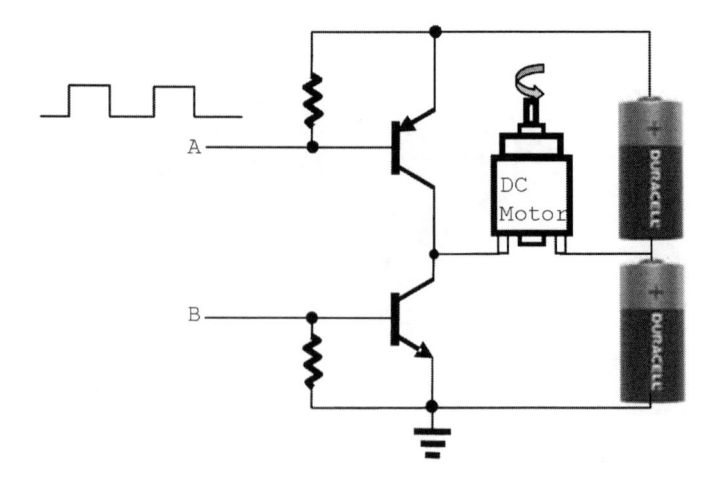

FIGURE 4.37

Motor speed control

and turn the bottom one (NPN) off, the current in the motor flows from left to right in the figure. On the other hand, when we turn the top transistor off and turn the bottom one on, the current in the motor flows from right to left in the figure. Both transistors should not be on at the same time, because that would create a short circuit. The input A turns on the top transistor with logic low and the input B turns on the bottom transistor with logic high. We can use 15K Ohms resistors. To control speed in one direction, we can keep one transistor off, and use a PWM (Pulse Width Modulated) signal on the other. This is shown in Figure 4.37. You can decide on some appropriate time period for a repetitive cycle, e.g., 100 Hz, and then if we keep the duty cycle of the PWM signal at 50%, we are keeping the transistor on for 50% of the time during each cycle. At 0% duty cycle the transistor is off at all times, and therefore will not move. At 100% duty cycle, we give the full voltage to the motor, which will produce the highest speed. At 50% it will be less. Therefore we can control the duty cycle of the PWM signal to control the speed.

Now, for the robotic arm, we have to control five motors. We have only 8 data pins on the parallel port. We can control five transistors directly from five ports and then use a 3 to 8 line decoder to get five outputs from only three pins. We can get eight outputs, but we need only five. Another way would be to use a serial to parallel converter chip. That method we will not pursue further here.

From the schematic shown in Figure 4.38, pins 2, 3, and 4 on DB25 connector are connected to the 3-to-8 decoder. These pins are D0, D1, and D2 for the data register of the parallel port. Let us study how a 3-to-8 decoder works.

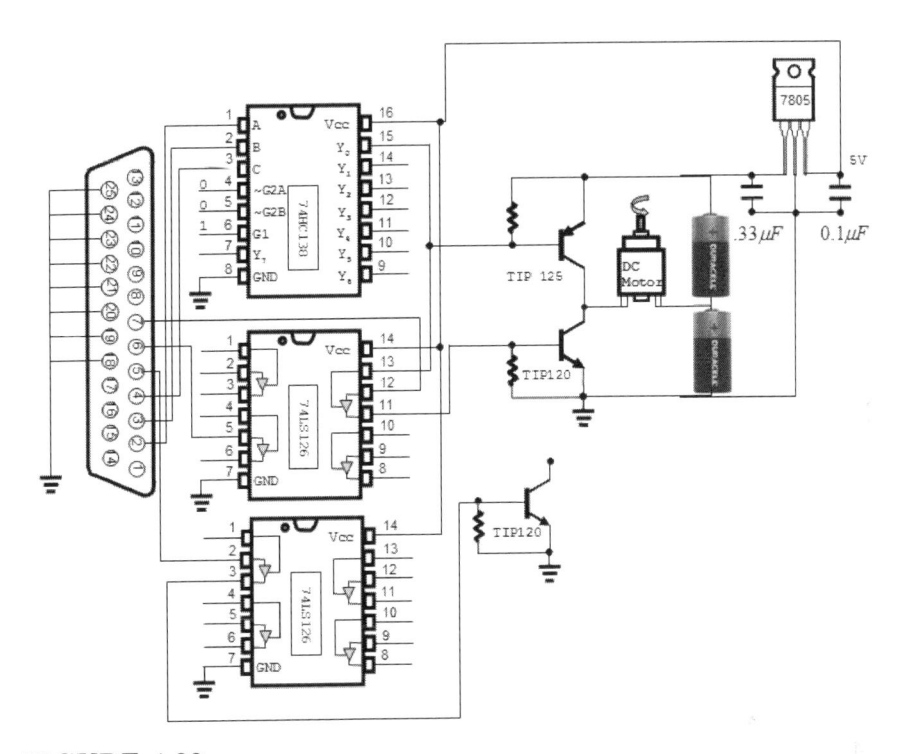

FIGURE 4.38 Schematic of motor control

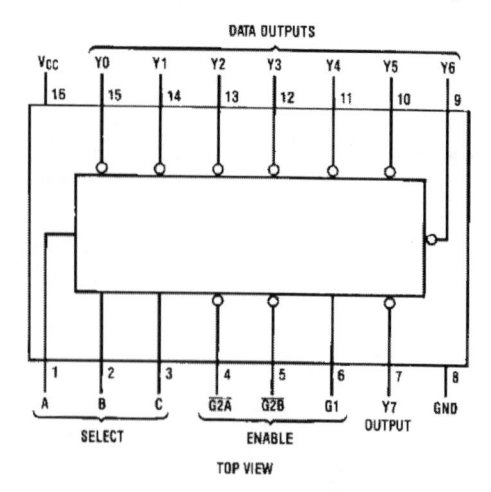

FIGURE 4.39

Pin description for 3 to 8 decoder (74LS138) (Datasheet used with permission, Fairchild Semiconductor)

TABLE 4.1
Pin description

Pin Names	Pin Description
$A_0 - A_2$	Address Inputs
$E_1 - E_2$	Enable Inputs
E_3	Enable Input
$O_0 - O_7$	Outputs

The following is taken from the data sheet on a 3 to 8 decoder from Fairchild Semiconductors. The connection diagram for the decoder chip is shown in Figure 4.39.

The pin description of the pins is shown in the Table 4.1.

The logic diagram of the decoder is shown Figure 4.40.

The truth table for the decoder is shown in Figure 4.41.

In the truth table H means logic 1, L means logic 0, and X means do not care. Either logic 0 or logic 1 are accepted for do not cares. The truth table and the logic diagram show that when the input on the three input pins is 000, then the output on the pin is 0, and the output on all other output pins is high. We connect five outputs to PNP transistors. When the output goes low, it turns on the PNP transistor. We want to make sure that when any PNP transistor is on, its corresponding NPN transistor is off. To accomplish that, we use a 74LS126 chip. It is a 3-state buffer chip. Its description below is taken from the data sheet from Fairchild Semiconductor. The connection diagram for the chip is shown in Figure 4.42.

The truth table for the buffer is given in Figure 4.43.

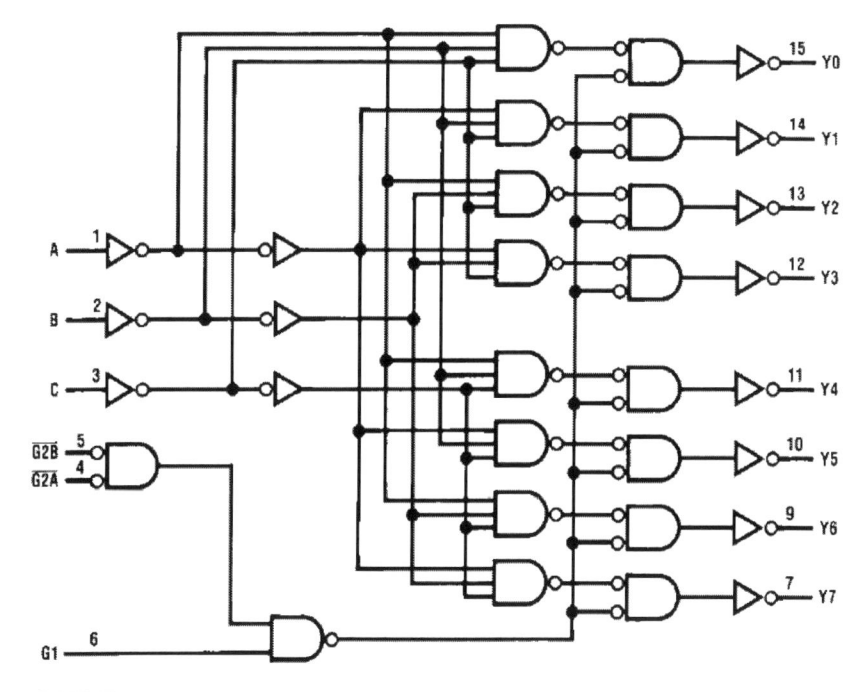

Robotic Arm

FIGURE 4.40
Logic diagram for 3 to 8 decoder (74LS138) (Datasheet used with permission, Fairchild Semiconductor)

Inputs					Outputs							
Enable		Select									
G1	$\overline{G2}$ (Note 1)	C	B	A	Y0	Y1	Y2	Y3	Y4	Y5	Y6	Y7
X	H	X	X	X	H	H	H	H	H	H	H	H
L	X	X	X	X	H	H	H	H	H	H	H	H
H	L	L	L	L	L	H	H	H	H	H	H	H
H	L	L	L	H	H	L	H	H	H	H	H	H
H	L	L	H	L	H	H	L	H	H	H	H	H
H	L	L	H	H	H	H	H	L	H	H	H	H
H	L	H	L	L	H	H	H	H	L	H	H	H
H	L	H	L	H	H	H	H	H	H	L	H	H
H	L	H	H	L	H	H	H	H	H	H	L	H
H	L	H	H	H	H	H	H	H	H	H	H	L

H = HIGH Level, L = LOW Level, X = don't care

Note 1: $\overline{G2}$ = G2A+G2B

FIGURE 4.41
Truth table for 3 to 8 decoder (74LS138) (Datasheet used with permission, Fairchild Semiconductor)

FIGURE 4.42
   Pin description for 3-state 74LS126 buffer

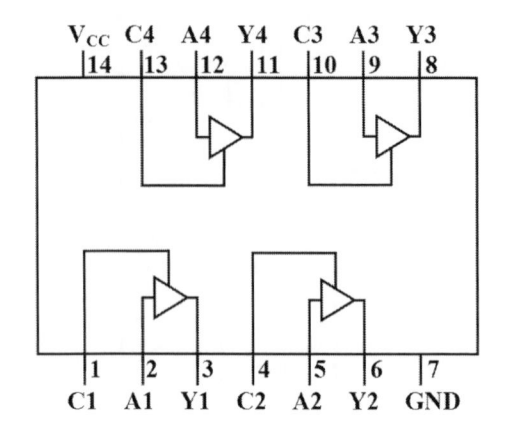

Inputs		Output
A	C	Y
H	H	H
L	H	L
X	L	Z

FIGURE 4.43
   Truth table for 3-state 74LS126 buffer

The truth table for the buffer shows that when we keep the control pin low (logic 0), the output is disconnected from the input. This is noted by Hi-Z, high impedance. Therefore, we can take the output of the 3 to 8 decoder that turns the PNP transistor on and use it to control the input to the corresponding input to the NPNP transistor. This is shown in Figure 4.38. Although it is not shown in that figure, the outputs $\bar{O}_0$ to $\bar{O}_4$ should be connected to five PNP transistors, each one controlling a corresponding NPN transistor. When we have to turn on any PNP transistor, we send a command to the 3 to 8 decoder for getting the corresponding output low, and that PNP transistor turns on, and the connected NPN transistor gets turned off. However, if we want to turn some NPN transistor on, we have to keep the corresponding PNP transistor off. To accomplish that, we can give a command like 111 to the 3 to 8 decoder which would keep all the PNP transistors turned off so that they do not interfere with the NPN operations because all the buffers would be enabled. To turn off all motors, we just send 111 to the decoder and send 0's to the NPN transistor.

Now let us modify the code written previously to control the arm using the circuit with transistors. Notice the commands to the parallel port for various motor control actions.

4.5.1 Code for PC Robotic Arm Control

```
#include <stdio.h>/* standard I/O Library */
#define DATA 0x0378     /* Value to access the data pins of
   the parallel port     */

void main()
{
/* Declare all variables                                    */
int GrpCl,GrpOp,ElbDn,ElbUp,BasCW,BasCCW;
int ShdUp,ShdDn,WstCCW,WstCW,Stop,finish;
char comd;

/* Initialize the variables */
finish =0;   /* Boolean value to end the program            */
comd ='n';   /* Initialize the command variable             */

GrpCl =0x00;   /* send 0000 0000; to close gripper          */
GrpOp =0x0f;   /* send 0000 1111; to open gripper           */
ElbDn =0x01;   /* send 0000 0001; move elbow down           */
ElbUp =0x17;   /* send 0001 0111; move elbow up             */
BasCCW =0x02;  /* send 0000 0010; rotate base CCW           */
BasCW =0x27;   /* send 0010 0111; rotate base CW            */
ShdUp =0x03;   /* send 0000 0011; move shoulder up          */
ShdDn =0x47;   /* send 0100 0111; move shoulder down        */
```

```c
WstCCW =0x04; /* send 0000 0100; rotate wrist CCW    */
WstCW  =0x87; /* send 1000 0111; rotate wrist CW     */
Stop   =0x07; /* send 0000 0111; stop all motors     */

/* Output message to user */
clrscr(); /* Clear the Screen */
printf("Please enter a command for the robot to execute");
printf("\n:"); /* Linefeed and colon */

/* Start the loop */
while(!finish)
{
comd = getch(); /* Get command from user           */
putch(comd); /* Echo the command to the screen     */

if (comd == 'r')  /* Close Gripper                 */
outportb(DATA,GrpCl); /* output data to
parallel port   */
else if (comd == 'f') /* Open Gripper              */
outportb(DATA,GrpOp);
else if (comd == 'a') /* Elbow Down                */
outportb(DATA,ElbDn);
else if (comd == 'q') /* Elbow Up                  */
outportb(DATA,ElbUp);
else if (comd == 'd') /* Base CCW                  */
outportb(DATA,BasCCW);
else if (comd == 's') /* Base CW                   */
outportb(DATA,BasCW);
else if (comd == 't') /* Shoulder Up               */
outportb(DATA,ShdUp);
else if (comd == 'g') /* Shoulder Down             */
outportb(DATA,ShdDn);
else if (comd == 'e') /* Wrist CCW                 */
outportb(DATA,WstCCW);
else if (comd == 'w') /* Wrist CW                  */
outportb(DATA,WstCW);
else if (comd == 'n') /* Stop all motors           */
outportb(DATA,Stop);
else if (comd == 'p') /* Exit the program          */
{
finish = 1; /* Set value as true           */
outportb(DATA,Stop); /* Stop all motors */
}
}
}
```

5

Robotic Arm Control

In this chapter we will cover various control tasks, robot modifications, and experiments you can design with the robotic arm from the previous chapter. You can make robot controllers for many tasks. One way to control the robot is via manual control using the hand controller. However, this chapter covers computer control.

Let us start by looking at the control of each joint. If you want to move a joint by some angle, one way to do it is to calculate by observing how fast the joint moves when its motor is on. Then to achieve some desired angle motion, we can keep that motor on for the proportional amount of time, given by equation 5.1, where we assume that motors start and stop immediately:

$$Time = \frac{desired angle movement}{joint speed} \quad (5.1)$$

If instead of moving a joint by some angle, you want the robot gripper to reach some specified point, you must think about how you know what that point is and how that information will be communicated to the robot.

5.1 Programmed Tasks

One way to teach the robot a point is to simply write a program to do a sequence of moves for each joint, spending a fixed amount of time for each move. Before starting the program, the robot must be positioned in the "home" position, which you specify. For example, one convenient home position for this robot is to turn all joints counterclockwise to their limits. Starting from this home position, you can run your programmed task multiple times, and you should be able to repeat the programmed moves every time. (This could be a pick and place operation described in the section "Automatic Control Using a Camera" later in this chapter.) There is one problem with this approach. When the robot's batteries run low, the robot will start moving slowly and will not be able to reach the exact specified points every time. Therefore, you need to use feedback to tell the robot when it has reached the desired point.

FIGURE 5.1

An EE-SX1042 switch (Copyright 2005 Omron Electronic Components LLC)

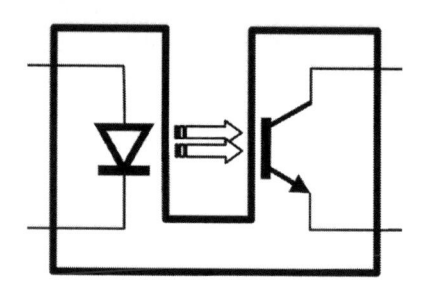

FIGURE 5.2

Internal circuitry of an EE-SX1042 switch

5.1.1 Encoder Feedback

You can use encoder feedback to know where the robot is. There are two types of encoders for this purpose: incremental encoders and absolute encoders. The incremental encoders give you information on relative motion (rotation) of a motor shaft (i.e., how much the shaft has rotated). Therefore, when you start the power, you can find out how much the joints have moved since the starting time. However, if you turn off the power and then turn it on again, you will not know how much the robot has moved. With absolute encoders, you can tell the robot's exact position as soon as you turn it on.

5.1.1.1 Incremental Encoders

You can build our own optical incremental encoder using an EE-SX1042 optical switch, shown in Figure 5.1.

Figure 5.2 shows the internal circuitry of the switch.

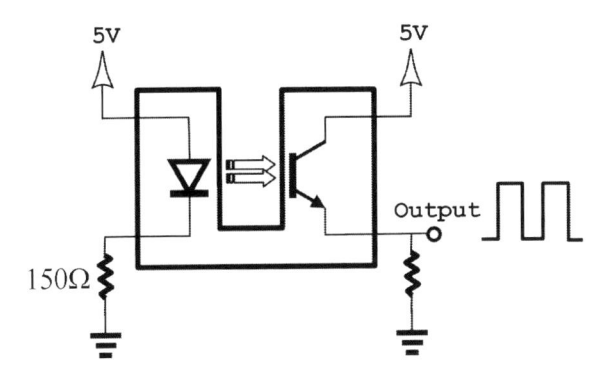

FIGURE 5.3
　　External interfacing of an EE-SX1042 switch

The switch contains an LED and a phototransistor. The LED can be powered from the input pins and is kept on all the time. If the light is interrupted so that it cannot reach the phototransistor, the phototransistor turns off; when the light can reach it, the phototransistor switch turns on. The external interfacing of the switch is shown in Figure 5.3.

When the phototransistor is on, the output is high ($\sim 5V$); when the phototransistor is off, the output is low ($\sim 0V$). Therefore, reading the output tells you if the light is being interrupted.

To the motor shaft, you can connect a round disk with slots through which light can pass. When the motor rotates, causing the disk to rotate, light passes through and is interrupted so that you get a pulse train at the output. By counting the pulses, you can find out how much the shaft has rotated, as shown in Figure 5.4.

You can design disks with many slots depending on the resolution that you require. Figure 5.5 shows a sample slotted disk that will give eight cycles per revolution of output.

By using a code wheel with the transmitter-receiver optical pair, you can find out how much a motor shaft is moving. To find out what direction it is moving, you can use two transmitter-receiver pairs placed so that when the code wheel rotates, the signals from the two receivers are 90 degrees out of phase, as shown in Figure 5.6. The Channel B signal will lag behind by 90 degrees if the wheel moves counterclockwise and will lead by 90 degrees if the wheel moves clockwise. This pair of signals is called a two-channel quadrature output.

Many incremental encoders come with one signal that emits one pulse per revolution. That allows the incremental encoder to give some absolute positioning information. You can design for that by adding a single slot per revolution on the disk at some distance from the center and having a separate transmitter-receiver pair for that. Moreover, instead of having the two

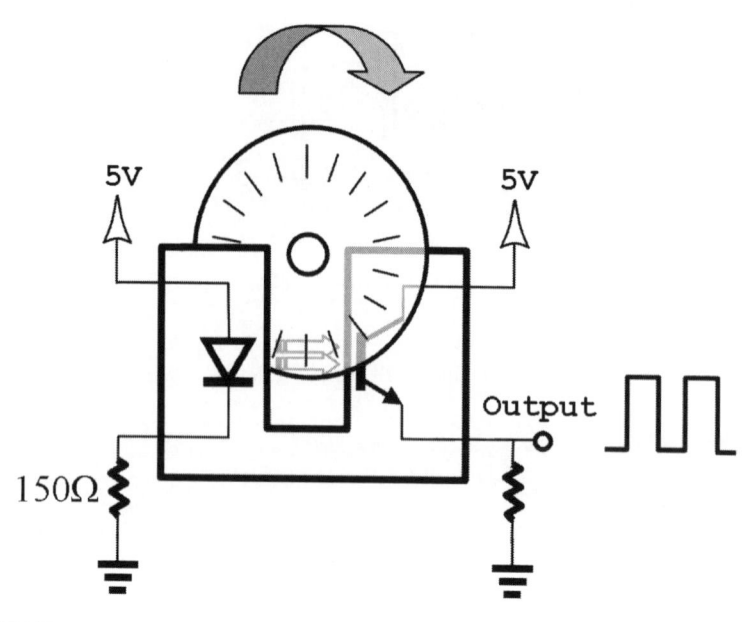

FIGURE 5.4

EE-SX1042 switch with a code wheel

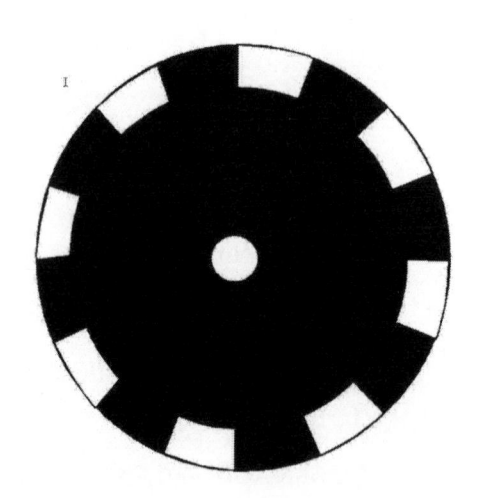

FIGURE 5.5

Code wheel for an incremental encoder

Robotic Arm Control   177

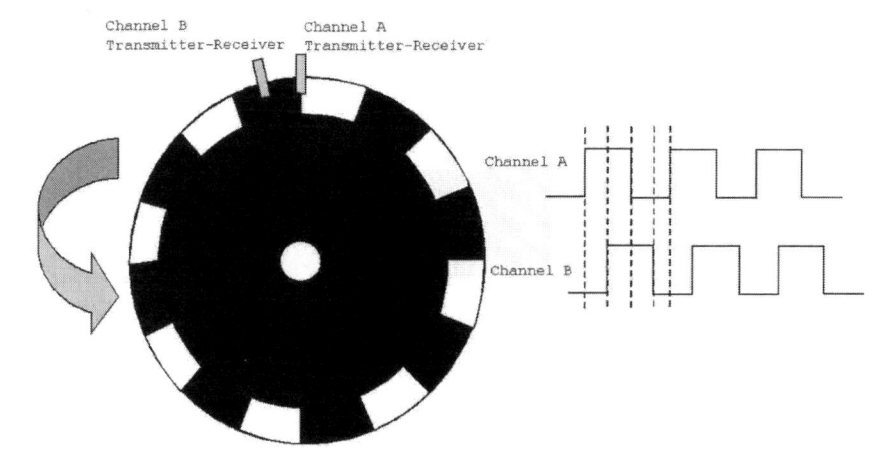

FIGURE 5.6
Code wheel for an incremental encoder with Channel A and Channel B pairs

transmitter-receiver pairs at the same distance from the center, you could have them at two different distances and have two sets of concentric slots, with one of the slots shifted to give the quadrature output. Figure 5.7 shows a code-wheel design that gives quadrature output as well as a zero marker (index marker) for one pulse per revolution.

You do not always have to build your own incremental encoder; you can buy commercial ones. Two examples are the HEDS-9000 and the HEDS-9100. These come with two transmitter-receiver pairs and a lens with the LED source. Figure 5.8 shows what these encoders look like.

Figure 5.9 presents a functional block diagram for the encoders.

The quadrature output for these encoders is shown in Figure 5.10.

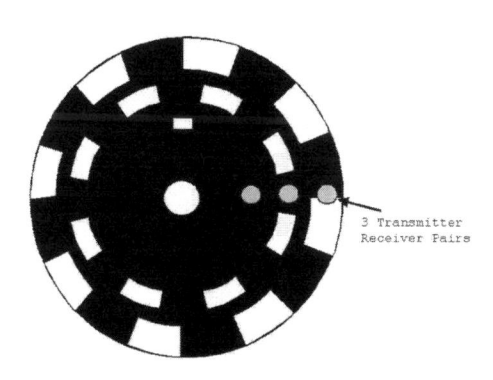

FIGURE 5.7
Code wheel for quadrature and index

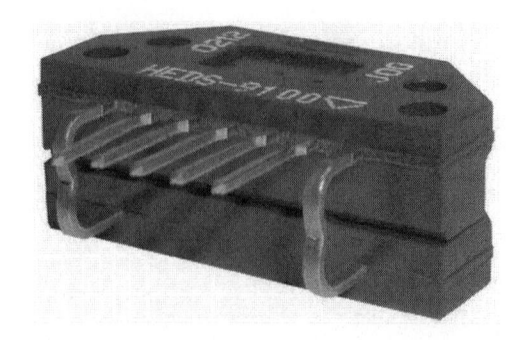

FIGURE 5.8
The HEDS-9000 or HEDS-9100 encoder (Copyright 2006 US Digital)

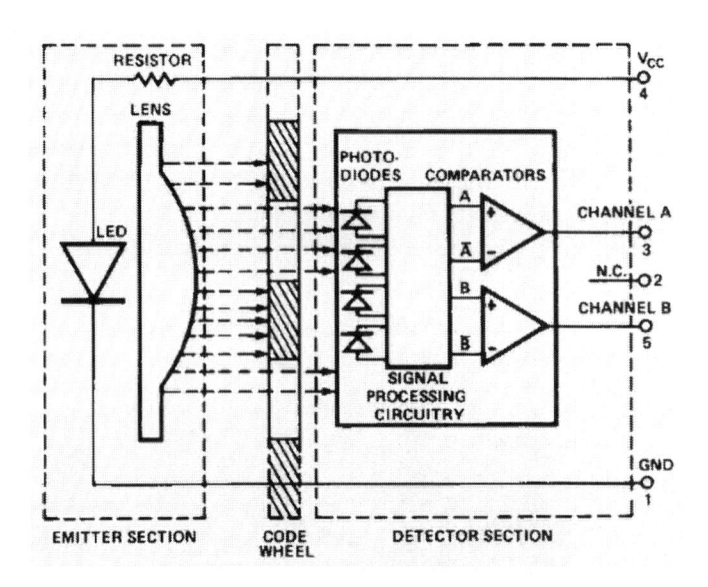

FIGURE 5.9
Functional block for HEDS-9000 and HEDS-9100 (Copyright 2006 Avago Technologies)

FIGURE 5.10
Quadrature output for HEDS-9000 and HEDS-9100 (Copyright 2006 Avago Technologies)

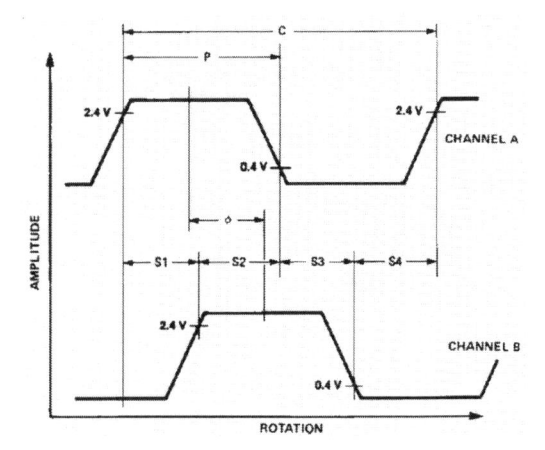

The mounting of a code wheel to this encoder is illustrated in Figure 5.11.

5.1.1.2 Absolute Encoders

Depending on the required resolution, absolute encoders use multiple transmitter-receiver pairs. The code wheels are designed using some coding scheme so that as soon as the power turns on, you can read the output from the multiple receivers and know within some resolution the angular position of the shaft. Figure 5.12 shows a sample absolute encoder code wheel.

Absolute encoders have multiple transmitter-receiver pairs at different points on the same radius of the wheel, as shown in Figure 5.13.

Most encoders use the natural binary coding of the output bits, or Gray coding. In natural binary coding, when the shaft moves, the bits follow the natural binary increasing sequence (e.g., 000, 001, 010, etc.). As you can see, when the bits change from 001 and 010, two bits change their values simultaneously for a single increment of the shaft angle. Gray coding changes only 1 bit per shaft-angle increment and reduces noise problems because of that.

You can make your own absolute encoder by cutting your own pattern on a disk and then using LED transmitter and phototransistor receiver pairs. A 3-bit absolute encoder using the natural binary coding can be designed on a wheel, as Figure 5.14 shows.

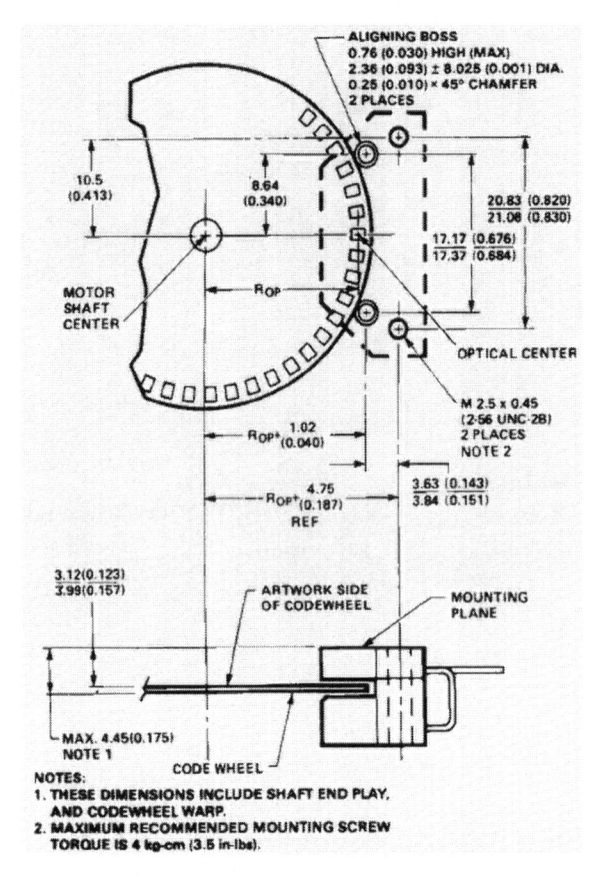

FIGURE 5.11
Code-wheel mounting for HEDS-9000 and HEDS-9100 (Copyright 2006 Avago Technologies)

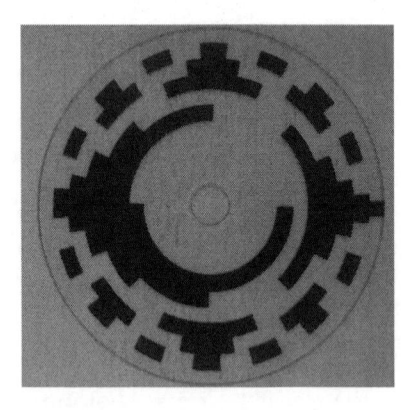

FIGURE 5.12
Code wheel for an absolute encoder

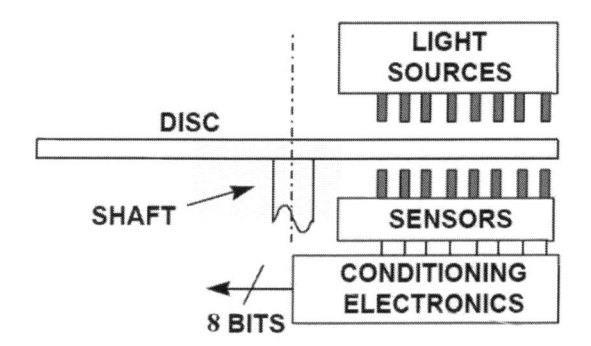

FIGURE 5.13

Absolute encoder internal construction

FIGURE 5.14

Code wheel for a 3-bit absolute encoder

FIGURE 5.15

Linear potentiometer

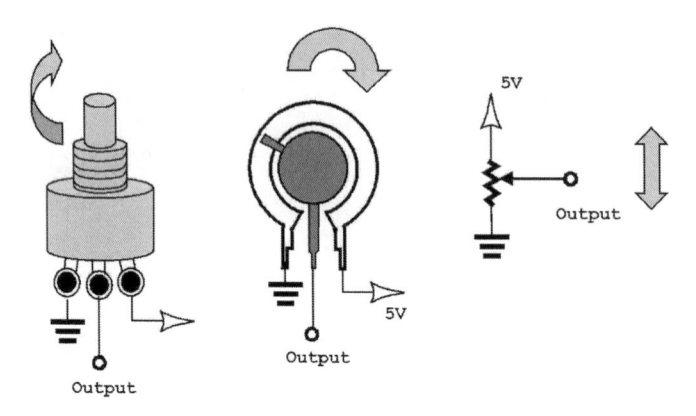

FIGURE 5.16

Potentiometer interfacing

5.1.2 Potentiometer Feedback

One easy option for sensing angular position of the motor and joints is to use potentiometers (often called pots). They give absolute angle information and are inexpensive. Figure 5.15 shows a potentiometer that you can get from RadioShack. It is a 50 ohms linear potentiometer.

To use the pot, attach it to the motor shaft of the joint. Figure 5.16 shows the pot's electronic interfacing.

One problem with using potentiometers with the robotic arm is the mechanical interfacing with the joints. String pots offer a nice way to solve that problem. One manufacturer of string pots is Celesco Transducer Products Inc. String pots have a spring-loaded string connected to a pot. You can tie the string around the robotic arms to interface them with the sensors to measure the joint angles. Figure 5.17 depicts the operation of these sensors.

5.1.3 Joint Control

Now, let us think of a way to perform a programmed task for the robotic arm using encoders. You can move the robot joints to specific positions and then

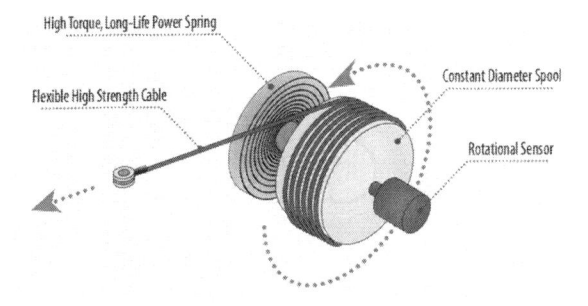

FIGURE 5.17
String potentiometer principle (Copyright 2004 Celesco Transducer Products Inc.)

record the positions (joint angles). After the recording is done, you can make the robot do a task following those recorded points. You can have a PC or a microcontroller as the brain of the robotic arm. You would have a switch to record points. You can move all the joints to extreme counterclockwise positions to go to the home position. After that you can use the incremental encoders for each joint to keep track of how much you want each joint to move. However, if you are using absolute encoders for each joint, there is no need for a homing sequence. You can make each joint move to the appropriate absolute joint angles.

5.2 Automatic Control Using a Camera

Sometimes you need to program a robot to do automatic tasks. For example, you might use a camera to see where on a flat surface some object is. Consider a square object. The camera can find where the object is, and then you can have the robot controller pick the object up and move it somewhere else. This is called a pick and place operation. Let us look at the details of a camera that you can set up easily with a PC or a microcontroller to recognize objects so the robotic arm can pick up an object from one location and place it at another.

We will use the CMUcam2+ camera, which is easy to use and set up and is a good candidate for interfacing with the robotic arm. 5.18 shows a front view of the camera with its board. Figure 5.19 shows a side view.

The block diagram shows that you can control and communicate with the camera using serial communication (RS-232). You can also interface servos directly with the camera control board. The camera can track an object by its color. Then the controller can get the coordinate of that object using serial

FIGURE 5.18
Front view of the CMUcam2+ (Copyright 1994-2006 Acroname, Inc.)

FIGURE 5.19
Side view of the CMUcam2+ (Copyright 1994-2006 Acroname, Inc.)

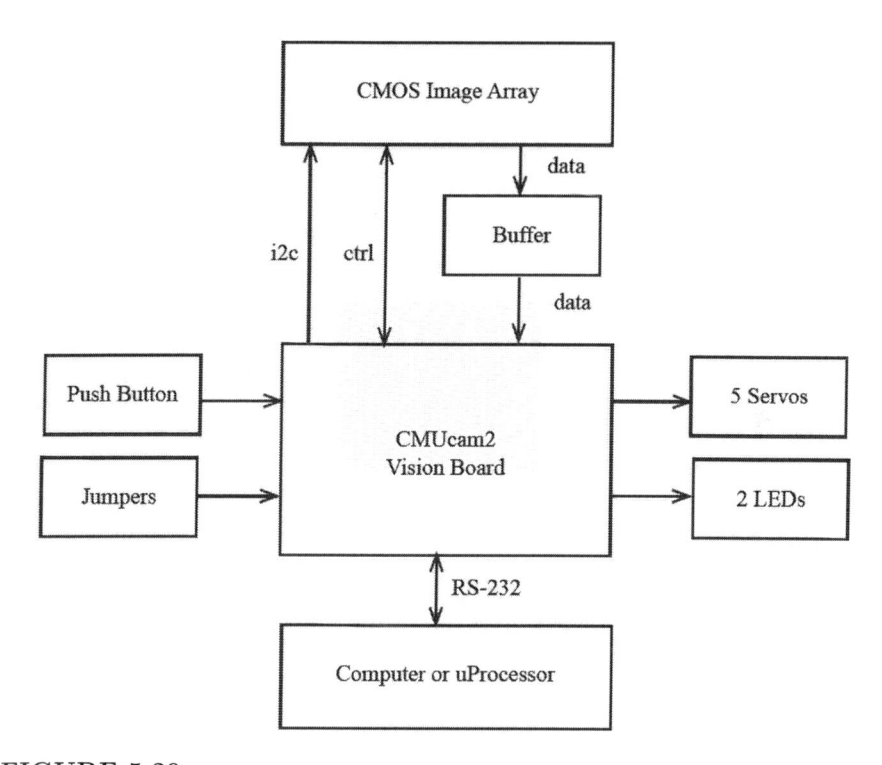

FIGURE 5.20
Block diagram for the CMUcam2+ (Copyright 2003 Anthony Rowe and Carnegie Mellon University)

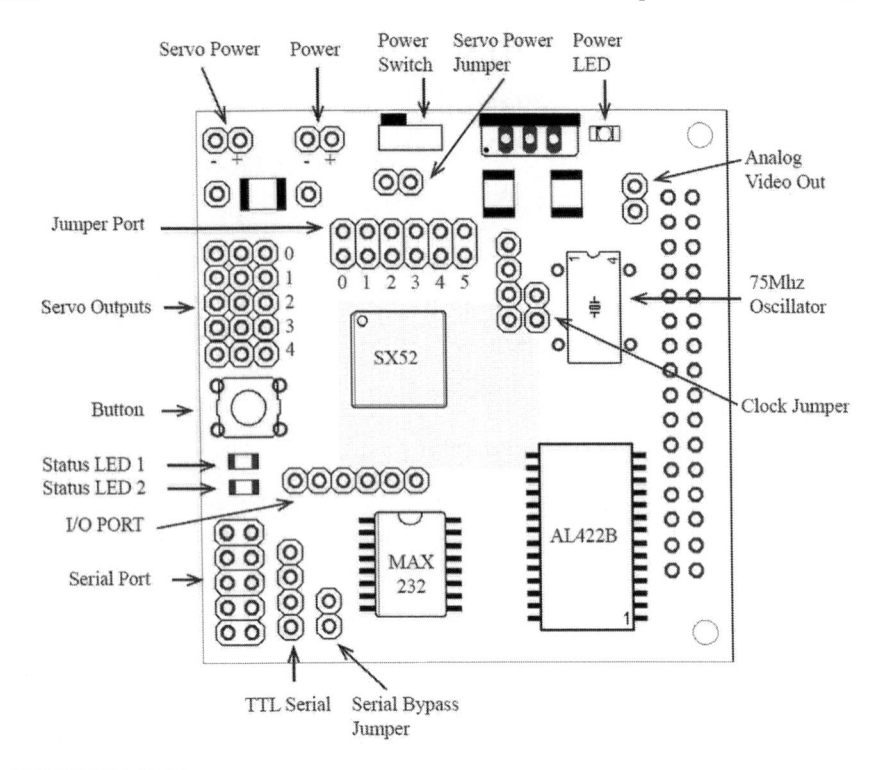

FIGURE 5.21

CMUcam board layout (Copyright 2003 Anthony Rowe and Carnegie Mellon University)

communication, and make the robot pick up that object. The board layout is shown in Figure 5.21.

The camera board has the MAX232 chip that converts transistor-transistor logic (TTL) into RS-232 logic. The board also has TTL for serial communication with microcontrollers. The communication parameters of the board are as follows:

- 8 data bits
- 1 stop bit
- No parity
- No flow control

The baud rate is configurable using jumpers on the board. When you connect the camera to a computer using a serial cable and turn the power on, and use a terminal program on the PC, you see the lines on the program:

> **TC [Rmin Rmax Gmin Gmax Bmin Bmax] \r**
>
> This command begins to Track a Color. It takes in the minimum and maximum RGB (CrYCb) values and outputs a type T packet. This packet by default returns the middle mass x and y coordinates, the bounding box, the number of pixels tracked, and a confidence value. The packet can be masked using the **OM** output mask function. Remember that the color values from the CMOS camera will range from between 16 and 240. If TC is called with no arguments it will track with the previous set of tracking parameters.
>
> Default Type T packet
> *T mx my x1 y1 x2 y2 pixels confidence\r*
>
> Example of how to Track a Color with the default mode parameters:
>
> ```
> :TC 130 255 0 0 30 30
> ACK
> T 50 80 38 82 53 128 35 98
> T 52 81 38 82 53 128 35 98
> ```

FIGURE 5.22
TC command from the CMUcam2GUI user guide (Copyright 2003 Anthony Rowe and Carnegie Mellon University)

```
CMUcam2 v1.0 c6
:
```

The camera comes with a Java program called CMUcam2GUI that allows the camera to be focused. After the camera is focused, you can employ it to track an object using color. The command that the PC or a microcontroller needs to send to the camera to obtain the coordinates of the object is TC, which is short for Track Color. Figure 5.22 provides the details of this command from the camera user guide.

After you get the coordinates of the object using the camera, you need to find the angle each joint should be at to reach the target.

5.3 Robot Kinematics

For the robot to work with a camera or any sensor that senses the "outside" world, you need to be able to express coordinates of objects in world coordinate systems and joint coordinate systems (in terms of the robot joint angles). This section shows how to do that for the robotic arm kit we are working with. The first step in deriving the formula for transforming one

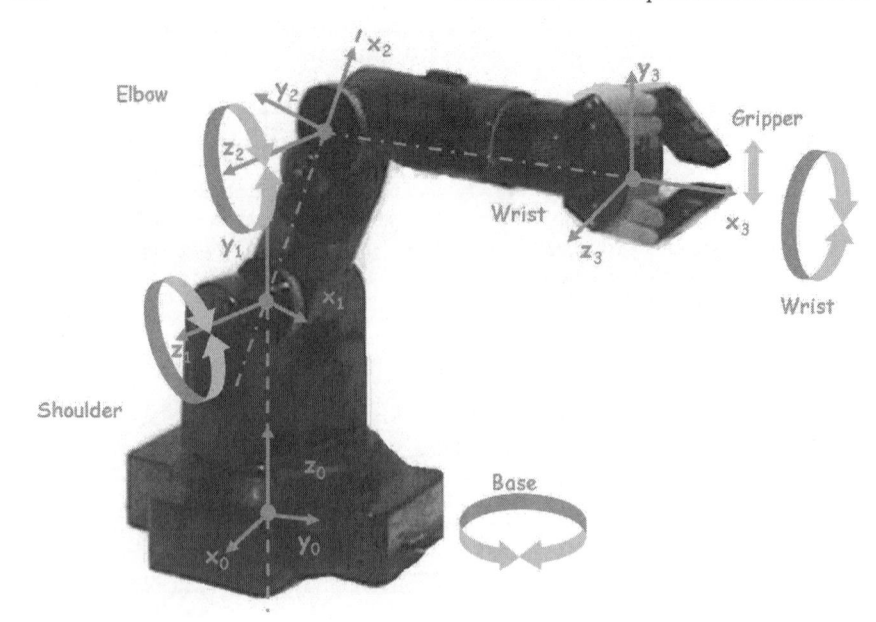

FIGURE 5.23
Coordinates for the robotic arm (Copyright 2006 OWI ROBOTS

coordinate system to the other is to use different coordinate systems for each joint of the robot. The joint angles will serve as variables, and you will use Denavit-Hartenberg (DH) representation for coordinate transformations. Derivations and details of the coordinate transformations are given in many robotics university textbooks. Here we will just apply the steps to the robotic arm. The DH transformations are based on the principle that coordinates for the robot joints are designed so that between any two consecutive frames you have the following two constraints:

- Constraint 1: Axis xi is perpendicular to axis z_{i-1}
- Constraint 2: Axis xi intersects axis z_{i-1}.

Because of these two constraints, instead of the two consecutive coordinates having six degrees of freedom between them, they have only four. Consider Figure 5.23, which shows various coordinate systems for the different joints of the robot. We will call o_i the origin point of each coordinate system. The base frame starts at o_0, the shoulder frame starts at o_1, the elbow frame starts at o_2, and the wrist frame starts at o_3. Note that the end effectors (in this case, the gripper) are not counted in the robot's degrees of freedom. Thus, the gripper's frame does not show.

The technique for building the DH table is presented in the box below titled "DH Coordinate System Buildup".

DH Coordinate System Buildup

1. *Base joint axes*: Identify the first joint axis and label it as z_0. The first joint axis for the robot is the axis about which the base joint is rotating. The base frame is located at the base at a point about which the shoulder rotates, as in Figure 5.23.

2. *Base frame*: Choose the x_0 and y_0 axes so that they make a right-handed coordinate system with z_0, as in Figure 5.23.

3. *Shoulder axes*: Choose the x_1 axis of the shoulder frame so that it is perpendicular to z_0 axis and intersects it. Locate the frame at the shoulder rotating axis.

4. *Shoulder frame*: Choose the y_1 and z_1 axes so that they make a right-handed coordinate system with x_1, as in Figure 5.23. Take z_1 to be the joint-rotation axis.

5. *Elbow axes*: Choose the x_2 axis of the elbow frame so that it is perpendicular to the z_1 axis and intersects it. Locate the frame at the elbow-rotating axis.

6. *Elbow frame*: Choose the y_2 and z_2 axes so that they make a right-handed coordinate system with x_2 as in Figure 5.23. Take z_2 to be the joint-rotation axis.

7. *Wrist axes*: Choose the x_3 axis of the elbow frame so that it is perpendicular to the z_2 axis and intersects it. Locate the frame at the wrist-rotating axis.

8. *Wrist frame*: Choose the y_3 and z_3 axes so that they make a right-handed coordinate system with x_3, as in Figure 5.23.

There are three links in this robot:

- Base link: This is the link between the base and the shoulder.

- Arm: This is the link between the shoulder and the elbow joints.

- Forearm: This is the link between the elbow and the wrist joints.

Each link has four parameters that show how the joints on each side of the links move or rotate. These parameters are listed in Table 5.1.

Using the information in Table 5.1, you can make the DH table for the robotic arm (Table 5.2).

If you know the values of the joint angles, you can calculate the position of the robot hand in world coordinates. This is called *forward kinematics*. To calculate the hand's position, you must calculate a transformation matrix by

TABLE 5.1
DH link parameters

a_i	Distance from the intersection of x_i and z_{i-1} to o_i; the distance taken along the x_i axis
α_i	The angle between z_{i-1} and z_i axes; the rotation taken about the x_i axis
d_i	Distance from o_{i-1} to the intersection of x_i and z_{i-1} axes; the distance taken along the z_{i-1} axis
θ_i	The angle between x_{i-1} and x_i axes; the rotation taken about the z_{i-1} axis

TABLE 5.2
DH table for the robotic arm

Link	a_i	d_i	α_i	θ_i
1	0	d_1	90°	θ_1
2	a_2	0	0°	θ_2
3	a_3	0	α_3	θ_3

multiplying three matrices, one for each link of the robot. The transformation matrix for each link is give by equation 5.2.

$$A_i = \begin{pmatrix} \cos\theta_i & -\sin\theta_i \cos\alpha_i & \sin\theta_i \sin\alpha_i & a_i \cos\theta_i \\ \sin\theta_i & \cos\theta_i \cos\alpha_i & \cos\theta_i \sin\alpha_i & a_i \sin\theta_i \\ 0 & \sin\alpha_i & \cos\alpha_i & d_i \\ 0 & 0 & 0 & 1 \end{pmatrix} \quad (5.2)$$

There are four transformations (two rotation and two translations) between each consecutive pair of joint coordinate systems to transform one coordinate into another. The four transformations make up the four link parameters of the DH table. The matrix A_i is derived from equation 5.3:

$$A_i = Rot_{z,\theta_i} Trans_{z,d_i} Trans_{x,a_i} Rot_{x,\alpha_i} \quad (5.3)$$

Here, Rot stands for rotation, and Trans stands for translation. This equation means that if you take frame i and perform the following four operations, it will become frame $i-1$.

- *Operation 1*: Rotate frame i about its x axis by an angle of α_i.

- *Operation 2*: Take the new frame obtained in operation 1 and translate it by a distance of a_i in its x direction.

- *Operation 3*: Take the new frame obtained in operation 2 and translate it by a distance of d_i in its z direction.

- *Operation 4*: Take the new frame obtained in operation 3 and rotate it about its z axis by an angle of θ_i.

Formula 5.3 can be expanded to derive equation 5.2 by using the following matrix multiplication. (The details of these can be obtained from any university book on robotics (see [16] or [6]).

$$A_i = \begin{pmatrix} \cos\theta_i & -\sin\theta_i & 0 & 0 \\ \sin\theta_i & \cos\theta_i & 0 & 0 \\ 0 & 0 & 1 & 0 \\ 0 & 0 & 0 & 1 \end{pmatrix}$$

$$\begin{pmatrix} 1 & 0 & 0 & 0 \\ 0 & 1 & 0 & 0 \\ 0 & 0 & 1 & d_i \\ 0 & 0 & 0 & 1 \end{pmatrix} \begin{pmatrix} 1 & 0 & 0 & a_i \\ 0 & 1 & 0 & 0 \\ 0 & 0 & 1 & 0 \\ 0 & 0 & 0 & 1 \end{pmatrix} \begin{pmatrix} 1 & 0 & 0 & 0 \\ 0 & \cos\alpha_i & \sin\alpha_i & 0 \\ 0 & \sin\alpha_i & \cos\alpha_i & 0 \\ 0 & 0 & 0 & 1 \end{pmatrix} \quad (5.4)$$

This is based on the fact that

$$Rot_{z,\theta_i} = \begin{pmatrix} \cos\theta_i & -\sin\theta_i & 0 & 0 \\ \sin\theta_i & \cos\theta_i & 0 & 0 \\ 0 & 0 & 1 & 0 \\ 0 & 0 & 0 & 1 \end{pmatrix} \quad (5.5)$$

$$Trans_{z,d_i} = \begin{pmatrix} 1 & 0 & 0 & 0 \\ 0 & 1 & 0 & 0 \\ 0 & 0 & 1 & d_i \\ 0 & 0 & 0 & 1 \end{pmatrix} \quad (5.6)$$

$$Rot_{x,a_i} = \begin{pmatrix} 1 & 0 & 0 & a_i \\ 0 & 1 & 0 & 0 \\ 0 & 0 & 1 & 0 \\ 0 & 0 & 0 & 1 \end{pmatrix} \quad (5.7)$$

$$Trans_{z,d_i} = \begin{pmatrix} 1 & 0 & 0 & 0 \\ 0 & \cos\alpha_i & \sin\alpha_i & 0 \\ 0 & \sin\alpha_i & \cos\alpha_i & 0 \\ 0 & 0 & 0 & 1 \end{pmatrix} \quad (5.8)$$

Using equation 5.2 and the DH table parameters for each link, you can calculate the total transformation from the wrist frame to the base frame:

$$T(\theta_1, \theta_2, \theta_3, \alpha_3) = A_1(\theta_1)A_2(\theta_2)A_3(\theta_3)A_4(\theta_4) \quad (5.9)$$

Equation 5.10 provides the position of the robot hand in world coordinates (i.e., in frame 0).

$$\begin{bmatrix} x \\ y \\ z \\ 1 \end{bmatrix} = T(\theta_1, \theta_2, \theta_3, \alpha_3) \begin{bmatrix} 0 \\ 0 \\ 0 \\ 1 \end{bmatrix} \qquad (5.10)$$

From equation 5.10, given the joint angles, you can calculate the x, y, and z values of the location of the robot hand. For the pick and place problem, you are actually more interested in the opposite problem: given the x, y, and z values of the object location (obtained from the camera and calculated in terms of the frame 0) find the desired joint angles. The equation for these is more involved and can be obtained from university-level robotics text books. Once you know the desired joint angles, the robot controllers can move the joint angles to those values. To obtain the inverse kinematics, you must first find the forward kinematics to discover the world coordinates of the gripper in terms of the joint angles; the world coordinates are functions of joint coordinates. Equation 5.10 provides the function. To obtain the inverse kinematics you need to find the inverse relationship of the function shown in equation 5.11.

$$\begin{bmatrix} x \\ y \\ z \end{bmatrix} = f(\theta_1, \theta_2, \theta_3, \alpha_3) \qquad (5.11)$$

The inverse kinematics relationships are obtained as shown in equation 5.12.

$$\begin{bmatrix} \theta_1 \\ \theta_2 \\ \theta_3 \\ \alpha_3 \end{bmatrix} = f^{-1}(x, y, z) \qquad (5.12)$$

5.3.1 Velocity Kinematics

If you want to control the arm to not only go to a specified position, but to move at a certain speed (which does not have to be constant), you need velocity relationships between world and joint coordinates. If you know the velocities of the joints, you can find the velocity of the gripper. This relationship can be obtained by differentiating equation 5.11 with respect to time.

$$\frac{d}{dt}\begin{bmatrix} x \\ y \\ z \end{bmatrix} = \frac{\partial f(\theta_1, \theta_2, \theta_3, \alpha_3)}{\partial(\theta_1, \theta_2, \theta_3, \alpha_3)} \begin{bmatrix} \dot{\theta}_1 \\ \dot{\theta}_2 \\ \dot{\theta}_3 \\ \dot{\alpha}_3 \end{bmatrix} \qquad (5.13)$$

In equation 5.13 $frac\partial f(\theta_1, \theta_2, \theta_3, \alpha_3)\partial(\theta_1, \theta_2, \theta_3, \alpha_3)$ is a 3×4 marix called the Jacobian. The matrix is time-varying and depends on the values of the joint variables at any given time. You can obtain the inverse relationship by solving equation 5.14.

$$\begin{bmatrix} \dot{\theta}_1 \\ \dot{\theta}_2 \\ \dot{\theta}_3 \\ \dot{\alpha}_3 \end{bmatrix} = \begin{bmatrix} \frac{\partial f(\theta_1, \theta_2, \theta_3, \alpha_3)}{\partial (\theta_1, \theta_2, \theta_3, \alpha_3)} \end{bmatrix}^{-1} \begin{bmatrix} \dot{x} \\ \dot{y} \\ \dot{z} \end{bmatrix} \quad (5.14)$$

Similarly, you can find relationships between the joint and world accelerations.

5.4 Dynamics and Control

To make the robot move along a certain trajectory, you need to apply different voltages to the motors (PWM commands to the motors) at different times. How do you know how much voltage to apply at different times to each motor? This is computed using feedback controllers that use sensors to detect the joint angles and/or the gripper position and then calculate the desired voltages. For details, consult the books [16] or [6]. The control designs are done based on the dynamics of the robot that can be derived using the Lagrangian approach or the Euler iterative method. Once you've derived the dynamics that show how the voltages dictate the accelerations, feedback controllers can be designed to control the robots for specific tasks.

5.5 Some Experiments

You can do many more experiments with the robot. Some of them are listed here.

- Design a complete path for the robot hand in world coordinates and then make the robot hand follow the path.

- Perform speed control of the robot, having it follow not just a path but a time-based trajectory.

- Add a force sensor to the gripper so that the robot can do force feedback control. Force feedback can ensure that you apply only a desired force. For instance, you can make the gripper hold eggs without breaking them.

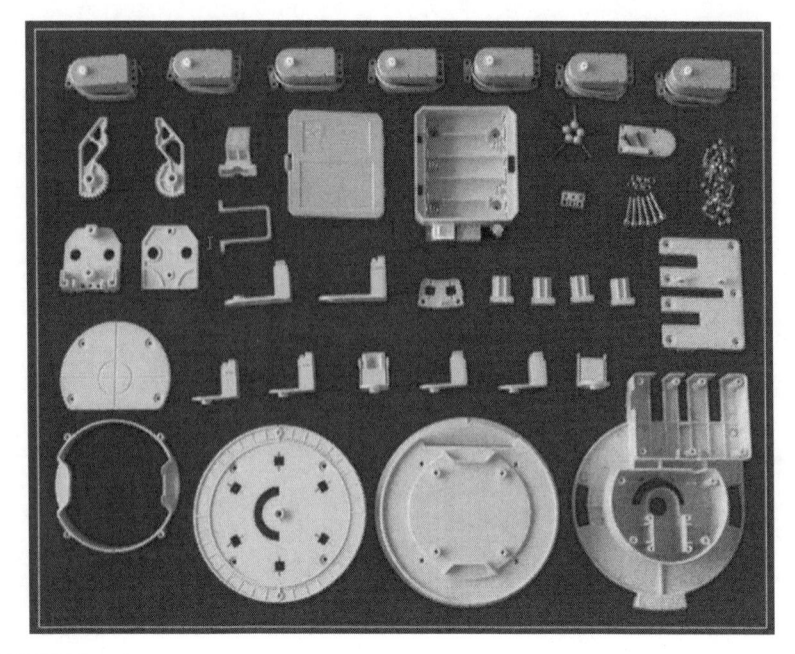

FIGURE 5.24
Necessary parts to construct the six-DOF arm robot (Copyright 2002 MCII Corporation)

5.6 Control of a Six-Degrees-of-Freedom Arm Robot

In this section, we will explore a servomotor-based six-degrees-of-freedom (DOF) arm robot by MCII Robot. We will briefly explore the mechanical construction and the Mini Servo Explorer software that controls the robot. Then we will derive the DH table and transformations from the robot's base to the hand. Finally, we will explore forward kinematics and inverse kinematics with some examples. Through careful programming, you can accurately position the robotic arm to perform various complicated movements in sequence.

5.6.1 Mechanical Construction

The six-DOF arm robot has seven mini servomotors. These motors provide one degree of freedom each to the base, the arm, the forearm, the forewrist, the back wrist, and the clamp, and enable them to move within 80 degrees of freedom. Figure 5.24 shows all the parts used to construct the six-DOF arm robot. Figure 5.25 shows the completed robot.

As mentioned earlier, the six-DOF arm robot has seven servos, including

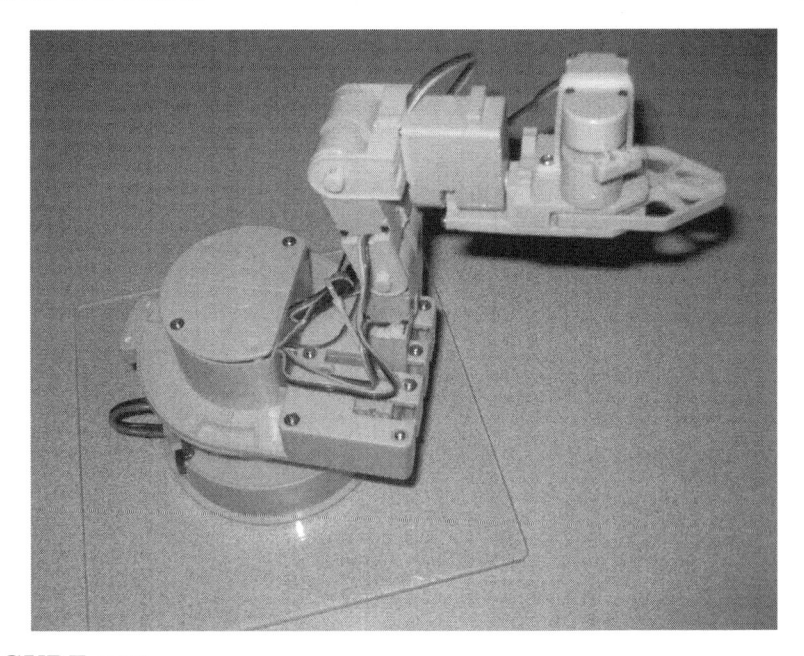

Robotic Arm Control 195

FIGURE 5.25
     Completed six-DOF arm robot

the servo in the gripper. The servos from the base to the gripper generate six degrees of freedom since each of them creates a degree of freedom. Figure 5.26 is a typical assembly drawing showing all the servos and major components of the robot. For details of the construction and parts, you can download the six-DOF arm robot user manual at MCII Corporation web site [44].

5.6.1.1 Hardware Components

The six-DOF arm robot has two major components: mini servomotors and JM-SSC16 mini servomotor controller board.

5.6.1.1.1 Mini Servomotors
Hobbyists in robotics and other fields use mini servomotors widely. A servomotor consists of a DC motor, a gear box, a variable resistor for feedback, and an electronic control board. Figure 5.27 presents internal components of a mini servomotor. The angle that a servomotor rotates is controlled through a feedback system. First a periodic pulse is applied to control circuitry. The difference between the input voltage and the voltage fed back from the variable resistor is applied to the DC motor. Then the DC motor rotates the gear box and the variable resistor attached to the gear box. The variable resistor has a voltage based on how much the gears rotate it. This voltage is fed back to the control circuitry and compared with the reference voltage applied to the control circuitry. This reference voltage

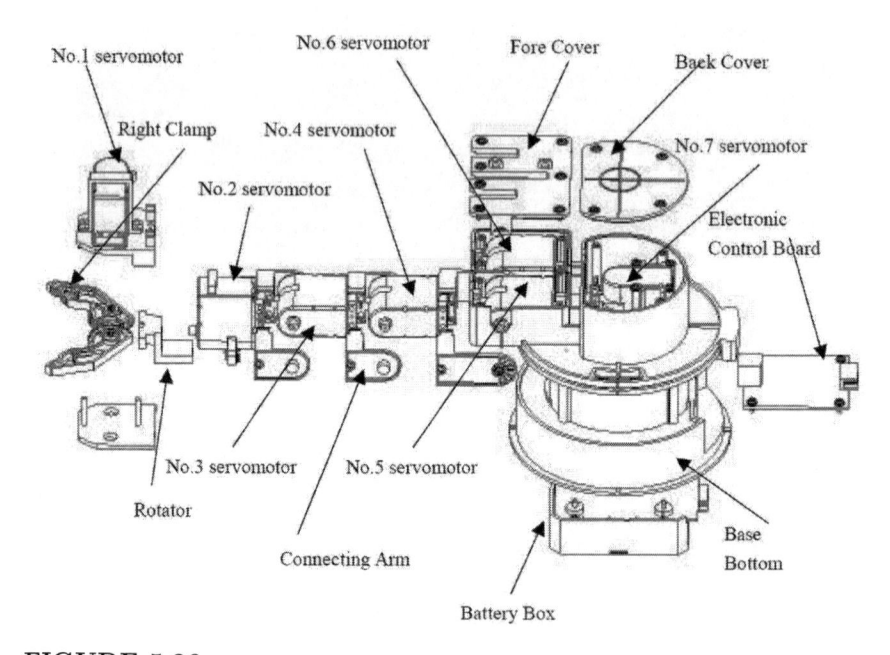

FIGURE 5.26
Major mechanical components of the six-DOF arm robot (Copyright 2002 MCII Corporation)

Robotic Arm Control

FIGURE 5.27
Internal components of a mini servomotor (Copyright 2002 MCII Corporation)

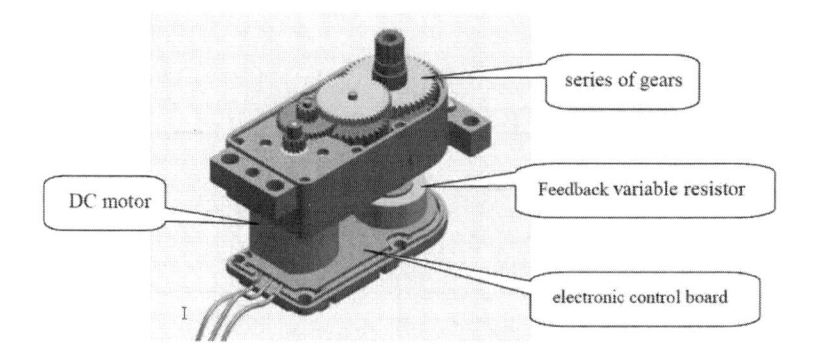

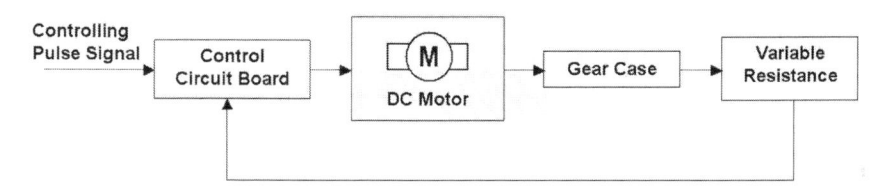

FIGURE 5.28
Internal components of a mini servomotor

corresponds to the desired angle. The difference in voltage will be large when the pulse is applied, and will gradually decrease, since the voltage on the variable resistor will increase after the gears turn the resistor. The difference will be zero when the servomotor reaches the target angle.

Figure 5.28 shows the block diagram of a servomotor.

A mini servomotor has three wires: power (red), ground (black), and control (yellow). The power and ground wires power the control circuitry and the DC motor. The power should be about 4 to 6 volts and be separated from the control line to suppress the noise the servomotor creates. The mini servomotor requires a pulse of 1 to 2 ms in a period of 20 ms. Since the mini servomotors are designed to have 80 degrees of freedom, the pulse width will vary approximately from 0.5 ms (-45 degrees) to 1.5 ms (45 degrees). However, the user does not need to think about the pulse width; the software and the mini servomotor controller create the necessary pulse width. In the software, the angle is represented as a number between 0 and 253, where the midpoint of the servomotor is 127.

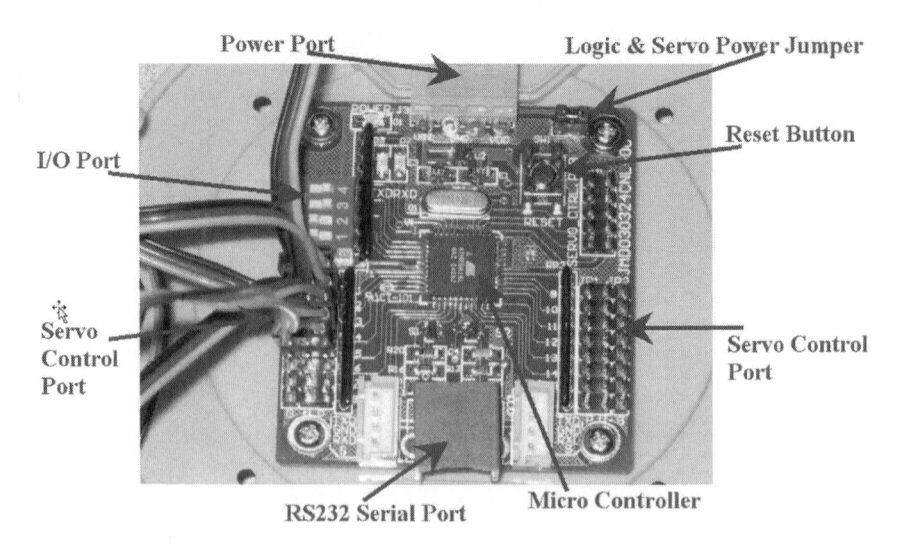

FIGURE 5.29
JM-SSC16 mini servomotor controller

5.6.2 JM-SSC16 Mini Servomotor Controller Board

The JM-SSC16 is a general purpose servomotor controller that can control up to 16 servomotors through the serial port of a PC. Figure 5.29 shows the JM-SSC16 servomotor controller. Detailed pinouts of the JM-SSC16 servomotor controller and how to connect it to a PC are explored in the user manual for the six-DOF arm robot.

5.6.3 Control Software

There are two communication protocols to control the robot using the serial port: instant control protocol and download protocol. In the instant control protocol, a control command (shown in Table 5.3) is sent to the servomotor controller through the serial port. The first byte is 255 (0xFF in hexadecimal form). The second byte is the target servomotor. The third byte is the value for the servomotor, which determines the angle of the motor. It takes values from 0 to 255 representing -45 degrees and 45 degrees, respectively. The value 127 represents the 0 degree angle.

In the download protocol, software called Mini Servo Explorer is used to program a series of angle values for all the servomotors in the robot and to send them to the servomotor controller sequentially or continuously. The following list presents the important features of the software.

- The upper and lower limits of servomotors can be set separately according to the parameters and environmental conditions of each servomotor.

Robotic Arm Control 199

TABLE 5.3
Meaning of the bytes in the instant control protocol

Byte Number	Meaning
1	0xFF
2	Servomotor number (0-16)
3	Servomotor parameter (0-253)

- Servomotors can be controlled individually or simultaneously.

- Servomotors can be moved to a target position with the highest speed, or a speed can be set for the operation.

- Servomotors can be controlled by inputting values of the positions or drawing a scroll bar on the interface.

- The values of the servomotors can be changed while you are running the robot. It is also possible to edit the values without connecting to the robot.

- The actions can be saved in a file for future usage.

- It is possible to drive two servomotors in a master-slave architecture where the slave servomotor follows the master.

The servomotor settings in the Mini Servo Explorer software are presented in Figure 5.30.

Figure 5.31 presents a sample program to move the robot up and down.

5.6.4 Forward and Inverse Kinematics

As stated earlier, it is possible to control the robot using the instant control protocol through the serial port. If the servomotor parameters are sent to the JM-SSC16 controller in the format specified in Table 5.3, the robot will move to the servomotor specified in the command to the target position based on the parameter. To move the robot on a path, you need to send a series of commands to move the robot to specified positions. In reality, if you want to move the robot hand in a linear path in Cartesian coordinates, you must generate the points (positions) by doing a series of geometric calculations. After you generate the points you want the robot to follow, you must find the angles that each servomotor should be at to place the robot hand at the desired location. You can create the DH table of the robot and find the transformation from the robot's base to its hand through forward kinematics equations. However, to find the desired angles (parameters) of the servomotors, you need to find equations for these angles in terms of the position and the orientation of the hand. This is called inverse kinematics, and it is not an easy task for a six-DOF arm robot.

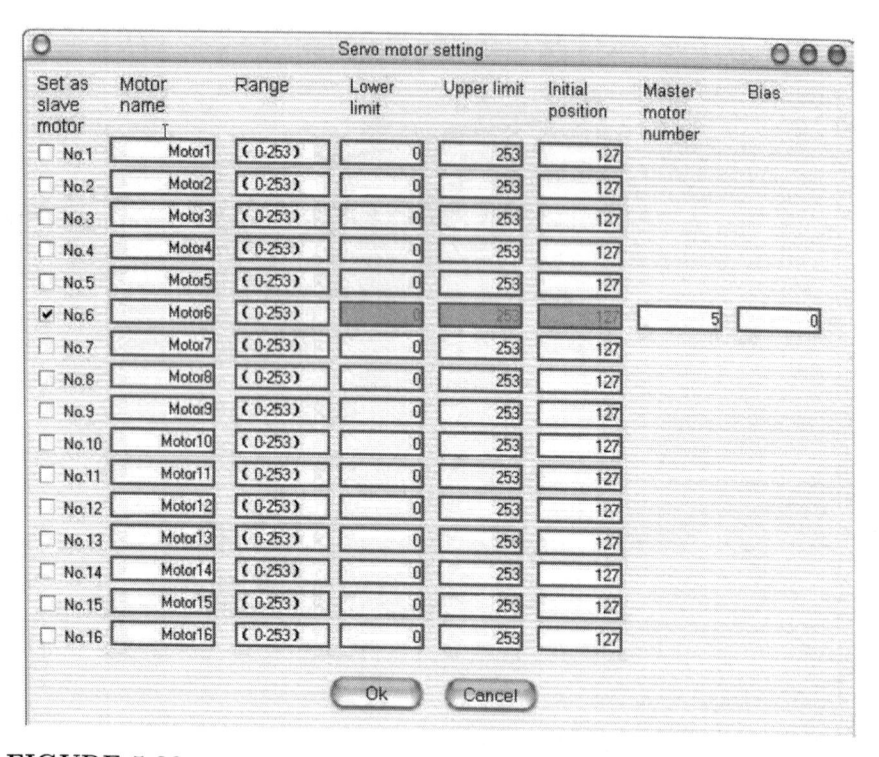

FIGURE 5.30

Servomotor settings in Mini Servo Explorer

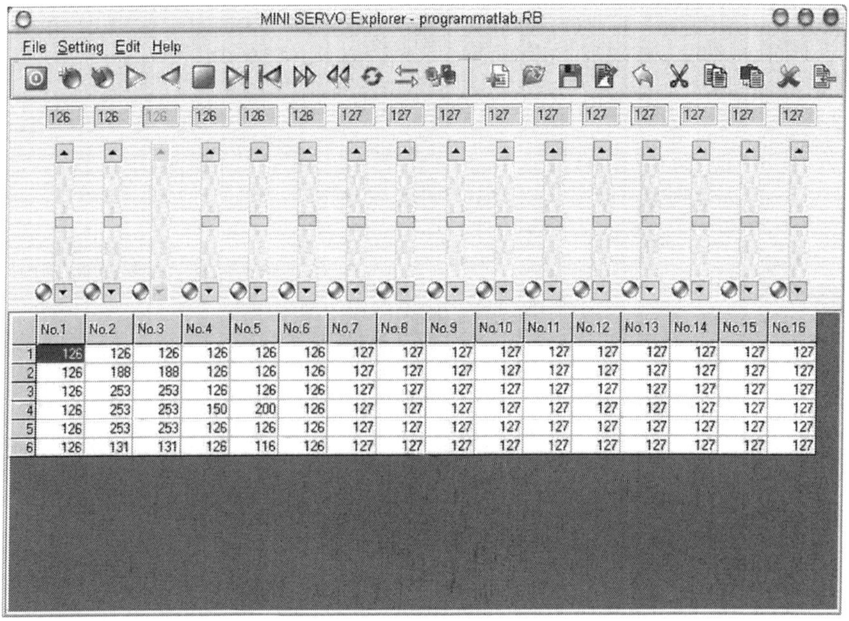

Robotic Arm Control

FIGURE 5.31
An example program to move the robot up and down

5.6.4.1 Forward Kinematics

To come up with inverse kinematics equations, you need to first find the forward kinematics equations of the robot. Figure 5.32 shows a picture of the robot with the joint frames. We put the frame 0 at the base of the robot and placed other frames based on the DH table rules. These rules are mentioned when we discuss the control of the five-DOF arm robot early in the chapter.

Table 5.4 is the DH table of the six-DOF arm robot. Some of the frames are specially placed to ease the calculation of the inverse and forward kinematics equations.

The parameters a_i and d_i are obtained from the robot's physical specifica-

TABLE 5.4
The DH table of the six-DOF arm robot

Link	a_i	d_i	α_i	θ_i
1	a_1	d_1	$90°$	θ_1
2	a_2	0	$0°$	θ_2
3	a_3	0	0	θ_3
4	a_4	0	$90°$	θ_4
5	0	0	0	θ_5

FIGURE 5.32

Six-DOF arm robot with joint-coordinate frames

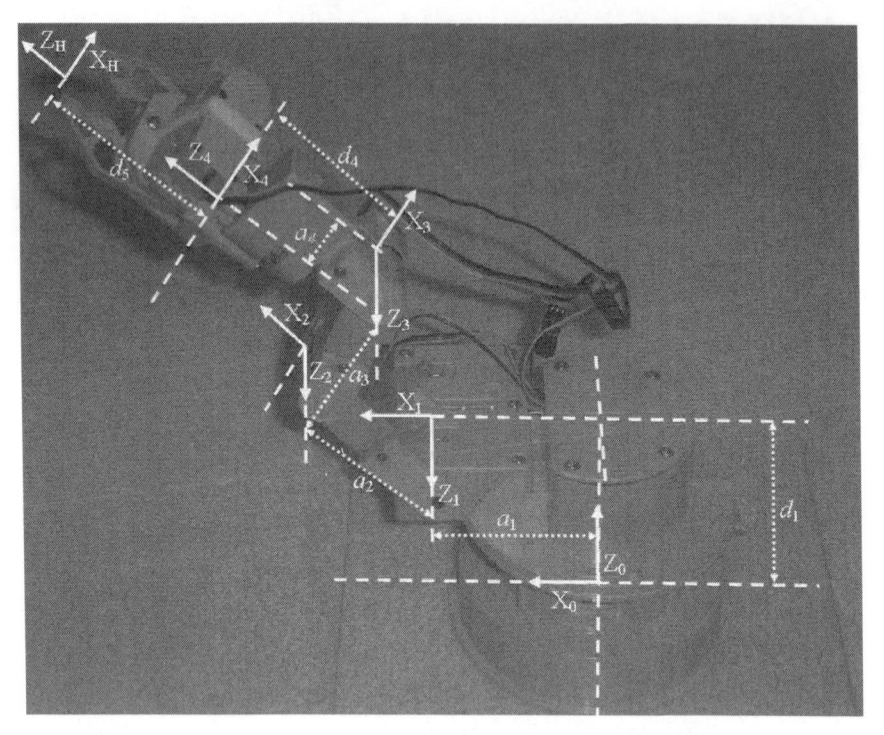

tions or from direct measurements of the robot parts. The parameter α_i is obtained based on the rules from the DH-table-generation technique shown in Table 5.1. The values for the measured parameters are $d_1 = 89$, $d_4 = 55$, $d_5 = 65$, $a_1 = 67$, $a_2 = 66$, $a_3 = 52$, $a_4 = 20$, $\alpha_1 = 90$, and $\alpha_4 = 90$.

Using the DH table, you can calculate five homogenous frame transformations using equation 5.2. The homogenous transformation from the robot's base frame to the hand frame can be calculated by multiplying these homogenous transformations from the right. Using equation 5.2, you can calculate frame transformations for each link. Equations 5.15, 5.16, 5.17, 5.18, and 5.19 present the frame transformations for the robot transformation. Equation 5.20 presents the robot's homogenous transformations from the base frame to the hand frame.

$$A_1 = \begin{pmatrix} \cos\theta_1 & -\sin\theta_1 \cos\alpha_1 & \sin\theta_1 & a_1\cos\theta_1 \\ \sin\theta_1 & -\cos\theta_1 \cos\alpha_1 & \cos\theta_1 & a_1\sin\theta_1 \\ 0 & \sin\alpha_1 & \cos\alpha_1 & d_1 \\ 0 & 0 & 0 & 1 \end{pmatrix}$$

$$= \begin{pmatrix} \cos\theta_1 & 0 & \sin\theta_1 & 67\cos\theta_1 \\ \sin\theta_1 & 0 & \cos\theta_1 & 67\sin\theta_1 \\ 0 & \sin\alpha_1 & \cos\alpha_1 & 89 \\ 0 & 0 & 0 & 1 \end{pmatrix} \quad (5.15)$$

$$A_2 = \begin{pmatrix} \cos\theta_2 & -\sin\theta_2 & 0 & 66\cos\theta_2 \\ \sin\theta_2 & -\cos\theta_2 & 0 & 66\sin\theta_2 \\ 0 & 0 & 1 & 0 \\ 0 & 0 & 0 & 1 \end{pmatrix} \quad (5.16)$$

$$A_3 = \begin{pmatrix} \cos\theta_3 & -\sin\theta_3 & 0 & 52\cos\theta_3 \\ \sin\theta_3 & -\cos\theta_3 & 0 & 52\sin\theta_3 \\ 0 & 0 & 1 & 0 \\ 0 & 0 & 0 & 1 \end{pmatrix} \quad (5.17)$$

$$A_4 = \begin{pmatrix} -\sin\theta_4 & 0 & \cos\theta_4 & 0 \\ \cos\theta_4 & 0 & \sin\theta_4 & 0 \\ 0 & 1 & 0 & 0 \\ 0 & 0 & 0 & 1 \end{pmatrix} \quad (5.18)$$

$$A_5 = \begin{pmatrix} -\cos\theta_5 & -\sin\theta_5 & 0 & 0 \\ \sin\theta_5 & \cos\theta_5 & 0 & 0 \\ 0 & 0 & 1 & 120 \\ 0 & 0 & 0 & 1 \end{pmatrix} \quad (5.19)$$

$$R_{T_H} = A_1 \cdot A_2 \cdot A_3 \cdot A_4 \cdot A_5$$

$$= \begin{pmatrix} C_1C_5S_{234} + S_1S_5 & C_1S_5S_{234} + S_1C_5 & C_1C_{234} \\ -S_1C_5S_{234} - C_1S_5 & S_1S_5S_{234} - C_1C_5 & S_1C_{234} \\ C_5C_{234} & -S_5C_{234} & S_{234} \\ 0 & 0 & 0 \end{pmatrix} \quad (5.20)$$

$$\begin{pmatrix} C_1(120C_{234} + 52C_{23} + 66C_2 + 67) \\ S_1(120C_{234} + 52C_{23} + 66C_2 + 67) \\ 120S_{234} + 52S_{23} + 66S_2 + 67 \\ 1 \end{pmatrix}$$

When the robot is programmed to move the tip of its hand to a desired location, you need to include the transformation from the hand frame to the tip of the hand. The tip of the hand is 22.5 mm away from the origin of the hand frame in the -x direction. Thus, the transformation from the hand frame to the tip can be calculated by equation 5.21.

$$A_6 = \begin{pmatrix} 1 & 0 & 0 & -22.5 \\ 0 & 1 & 0 & 0 \\ 0 & 0 & 1 & 0 \\ 0 & 0 & 0 & 1 \end{pmatrix} \quad (5.21)$$

Equation 5.22 shows the transformation from the robot's base frame to the tip of the hand.

$$R_{T_H} = A_1 \cdot A_2 \cdot A_3 \cdot A_4 \cdot A_5 \cdot A_6$$

$$= \begin{pmatrix} C_1C_5S_{234} + S_1S_5 & C_1S_5S_{234} + S_1C_5 & C_1C_{234} \\ -S_1C_5S_{234} - C_1S_5 & S_1S_5S_{234} - C_1C_5 & S_1C_{234} \\ C_5C_{234} & -S_5C_{234} & S_{234} \\ 0 & 0 & 0 \end{pmatrix} \quad (5.22)$$

$$\begin{pmatrix} +52C_{23} + 66C_2 + 67 - 22.5(C_1C_5S_{234} + S_1S_5) \\ +52C_{23} + 66C_2 + 67 + 22.5(S_1C_5S_{234} + C_1S_5) \\ +67 - 22.5C_5C_{234} \\ 1 \end{pmatrix}$$

In equation 5.22, we have used trigonometric abbreviations to represent the cosine and sine of the angles, and short notation for combined angles. The abbreviations are listed in Table 5.5.

After obtaining the forward kinematics equations (transformations from the robot base to the tip of the hand), you can calculate the position of the tip of the hand and the orientation of the hand if the joint parameters (angles) are given. The forward kinematics of an arm robot are straightforward after you have obtained the robot's homogeneous transformations. If you do not know the joint parameters but you do know the position and orientation of the robot hand, you need to come up with inverse kinematics equations, which can

Robotic Arm Control   205

TABLE 5.5
DH link parameters trigonometric abbreviations used in the inverse and forward kinematics

Abbreviation	Meaning
S_i	$\sin \theta i$
C_i	$\cos \theta i$
S_{ij}	$\sin \theta i \cos \theta j + \cos \theta i \sin \theta j$
C_{ij}	$\cos \theta i \cos \theta j - \sin \theta i \sin \theta j$

lead you to joint parameters. In general, the complexity of inverse kinematics equations is higher than that of forward kinematics. The next section explores the inverse kinematics equations of the six-DOF arm robot in detail.

5.6.4.2 Inverse Kinematics

Inverse kinematics equations of an arm robot can be hard to obtain when the degrees of freedom are higher than three. The robot we discuss in this section has six degrees of freedom, so its inverse kinematics equations are not easy to solve. We will describe a standard way of calculating inverse kinematics of an arm robot.

To come up with inverse kinematics equations, you need to have the target position and orientation of the robot hand or the tip of the hand. Here, you will calculate the inverse kinematics for the tip of the robot. The target position and the orientation of the tip of the hand can be given as a homogeneous transformation, as in equation 5.23.

$$A_T = \begin{pmatrix} n_x & o_x & a_x & p_x \\ n_y & o_y & a_y & p_y \\ n_z & o_z & a_z & p_z \\ 0 & 0 & 0 & 1 \end{pmatrix} \quad (5.23)$$

The matrix in equation 5.23 equals the transformation from the base frame to the tip of the robot, given in equation 5.22. This equality is presented in equation 5.24:

$$R_{T_H} = A_1 \cdot A_2 \cdot A_3 \cdot A_4 \cdot A_5 \cdot A_6 \quad (5.24)$$

In equation 5.24, the left side represents the desired position and the orientation of the tip. The right side is a matrix representing transformations from the base frame to the tip of the robot. Thus, elements of this matrix are functions of joint parameters θ_i. As equation 5.22 shows, you cannot determine the joint parameters by matching elements in the matrices on the left and right. The number of unknowns is six, but the variables are coupled, which makes the solution harder. One solution to this coupling problem is to move one unknown to the left side by multiplying the inverse of the first matrix from the left. Then the equation becomes

$$A_1^{-1} A_T = A_2 \cdot A_3 \cdot A_4 \cdot A_5 \cdot A_6 \quad (5.25)$$

Equation 5.25 has one unknown on the left side and five unknowns on the right side. Thus, parameter θ_1 is decoupled and moved to the left side. Now you can match the elements in the same locations in the matrices to come up with an equation for θ_1. Equation 5.26 shows the equality.

$$\begin{pmatrix} C_1 n_x + S_1 n_y & C_1 o_x + S_1 o_y & C_1 a_x + S_1 a_y & C_1 p_x + S_1 p_y - 67 \\ n_z & o_z & a_z & -80 + p_z \\ S_1 n_x - C_1 n_y & S_1 o_x - C_1 o_y & S_1 a_x - C_1 a_y & S_1 p_x - C_1 p_y \\ 0 & 0 & 0 & 1 \end{pmatrix}$$

$$= \begin{pmatrix} -S_{234} C_5 & S_{234} S_5 & C_{234} & 22.5 S_{234} C_5 + 120 C_{234} = 52 C_{23} + 66 C_2 \\ C_{234} C_5 & -C_{234} S_5 & S_{234} & -22.5 C_{234} C_5 + 120 S_{234} = 52 S_{23} + 66 S_2 \\ S_5 & C_5 & 0 & -22.5 S_5 \\ 0 & 0 & 0 & 1 \end{pmatrix}$$

(5.26)

Now you can match the corresponding elements in the matrices to create an equation for parameter θ_1. By equating the elements at the third row and the third column, we get equation 5.27.

$$0 = S_1 a_x - C_1 a_y \quad (5.27)$$

Using equation 5.27, we can calculate parameter θ_1 as follows:

$$S_1 a_x = C_1 a_y \quad (5.28)$$

$$\frac{S_1}{C_1} = \frac{a_y}{a_x} \quad (5.29)$$

$$\theta_1 = ATAN2(a_y, a_x) \quad (5.30)$$

In equation 5.30, a special function called ATAN2 is used to find the angle by determining in which quadrant the angle is located. If you were to use the regular ATAN function, you could not differentiate some angles. For instance, (-1, -1) and (1, 1) will generate 45 degrees if you use ATAN, but ATAN2 will give 215 degrees and 45 degrees, respectively. Thus, to discover an angle you need both the cosine and the sine of the angle for the ATAN2 function.

In general, you repeat this process for every angle you would like to discover. However, sometimes you may be able to discover more than one angle in a step. In fact, you can match the elements at the third row and the first column, and the elements at the third row and the second column. The first

Robotic Arm Control

match gives you S_5 and the second match gives you C_5. Therefore, you can calculate the angle θ_5. The following three equations lay out this process.

$$S_5 = S_1\eta_x - C_1\eta_y \tag{5.31}$$

$$C_5 = S_1 o_x - C_1 o_y \tag{5.32}$$

$$\theta_5 = ATAN2(S_1\eta_x - C_1\eta_y, S_1 O_x - C_1 O_y) \tag{5.33}$$

At the end of the first step you have discovered the angles θ_1 and θ_5. In the second step, you will decouple the angle θ_2 by multiplying the inverse of the transformation representing this angle. Equation 5.34 shows both sides of the equation.

$$A_2^{-1} A_1^{-1} A_T = A_3 \cdot A_4 \cdot A_5 \cdot A_6 \tag{5.34}$$

After going through the multiplication, you have one unknown on the left side and two unknowns on the right side, as given in equation 5.35.

$$\begin{pmatrix} C_2(C_1 n_x & C_2(C_1 o_x & C_2(C_1 a_x & C_2(C_1 p_x + S_1 p_y) \\ +S_1 n_y) + S_2 n_z & +S_1 o_y) + S_2 o_z & +S_1 a_y) + S_2 a_z & +S_2 p_z - 67 C_2 \\ & & & -89 S_2 - 66 \\ -S_2(C_1 n_x & -S_2(C_1 o_x & -S_2(C_1 a_x & -S_2(C_1 p_x + S_1 p_y) \\ +S_1 n_y) + C_2 n_z & +S_1 o_y) + C_2 o_z & +S_1 a_y) + C_2 a_z & +C_2 p_z + 67 S_2 \\ & & & -89 C_2 \\ S_1 n_x - C_1 n_y & S_1 o_x - C_1 o_y & S_1 a_x - C_1 a_y & S_1 p_x - C_1 p_y \\ 0 & 0 & 0 & 1 \end{pmatrix}$$
$$= \begin{pmatrix} -S_{34} C_5 & S_{34} S_5 & C_{34} & 22.5 S_{34} C_5 + 120 C_{34} + 52 C_3 \\ C_{34} C_5 & -C_{34} S_5 & S_{34} & -22.5 C_{34} C_5 + 120 S_{34} + 52 S_3 \\ S_5 & C_5 & 0 & -22.5 S_5 \\ 0 & 0 & 0 & 1 \end{pmatrix}$$
$$\tag{5.35}$$

By matching elements at the first row and the first column, you get

$$-S_{34} C_5 = C_2(C_1 n_x + S_1 n_y) + S_2 n_z \tag{5.36}$$

$$S_{34} = -[C_2(C_1 n_x + S_1 n_y) + S_2 n_z]/C_5 \tag{5.37}$$

Similarly, if you match the elements at the first row and the second column, you get

$$S_{34} S_5 = C_2(C_1 o_x + S_1 o_y) + S_2 o_z \tag{5.38}$$

$$S_{34} = [C_2(C_1 o_x + S_1 o_y) + S_2 o_z]/S_5 \tag{5.39}$$

By combining equation 5.39 and equation 5.37, you get the following:

$$\frac{[C_2(C_1 o_x + S_1 o_y) + S_2 o_z]}{S_5} = -\frac{[C_2(C_1 n_x + S_1 n_y) + S_2 n_z]}{C_5}$$
$$[C_2(C_1 o_x + S_1 o_y) + S_2 o_z] C_5 = -[C_2(C_1 n_x + S_1 n_y) + S_2 n_z] S_5$$
$$C_2(C_1 o_x + S_1 o_y)C_5 + C_2(C_1 n_x + S_1 n_y)S_5 = -S_2 n_z S_5 + S_2 o_z C_5$$
$$C_2[(C_1 o_x + S_1 o_y)C_5 + (C_1 n_x + S_1 n_y)S_5] = S_2[o_z C_5 - n_z S_5] \quad (5.40)$$
$$\frac{S_2}{C_2} = \frac{[(C_1 o_x + S_1 o_y)C_5 + (C_1 n_x + S_1 n_y)S_5]}{[o_z C_5 - n_z S_5]}$$
$$\theta_2 = ATAN2((C_1 o_x + S_1 o_y)C_5 + (C_1 n_x + S_1 n_y)S_5, o_z C_5 - n_z S_5)$$

So far, you have inverse kinematics equations for parameters (angles) θ_1, θ_2, and θ_5. You may repeat the process for the remaining parameters, θ_3 and θ_4, to decouple them. However, the parameters are decoupled enough that you can first determine θ_{34} and then you can determine θ_3.

The sine and cosine of θ_{34} appear in the right side of the equation at the third column. The same locations on the left side are filled with equations that have known parameters. Equations 5.41 and 5.42 reveal the sine and cosine of θ_{34} by matching the elements in these locations in the equations.

$$S_{34} = -S_2(C_1 a_x + S_1 a_y) + C_2 a_z \quad (5.41)$$

$$C_{34} = C_2(C_1 a_x + S_1 a_y) + S_2 a_z \quad (5.42)$$

Using the sine and cosine of θ_{34}, you can calculate its value by the ATAN2 function, as shown in equation 5.43.

$$\theta_{34} = ATAN2(-S_2(C_1 a_x + S_1 a_y) + C_2 a_z, C_2(C_1 a_x + S_1 a_y) + S_2 a_z) \quad (5.43)$$

Similarly, if you match the elements in the first row and the fourth column, you can find an equation for the cosine of θ_3, as shown in equation 5.44.

$$C_5 + 120C_{34} + 52C_3 = C_2(C_1 p_x + S_1 p_y) + S_2 p_z - 67C_2 - 89S_2 - 66$$
$$52C_3 = C_2(C_1 p_x + S_1 p_y) + S_2 p_z - 67C_2 - 89S_2 - 66 - 22.5S_{34}C_5 - 120C_{34}$$
$$(5.44)$$

You can find the sine of θ_3 by matching the elements in the second row and the fourth column, as given by equation 5.45.

$$22.5S_{34} - 22.5C_{34}C_5 + 120S_{34} + 52S_3 =$$
$$-S_2(C_1 p_x + S_1 p_y) + C_2 p_z + 67S_2 - 89C_2$$
$$52S_3 = -S_2(C_1 p_x + S_1 p_y) + C_2 p_z + 67S_2 - 89C_2 + 22.5C_{34}C_5 - 120S_{34}$$
$$(5.45)$$

Using equations 5.44 and 5.45, you can calculate θ_3 by the ATAN2 function, as shown in equation 5.46.

$$\theta_3 = ATAN2(-S_2(C_1p_x + S_1p_y) + C_2p_z + 67S_2 - 89C_2 + \\ 22.5C_{34}C_5 - 120S_{34}, \\ C_2(C_1p_x + S_1p_y) + S_2p_z - 67C_2 - 89S_2 - 66 - 22.5S_{34}C_5 - 120C_{34}) \tag{5.46}$$

Since you determined θ_{34} and θ_3, you can determine θ_4 by simple subtraction, as shown in equation 5.47.

$$\theta_4 = \theta_{34} - \theta_3 \tag{5.47}$$

Equation 5.47 completes the inverse kinematics analysis of the six-DOF robot discussed in this chapter. The next section presents MATLAB M-files used to create symbolic matrices for forward and inverse kinematics. It also shows how your MATLAB program can create values instead of angles for the servomotors on the robots.

5.7 Examples and MATLAB Programs

In this section we present six MATLAB M-files for forward kinematics, inverse kinematics, and linear path planning. Table 5.6 lists the M-files. The MATLAB files do not require any of the MATLAB toolboxes. If you have a student version of MATLAB or Octave (open-source MATLAB-like software), you can run these files.

Subsection 5.7.1 is the M-file for calculating the homogenous transformation of a row of a DH table.

5.7.1 DH.m: M-File for a Homogenous Transformation of a Row of a DH Table

```
% T: Joint angle Theta in degrees
% a: Distance between Zi and Zi-1
% d: Distance between Xi and Xi-1
% al: Angle Alfa about Xi from Zi-1 to Zi

function A = DH(T, a, d, al)

A = [cosd(T) -sind(T)*cosd(al) sind(T)*sind(al) a*cosd(T);
     sind(T)  cosd(T)*cosd(al) -cosd(T)*sind(al) a*sind(T);
     0        sind(al)          cosd(al)          d;
     0        0                 0                 1];
```

TABLE 5.6
MATLAB M-files for aix-DOF arm robot

M-File	Description
DH.m	Calculates the homogenous transformation for a row of a DH table. This is used as DH(θ, a, d, α). The angles are in degrees, and lengths are in millimeters.
RobotTF.m	Calculates the robot transfer function for a given set of joint angles θ_1, θ_2, θ_3, θ_4, and θ_5. This can be used for the forward kinematics calculations.
RobotSym.m	Generates symbolic transformation matrices for forward and inverse kinematics components. You can see the transformations from any frame to another frame by simply multiplying the matrices this M-file creates. The results of the multiplications will be in symbolic form.
InverseKin.m	Takes a matrix for the desired position and orientation of the hand as a parameter. Then it calculates the angles θ_1, θ_2, θ_3, θ_4, and θ_5 required for the desired position and orientation by calculating the inverse kinematics equations.
Angle2Servo.m	Converts angle values calculated by InverseKin.m into corresponding servomotor values to move the robot properly. These values are applied to the mini servo controller either through the serial port or through the Mini Servo Explorer software.
Path.m	Generates series of values for the servomotors so that the robot moves from the starting position to the target position linearly. It writes these servomotor values into a file that Mini Servo Explorer can understand and execute.

Robotic Arm Control

Subsection 5.7.2 is the M-file for calculating the robot transformations from the base frame to the hand frame if the joint angles are provided.

5.7.2 RobotTF.m: M-File for Calculating the Transformation Matrix of the Robot

```
%6 DOF robot transfer function

% T1:    Angle for Joint 1, Theta 1;
% T2:    Angle for Joint 2, Theta 2;
% T3:    Angle for Joint 3, Theta 3;
% T4:    Angle for Joint 4, Theta 4;
% T5:    Angle for Joint 5, Theta 5;

function A=robotTF(T1, T2,T3,T4,T5)

a1= 67;
a2 =66;
a3 =52;
a4 =0;
a5=0;

d1=89;
d2=0;
d3=0;
d4 =0;
d5 =55+65;   %75 till the tip of the gripper.

al1 = 90;
al2=0;
al5=0;
al3=0;
al4= 90;
al5 =0;

%All the angles must be between -85 and 85.

A1=DH(T1,a1,d1,al1);
A2=DH(T2,a2,d2,al2);
A3=DH(T3,a3,d3,al3);
A4=DH(T4 + 90,a4,d4,al4);
A5=DH(T5,a5,d5,al5);
A6=DH(0,-22.5,0,0);

A= A1*A2*A3*A4*A5*A6;
```

Subsection 5.7.3 presents the symbolic analysis of the forward and inverse kinematics.

5.7.3 RobotSym.m: M-File for Symbolic Analysis of Inverse and Forward Kinematics

```
clear
t1 = sym('t1');
t2 = sym('t2');
t3 = sym('t3');
t4 = sym('t4');
t5 = sym('t5');
nx = sym('nx');
ny = sym('ny');
nz = sym('nz');
ox = sym('ox');
oy = sym('oy');
oz = sym('oz');
ax = sym('ax');
ay = sym('ay');
az = sym('az');
px = sym('px');
py = sym('py');
pz = sym('pz');


A1 = [cos(t1) 0 sin(t1) 67*cos(t1);
sin(t1) 0 -cos(t1) 67*sin(t1);
; 0 1 0 89; 0 0 0 1];

A2 = [cos(t2) -sin(t2) 0  66*cos(t2);
sin(t2) cos(t2) 0 66*sin(t2);
; 0 0 1 0; 0 0 0 1];

A3 =[cos(t3) -sin(t3) 0  52*cos(t3);
sin(t3) cos(t3) 0 52*sin(t3);
; 0 0 1 0; 0 0 0 1];

A4 = [-sin(t4) 0 cos(t4) 0;
cos(t4) 0 sin(t4) 0
; 0 1 0 0; 0 0 0 1];

A5 =[cos(t5) -sin(t5) 0  0;
sin(t5) cos(t5) 0 0;
```

```
    ; 0 0 1 120; 0 0 0 1];
A6 = [1 0 0 -22.5;
      0 1 0 0;
      0 0 1 0;
      0 0 0 1];

A = A1*A2*A3*A4*A5*A6;

Ar = [nx ox ax px;
      ny oy ay py;
      nz oz az pz;
      0 0 0 1;]

A11 =[ cos(t1),  sin(t1),          0,       -67;
        0,         0,             1,       -89;
       sin(t1), -cos(t1),          0,         0;
        0,         0,             0,         1;]

A21 = [ cos(t2),  sin(t2),         0,       -66;
       -sin(t2),  cos(t2),         0,         0;
       0,        0,               1,         0;
       0,        0,               0,         1];

A31 =[ cos(t3),  sin(t3),          0,       -52;
       -sin(t3), cos(t3),          0,         0;
        0,        0,              1,         0;
        0,        0,              0,         1];
A41 = [ -sin(t4), cos(t4),         0,         0;
        0,        0,              1,         0;
       cos(t4),  sin(t4),         0,         0;
       0,        0,         0,        1];

A51 = [ cos(t5),  sin(t5),         0,         0;
       -sin(t5),  cos(t5),         0,         0;
       0,        0,              1,       -120;
       0,        0,              0,         1];
```

Subsection 5.7.4 presents the M-file that calculates the required joint angles for a desired position and orientation of the robot tip by going through the inverse kinematics equations. The angles this M-file calculates need to be converted to values that the mini servo controller can understand. This conversion is done by Angle2Servo.m. Finally, we also present an M-file that calculates series of joint angles to move the robot tip from a starting point to a target point linearly. The program writes these angle values into a file that Mini Servo Explorer can understand.

5.7.4 InverseKin.m: M-File for Inverse Kinematics Equations

```
%Inverse kinematics for 6DOF robot
% A is the matrix that represents the desired position
% and the orientation of the robot hand.
% n is a vector that holds the values of the joint angles.
function [n] = invKin(A);

nx = A(1,1);
ny = A(2,1);
nz = A(3,1);
ox = A(1,2);
oy = A(2,2);
oz = A(3,2);
ax = A(1,3);
ay = A(2,3);
az = A(3,3);
px = A(1,4);
py = A(2,4);
pz = A(3,4);
A(4,1) = 0;
A(4,2) = 0;
A(4,3) = 0;
A(4,4) = 1;


t1 = atan2(ay, ax)
if (t1>pi/2)
    t1 = t1 -pi;
elseif t1<-pi/2
    t1 = t1 + pi;
end
t5 = atan2(nx*sin(t1) - ny*cos(t1), sin(t1)*ox - cos(t1)*oy)
if (t5>pi/2)
    t5 = t5 -pi;
elseif t5<-pi/2
    t5 = t5 + pi;
end
t2 = atan2(cos(t5)*(cos(t1)*ox + sin(t1)*oy) +
sin(t5)*(cos(t1)*nx + sin(t1)*ny),
oz*cos(t5)+az*sin(t5))
if (t2>pi/2)
    t2 = t2 -pi;
elseif t2<-pi/2
```

```
        t2 = t2 + pi;
end

t34 = atan2(-sin(t2)*(cos(t1)*ax+sin(t1)*ay) +cos(t2)*az,
cos(t2)*(cos(t1)*ax+sin(t1)*ay) + sin(t2)*az)
t3 = atan2(-sin(t2)*(cos(t1)*px +sin(t1)*py) + cos(t2)*pz +
67*sin(t2)
- 89*cos(t2)- 120*sin(t34) + 22.5*cos(t34)*cos(t5),
cos(t2)*(cos(t1)*px +sin(t1)*py) + sin(t2)*pz - 67*cos(t2)
- 89*sin(t2) -66 -120*cos(t34) - 22.5*sin(t34)*cos(t5))
if (t3>pi/2)
     t3 = t3 -pi;
elseif t3<-pi/2
     t3 = t3 + pi;
end

t4 = t34 - t3
if (t4>pi/2)
     t4 = t4 -pi;
elseif t4<-pi/2
     t4 = t4 + pi;
end

t1 = (180/pi)*t1;
t2 = (180/pi)*t2;
t3 = (180/pi)*t3;
t4 = (180/pi)*t4;
t5 = (180/pi)*t5;
n = [t1, t2, t3, t4, t5];
```

Subsection 5.7.1 is the M-file for calculating the homogenous transformation of a row of a DH table.

5.7.5 Angle2Servo.m: M-File for Converting Joint Angles to Servomotor Values

```
%Converts angles to servo numbers
% t1: Joint Angle
% n1: Corresponding Servomotor value
function n1 = angle2servo(t1)

n1 = 126 + t1*126/90;
n1 = fix(n1);
```

5.7.6 Path.m: M-File for Generating Joint Angles to Move the Robot Linearly

```
% Path for linear motion
% xs, ys, zs are starting position of the hand
% xe, ye, ze are ending position of the hand
function path = LinPath(xs, xe, ys, ye, zs, ze, st)

xs=8.5; xe=12; ys=0; ye=10; zs=24; ze=26; st=10;
f =fopen('prgmatlab.RB','w');
A=[0 0 1 0;
   0 -1 0 0;
   1 0 0 0;
   0 0 0 1];
slopy = (ye-ys)/(xe-xs);
slopz = (ze-zs)/(xe-xs);

fprintf(f, "16\n1\n203\n400\n"No.1",0,253,127,0,0\n");
fprintf(f, ""No.2",0,253,127,0,0\n"No.3",0,253,127,2,
0\n"No.4",0,253,127,0,0\n");
fprintf(f, ""No.5",0,253,127,0,0\n"No.6",0,253,127,0,
0\n"No.7",0,253,127,0,0\n");
fprintf(f, ""No.8",0,253,127,0,0\n"No.9",0,253,127,0,
0\n"No.10",0,253,127,0,0\n");
fprintf(f, ""No.11",0,253,127,0,0\n"No.12",0,253,127,0,
0\n"No.13",0,253,127,0,0\n");
fprintf(f, ""No.14",0,253,127,0,0\n"No.15",0,253,127,0,
0\n"No.16", ~CCC0,253,127,0,0\n10\n");

x =0;
for i=1:st
    x =  (xe-xs)*i-1;
    y = ys + slopy*x;
    z = zs + slopz*x;
    A(:,4) = [xs+x, y, z, 1]'
    path(i,:) = invKin(A);
    fprintf(f,'%d,%d,%d,%d,%d,%d,127,127,127,
    127,127,127,127,127, ~CCC 127,127,\n',path(i,:))
end

fprintf(f, """\n0\n"EmptyMuisc"\n0\nf\n");
```

The M-file path.m writes the servomotor values into a file with RB format which Mini Servo Explorer can understand. Then this file can be opened and run by the Mini Servo Explorer. An example of such a file can be seen in subsection 5.7.7.

5.7.7 A Sample RB File Created by path.m for Mini Servo Explorer

```
16
1
203
400
"No.1",0,253,127,0,0
"No.2",0,253,127,0,0
"No.3",0,253,127,2,0
"No.4",0,253,127,0,0
"No.5",0,253,127,0,0
"No.6",0,253,127,0,0
"No.7",0,253,127,0,0
"No.8",0,253,127,0,0
"No.9",0,253,127,0,0
"No.10",0,253,127,0,0
"No.11",0,253,127,0,0
"No.12",0,253,127,0,0
"No.13",0,253,127,0,0
"No.14",0,253,127,0,0
"No.15",0,253,127,0,0
"No.16",0,253,127,0,0
10
126,126,126,139,112,126,127,127,127,127,127,127,127,127,127,127,
126,126,126,139,112,126,127,127,127,127,127,127,127,127,127,127,
126,126,126,138,113,126,127,127,127,127,127,127,127,127,127,127,
126,126,126,138,113,126,127,127,127,127,127,127,127,127,127,127,
126,126,126,137,114,126,127,127,127,127,127,127,127,127,127,127,
126,126,126,137,114,126,127,127,127,127,127,127,127,127,127,127,
126,126,126,136,115,126,127,127,127,127,127,127,127,127,127,127,
126,126,126,135,116,126,127,127,127,127,127,127,127,127,127,127,
126,126,126,135,116,126,127,127,127,127,127,127,127,127,127,127,
126,126,126,134,117,126,127,127,127,127,127,127,127,127,127,127,
""
0
"EmptyMusic"
0
```

6
Differential Drive Robot

In this chapter we will study a differential drive robot kit developed by Chaney Electronics that is driven by two wheels and comes with a breadboard so that you can conduct many experiments. First we will examine its mechanical construction, then we will look at the electronics to perform various tasks. Finally, we will cover some mathematical modeling for advanced concepts for wheeled robots. This robotic kit can be obtained from Chaney Electronics web site [27]. Figure 6.1 shows the completed car kit.

This robotic kit comes with three lessons and 31 experiments outlined in the instruction manual:

1. Assembling and Testing the Robot Motor System
2. Assembling the Robot Base and Power System
3. Assembling and Testing the Relay Control Board
4. Robot Forward Motion
5. Robot Backward Motion
6. Robot Forward Right Turn
7. Robot Backward Right Turn
8. Robot Hard Right Turn
9. Robot Forward Left Turn
10. Robot Backward Left Turn
11. Robot Hard Left Turn
12. Time-Controlled Motion I
13. Time-Controlled Motion II
14. Time-Controlled Hard Right Turn
15. Time-Controlled Hard Left Turn
16. Explorer I Robot

FIGURE 6.1
The robotic kit (Copyright 2006 Chaney Electronics. Reproduced with permission)

17. Explorer II Robot

18. The Unstoppable Robot

19. The Independent-Minded Robot

20. The Night Runner Robot

21. The Day Runner Robot

22. Infrared (IR) Vision

23. The IR-Eye Robot

24. Obedient I IR Remote-Controlled Robot

25. Obedient II IR Remote-Controlled Robot

26. Robot Pet

27. IR Obstacle-Avoiding Robot I

28. IR Obstacle-Avoiding Robot II

29. Sound Detection (Robot Ears)

30. Big Ears Robot

31. The Music-Dancer Robot

Differential Drive Robot 221

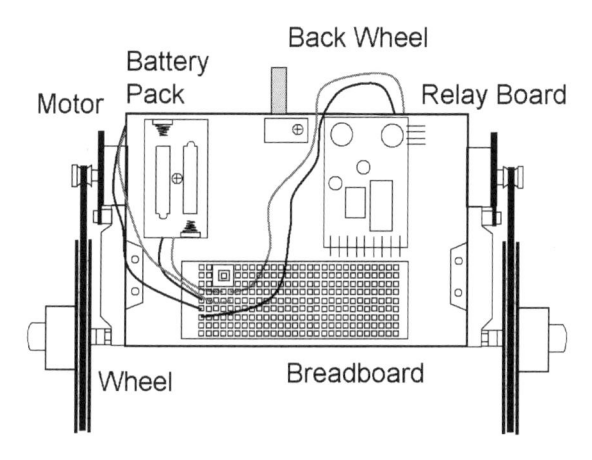

FIGURE 6.2
 Top view of the robot

FIGURE 6.3
 Breadboard used on the robot

6.1 Construction and Mechanics

The top view of the robot (Figure 6.2) shows its main mechanical components.

The robot's construction can be broken down into four parts: robot base with breadboard, traction system, power system, and relay board.

6.1.1 Robot Base with Breadboard

The robot base is Plexiglas and holds all the other parts of the robot together. The robot uses a solderless breadboard (Figure 6.3) for experimenting with electronics.

The breadboard needs to be attached to the Plexiglas base with two pieces of double-sided tape.

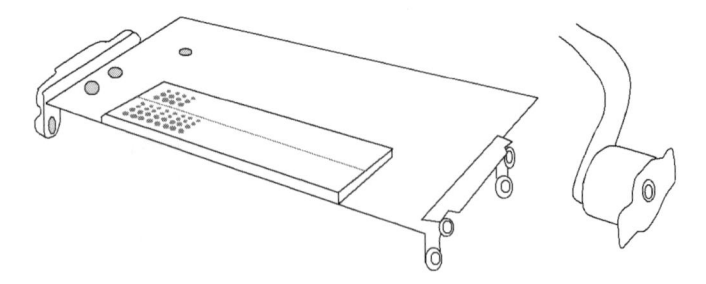

FIGURE 6.4
Board connecting to the motor using slide rails

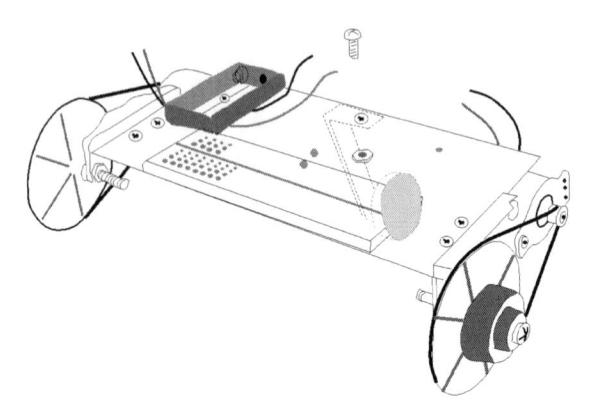

FIGURE 6.5
Connected board

6.1.2 Traction System

The robot's traction system essentially consists of two DC motors connected to two wheels using two rubber belts, creating a pulley system. The DC motors are attached to the base with the breadboard using slide rails as connectors between them (Figure 6.4).

The wheels are fastened to the board using the same slide rails. Rubber belts are installed between the pulley and the two front wheels. The back wheel moves freely and is just for support; it has no motor attached to it. Figure 6.5 depicts the connected robot.

6.1.2.1 Belt System

Figure 6.6 shows how the pulley system with the belt transfers torque and speed.

The motor shaft has radius R_1 and exerts a force F_1 on the belt at the point of contact between the belt and the motor shaft. The speed of the point

Differential Drive Robot

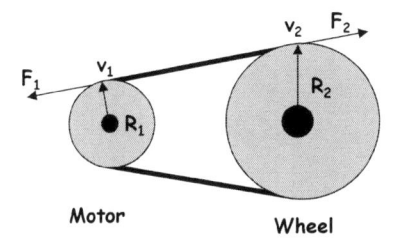

FIGURE 6.6

The robot pulley system

of contact is v_1. The driven wheel has radius R_2 and exerts a force F_2 on the belt at the point of contact between the belt and the wheel. The speed of the point of contact is v_2. The following two balance equations govern the pulley action. The first states that the speed of the belt is the same at the two points:

$$v_1 = v_2 \tag{6.1}$$

The second states that the two forces are equal:

$$F_1 = F_2 \tag{6.2}$$

Let ω_1 be the angular speed of the motor shaft and ω_2 be the angular speed of the driven wheel. Since angular speed is equal to the product of angular speed and radius, we can rewrite equation 6.1 as

$$r_1 \omega_1 = R_2 \omega_2 \tag{6.3}$$

We can use this equation as follows:

$$\frac{\omega_1}{\omega_2} = \frac{R_2}{R_1} \tag{6.4}$$

The ratio R_2/R_1 is the radius ratio. Torque, τ, is equal to the product of force and radius. In other words, force is the quotient of torque when it is divided by radius. Using this in equation 6.2 gives the following:

$$\frac{\tau_1}{R_1} = \frac{\tau_2}{R_2} \tag{6.5}$$

We can also write this as

$$\frac{\tau_1}{\tau_2} = \frac{R_1}{R_2} \tag{6.6}$$

Combining 6.4 and 6.6 and rearranging terms yields

$$\frac{\tau_1}{\omega_1} = \frac{\tau_2}{\omega_2} \tag{6.7}$$

FIGURE 6.7

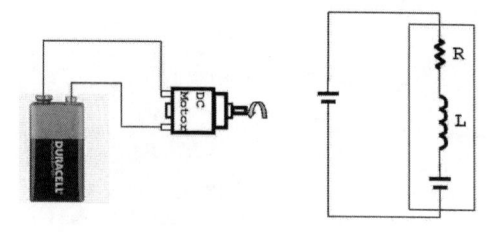

DC motor model

This shows that in an ideal pulley system, the input power is equal to the output power. Therefore, if we reduce the speed, the torque increases and vice versa. In a practical pulley system, there will be some slip between the wheel/shaft and the belt, which will cause a small reduction in the speed and torque transfer. To understand the advantage we gain by using the belt system, let's look at the dynamics of a DC motor.

6.1.2.2 DC Motor Dynamics

This section has been adapted with permission from Mobile Robotic Car Design by Pushkin Kachroo and Patricia Mellodge [7].

Let us consider a DC motor connected to a power supply. The motor has internal resistance and inductance of the internal coil. When a voltage is applied to a DC motor, the motor produces a voltage opposing the voltage applied. This is called the back emf. (Details of the back emf and electromagnetic theory associated with motors are covered in many textbooks on electromagnetism [11]). Figure 6.7 shows the motor connection and an equivalent electrical model.

The model shows the internal resistance and inductance of the motor, as well as a source for back emf. By applying Kirchhoff's voltage law in the circuit, we get

$$V = iR_a + L_a \frac{di}{dt} + E_a \quad (6.8)$$

The linear relationship between the back emf and the mechanical rotational velocity of the motor is as follows:

$$E_a = k_{emf}\omega_m \quad (6.9)$$

Here, we have k_{emf} as the back emf constant and ω_m as the motor shaft's angular velocity. Readjusting the terms of 6.8 and 6.9 gives us

$$\frac{di}{dt} = \frac{1}{L_a}[V - iR_a - k_{emf}\omega_m] \quad (6.10)$$

Assuming no friction on the motor shaft and taking the moment of inertia of the motor shaft and load combined as I_m, we can obtain the dynamic

Differential Drive Robot

equation for the mechanical motion of the motor by applying Newton's law for rotation. The result is

$$\frac{d\omega_m}{dt} = \frac{1}{I_m}\tau_m \qquad (6.11)$$

The torque is the product of a torque constant, k_{torque}, and the current flowing through the motor, i. This transforms equation 6.11 to

$$\frac{d\omega_m}{dt} = \frac{1}{I_m}k_{torque}i \qquad (6.12)$$

Equations 6.10 and 6.12 give the electrical and mechanical dynamics of the system. If we assume no loss of power, all the electrical power used in the motor equals the mechanical power generated. The electrical power consumed is the product of the back emf and the current:

$$P_{electrical} = E_a i = k_{emf}\omega_m i \qquad (6.13)$$

Equation 6.13 shows the mechanical power dissipated:

$$P_{mechanical} = \tau_m \omega_m = k_{torque} i \omega_m \qquad (6.14)$$

By equating the electrical power consumed with the mechanical power generated, we get

$$k_{emf} = k_{torque} = k \qquad (6.15)$$

We can make the model better by modeling friction in 6.12 that could include static and viscous friction terms. In the equation, the sign function is +1 if the parameter is positive; it is -1 if the parameter is negative.

$$\frac{d\omega_m}{dt} = \frac{1}{I_m}[ki - a\,sign(\omega_m) - b\omega_m] \qquad (6.16)$$

In this equation, a is the static friction coefficient, and b is the viscous friction coefficient.

6.1.2.3 DC Motor Steady State Analysis

If we apply a constant voltage to a DC motor, the motor will accelerate initially and then reach a steady state at some angular velocity. If we add a load torque to the shaft, the steady state angular velocity will decrease. We can increase the load torque so much that the motor stops rotating (for instance, by adding a big load on a rope). We can then plot the steady state angular velocity with respect to the load torque. These are called the torque-speed characteristics. We can derive these using dynamic equations. Our dynamic equations when we apply the load torque T and assume no static or viscous friction are as follows:

$$\frac{di}{dt} = \frac{1}{L_a}[V - iR_a - k\omega_m] \qquad (6.17)$$

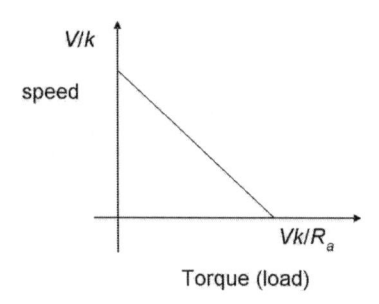

FIGURE 6.8

DC motor torque characteristic curve

and
$$\frac{d\omega_m}{dt} = \frac{1}{I_m}[ki - T] \tag{6.18}$$

In steady state, the rate of change of current and velocity become 0, and so we get the following:

$$0 = \frac{1}{L_a}[V - iR_a - k\omega_m] \tag{6.19}$$

and
$$0 = \frac{1}{I_m}[ki - T] \tag{6.20}$$

Equation 6.20 gives us the torque in terms of the current. That expression can be substituted in 6.19 to give us

$$\omega_m = \frac{V}{k} - \frac{R_a}{k^2}T \tag{6.21}$$

This is a straight line plot with a y-axis intercept of V/k and the x-axis intercept of $T = Vk/R_a$. This means when there is no load torque (the motor runs freely), by applying voltage V the steady state angular speed will be V/k, as shown in Figure 6.8. When we apply the minimum torque that makes the motor stop, the applied torque will be $T = Vk/R_a$.

6.1.2.4 Reason for Using the Belt

To understand the advantage of using the belt, imagine that we need to use a wheel of radius R_2. We can connect the wheel directly to the motor as shown in Figure 6.9.

The traction force from the ground is given by F. Let's say the torque (which is the product of F and R_2) is such that we are operating at the part of the curve from Figure 6.8 where the speed is 0. This means that when we connect the wheel as shown and turn on power to the motor, the wheel does not move because the motor is not able to produce the desired torque

Differential Drive Robot   227

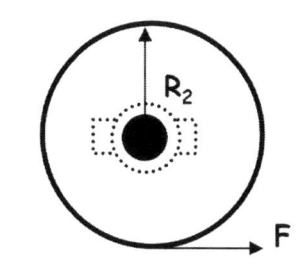

FIGURE 6.9
Direct wheel drive

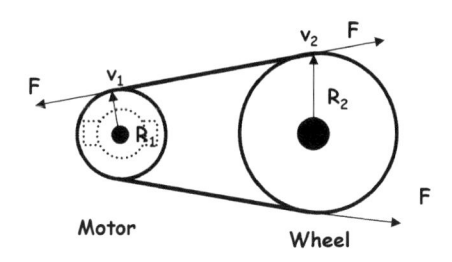

FIGURE 6.10
Belt drive

to counter the torque from the ground interaction of the wheel. Now we will connect the belt from the motor shaft to the wheel instead of connecting the wheel directly to the motor. Figure 6.10 shows this connection.

The torque on the motor is the product of F and R_1. If R_1 is half of R_2, then the torque on the motor will be half when the radius is R_1 as compared to when the radius is R_2. This would mean that the motor is operating at half the torque that produces zero speed in Figure 6.8. At this torque, the wheel will spin at a speed dictated by the curve in Figure 6.8. We can choose the radius of the shaft and the wheel to get the desired speed from the robot.

6.1.3 Power System

At present the robot frame has the breadboard and the traction system. We can add batteries so that when we connect the battery to the motor, the robot will start moving. Figure 6.11 shows how the battery holder is connected to the robot frame.

Now we can add a push-button switch between the battery and the motors so that when we turn the switch on, the wheels start moving; and when we switch it off, the wheels stop moving. We can test this by lifting the robot and checking out the simple circuit shown in Figure 6.12.

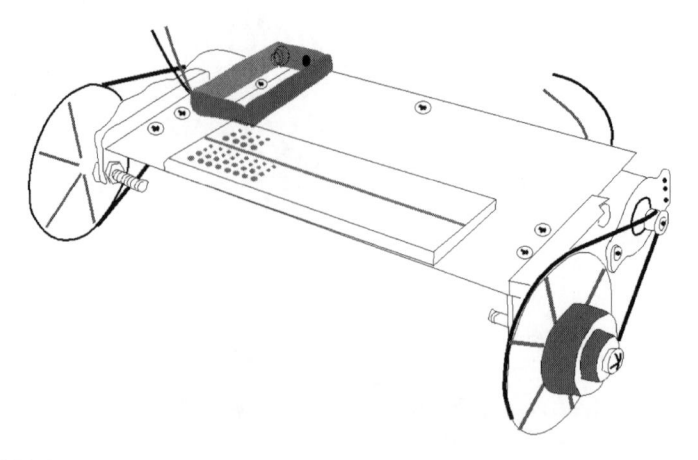

FIGURE 6.11

Motor power from a battery

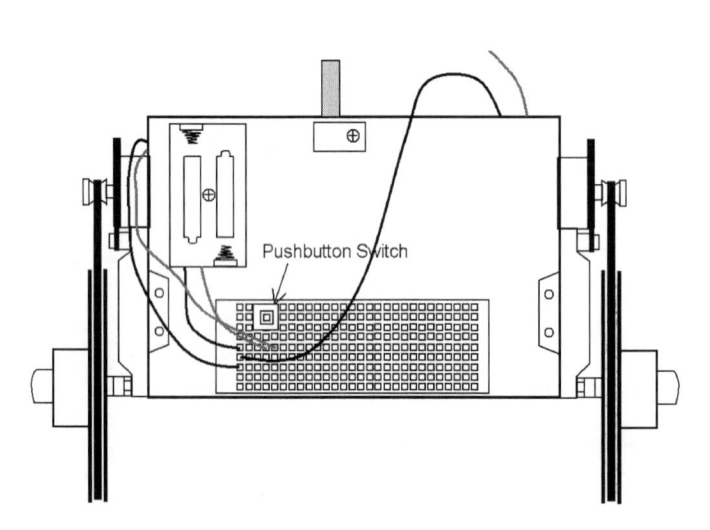

FIGURE 6.12

On-off control

Differential Drive Robot 229

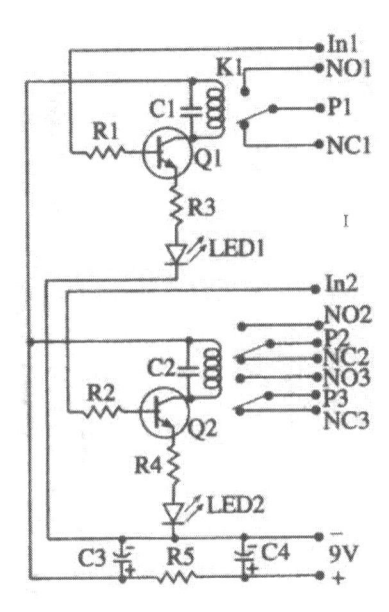

FIGURE 6.13
Relay board schematic (Copyright 2006 Chaney Electronics. Reproduced with permission)

6.1.4 Relay Board

The relay board provides electronic control of the motors so that you can give the robot controlled motion. Figure 6.13 shows the schematic of the relay board.

The relay board has two relays, each one controlled by an NPN transistor. Relay K1 is an SPDT (single-pole double-throw) 5 V relay; and relay K2 is a DPDT (double-pole double-throw). The relays are powered by the 9 V battery. The board has connections for the 9 V battery, input connections for the transistor inputs, and connections for the motors at the relay outputs. Figure 6.14 illustrates the parts layout of the board and the board placement on the robot.

Table 6.1 provides the list of parts used in the relay board.

6.1.5 Basic Robot Movements

By connecting the motors in different ways you can achieve different movements.

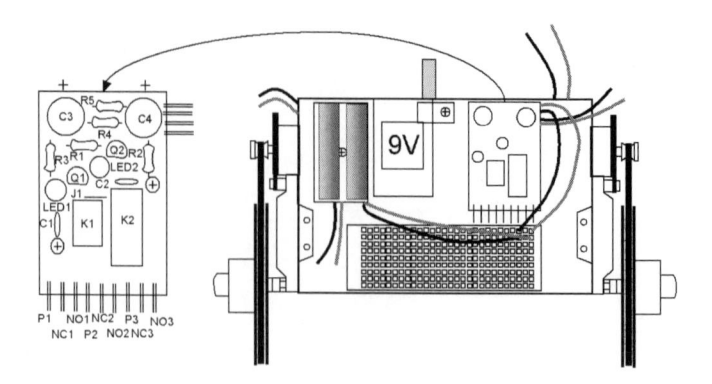

FIGURE 6.14

Parts layout and board placement

TABLE 6.1
Relay-board parts

Parts	Description
R1, R2	Resistor; 4.7K ohm (yellow, violet, red)
R3, R4	Resistor; 2.7 ohm (red, violet, gold)
R5	Resistor; 47 ohm (yellow, violet, black)
C1, C2	Disc capacitor; 0.01 microF (103)
C3, C4	Electrolytic capacitor; 100 microF
Q1, Q2	Transistor; 2N3904
LED1	Green LED
LED2	Red LED
K1	SPDT 5 V relay
K2	DPDT 5 V relay
Misc.	Circuit board, solid wire 3 3/4 inch long (four pieces), Solid wire 2 inch long (nine pieces)

Differential Drive Robot 231

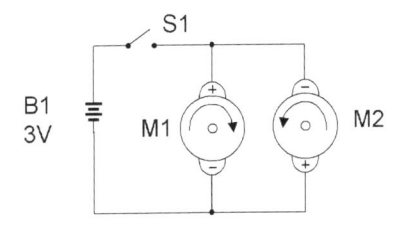

FIGURE 6.15
Forward motor motion

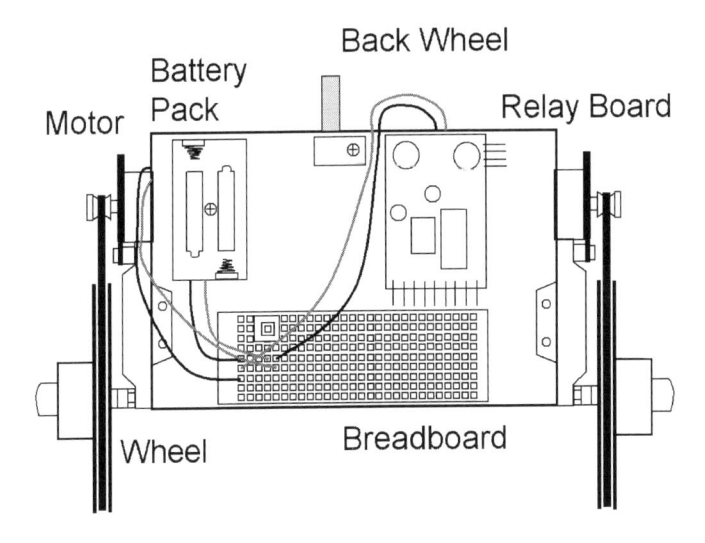

FIGURE 6.16
Breadboard connections for forward motion

6.1.5.1 Forward Motion

Forward motion is obtained by connecting the motors to power so that both motors make the wheels move forward. The motors are connected to the wheels of this robot so that the connections (shown in Figure 6.15) produce the forward motion.

Remember that motor M1 is connected to the power supply to turn the motor in the positive direction; motor M2 is connected to the power in reverse. Figure 6.16 shows the breadboard connections for this motion.

Figure 6.17 illustrates the motion of the robot with these connections.

6.1.5.2 Backward Motion

Backward motion is obtained by connecting the motors to power so that both motors make the wheels move backward. The motors are connected to the

FIGURE 6.17

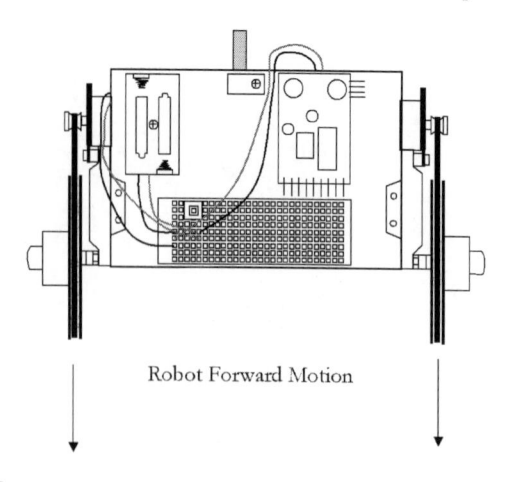

Robot forward motion

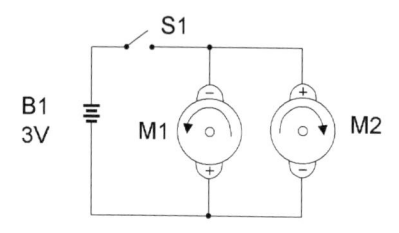

FIGURE 6.18

Backward motion

wheels of this robot so that the connections (shown in Figure 6.18) produce the backward motion.

6.1.5.3 Forward Right Turn

The forward right turn motion is obtained by disconnecting the right motor from the power and having the left one make the wheel move forward. The motors are connected to the wheels so that the connections in Figure 6.19 produce the forward right turn motion.

In this connection scheme, the right wheel does not turn and the left wheel makes the left side of the robot move forward. This way the robot turns about its right wheel, producing a forward right turn as in Figure 6.20.

6.1.5.4 Backward Right Turn

The backward right turn motion is obtained by disconnecting the left motor from power and having the right one make the wheel move backward. The motors are connected to the wheels so that the connections (Figure 6.21)

Differential Drive Robot 233

FIGURE 6.19

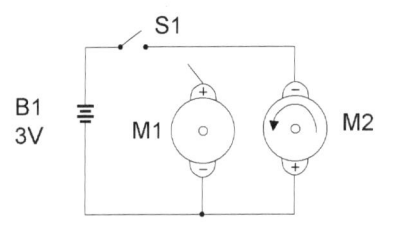

Forward right turn motor motion

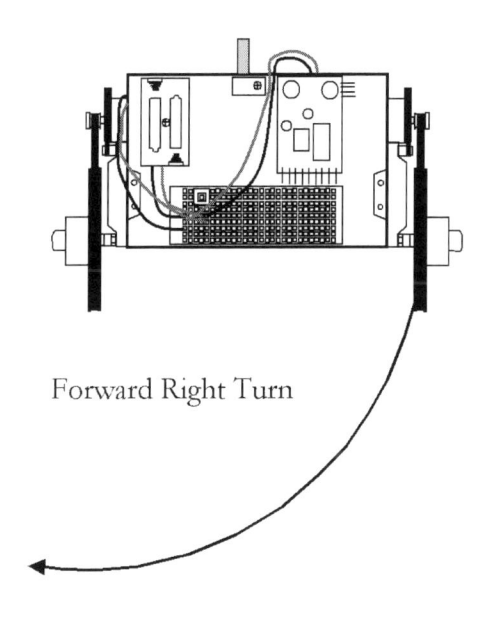

FIGURE 6.20

Forward-right-turn robot motion

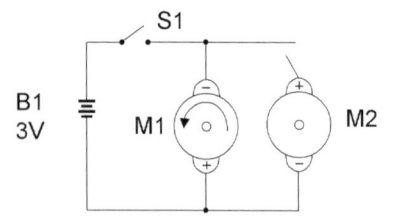

FIGURE 6.21
  Backward right-turn motor motion

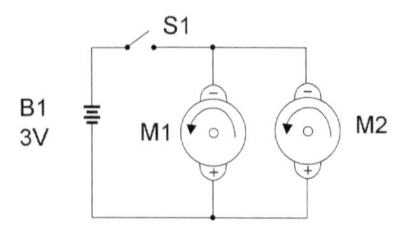

FIGURE 6.22
  Sharp forward right turn robot motion

produce the backward right turn motion.

6.1.5.5 Sharp Right Turn

The sharp forward right turn motion is obtained by connecting the motors to power so that the right motor makes the wheel move backward and the left one makes the wheel move forward. The motors are connected to the wheels so that the connections shown in Figure 6.22 produce the sharp forward right turn motion.

In this connection scheme, the right wheel turns backward and the left wheel makes the left side of the robot move forward. This way the robot turns about its center, producing a sharp forward right turn, as shown in Figure 6.23.

6.1.5.6 Summary of the Basic Movements

Table 6.2 summarized the robot's basic movements and shows which connections produce which motions.

6.1.6 Timed Movements

When you give a 5 V signal to the input of a relay on the relay board, that relay turns on; when you connect the input to the ground (0 V), the relay is off. You can control the motors by giving different signals (on or off) at the relay board inputs. You can use a 555 timer chip to produce pulses on the output, which can control the on-time and off-time of the motor. We will

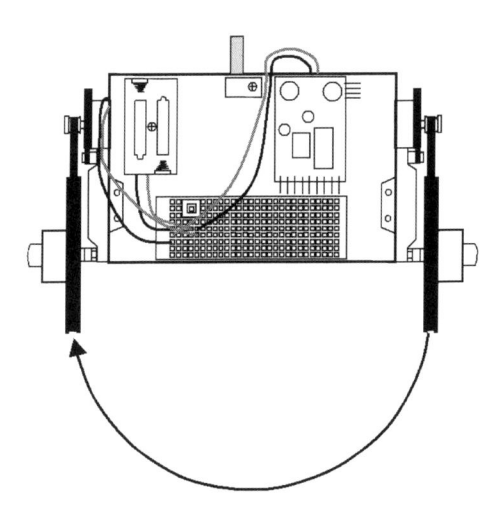

Sharp Forward Right Turn

FIGURE 6.23
Sharp forward right turn robot motion

TABLE 6.2
Basic robot movements

Movement	Left Motor	Right Motor
Robot forward motion	Forward	Forward
Robot backward motion	Backward	Backward
Robot forward right turn	Forward	Disconnected
Robot backward right turn	Disconnected	Backward
Robot sharp right turn	Forward	Backward
Robot forward left turn	Disconnected	Forward
Robot backward-left turn	Backward	Disconnected
Robot sharp left turn	Backward	Forward

FIGURE 6.24

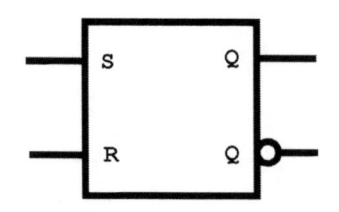

R-S flip-flop

TABLE 6.3
R-S flip-flop truth table

Input S	Input R	Present State (Q)	Next State (Q*)	Action
0	0	0	0	Hold
0	0	1	1	Hold
0	1	0	0	Reset
0	1	1	0	Reset
1	0	0	1	Set
1	0	1	1	Set
1	1	0	X	Not allowed
1	1	1	X	Not allowed

study the 555 timer chip and use it for robot control. But first we will look at how an R-S flip-flop works.

6.1.6.1 R-S Flip-Flop

Figure 6.24 is the symbol of an R-S flip-flop.

Table 6.3 is the truth table of the R-S flip-flop; it shows how the output of the flip-flop depends on the inputs and the previous state.

The table shows that when input R is 1 and input S is 0, the output Q resets to 0; when input R is 0, and input S is 1, the output Q sets to 1; when input R is 0 and input S is 0, the output Q remains the same as its previous value; the input combination of R at 1 and S at 1 is not allowed.

6.1.6.2 555 Timer

Figure 6.25 presents the pin numbers and their names in a 555 timer (also called *triple nickel*).

Figure 6.26 shows a simplified functional block diagram.

Let's look at how to use a 555 timer as an oscillator. We connect the timer as shown in Figure 6.27. This is called an *astable mode* of the timer, since it creates oscillations.

Differential Drive Robot

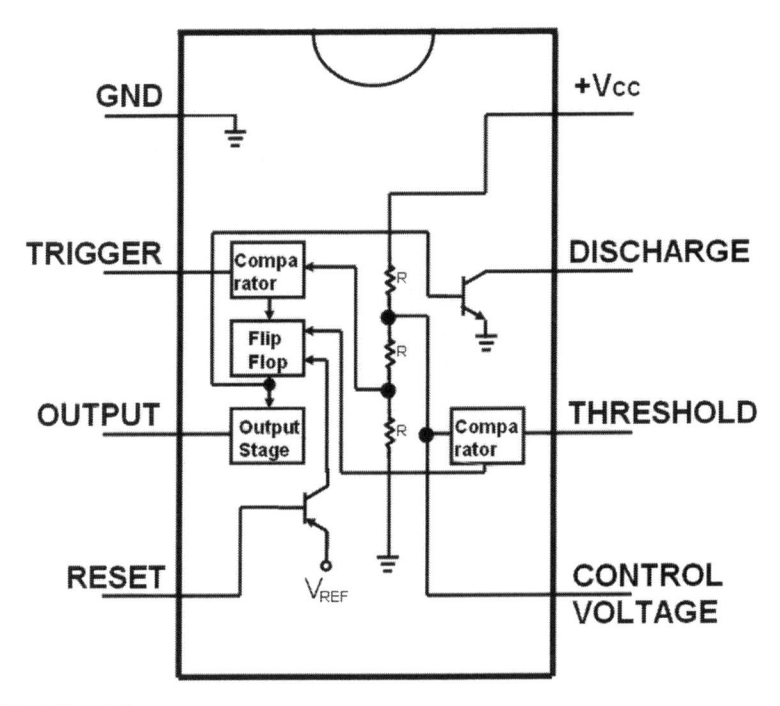

FIGURE 6.25

555 timer pins [47].

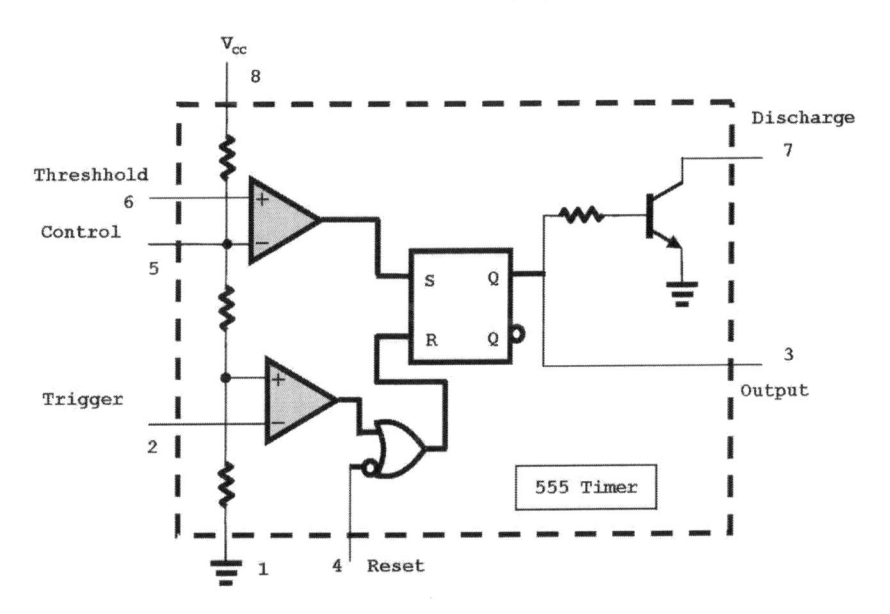

FIGURE 6.26

555 timer functional block diagram

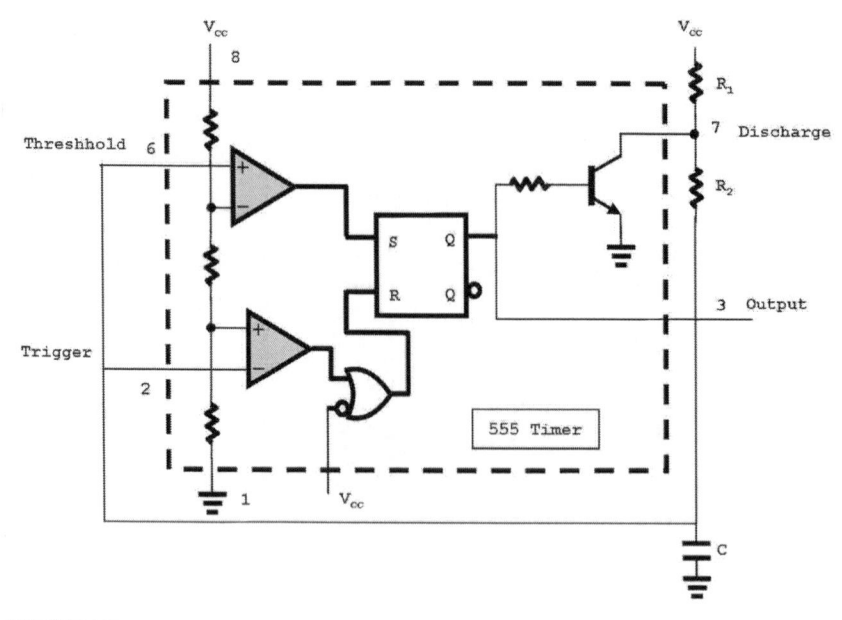

FIGURE 6.27

555 timer as an oscillator (astable mode)

Let us start the time when the capacitor has zero charge, and therefore zero voltage across it. This implies that the voltage at the threshold and trigger inputs is also 0. Because the comparator on top has 0 on the input shown with the positive sign and has $2/3V_{CC}$ voltage on the input shown with the negative sign, its output will be 0. However, because the comparator on the bottom has 0 on the input shown with the negative sign and has V_{CC} voltage on the input shown with the positive sign, its output will be 1. Checking from the operation of an R-S flip-flop, we see that the output Q of the flip-flop will be 0. This will give the output of the timer chip as 0. This 0 will turn the transistor off so that we have V_{CC} connected to the two resistors and the capacitor. This allows the capacitor whose charging time constant is $(R_1 + R_2)C$ to begin charging. It keeps charging until its voltage reaches just above $2/3V_{CC}$. Then the top comparator gives an output of 1 and the bottom comparator gives 0.

These inputs set the R-S flip-flop to give an output of 1, and therefore the timer output is 1. This output also turns on the transistor, which connects pin 7 to the ground. The capacitor starts discharging through resistor R_2 with a time constant of R_2C. The discharging continues until the capacitor voltage reaches just below $1/3V_{CC}$. At that time, the bottom comparator gives the output of 1 again, and the whole process continues cyclically to produce oscillatory output. The frequency of oscillation is given by the following

Differential Drive Robot

FIGURE 6.28

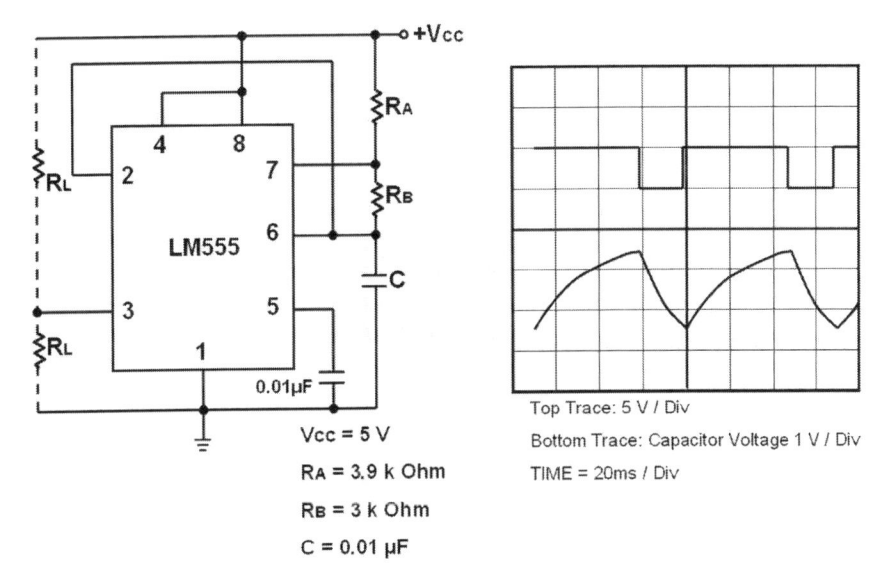

555 timer as an oscillator [47].

Vcc = 5 V
R_A = 3.9 k Ohm
R_B = 3 k Ohm
C = 0.01 μF

Top Trace: 5 V / Div
Bottom Trace: Capacitor Voltage 1 V / Div
TIME = 20ms / Div

equation:

$$f = \frac{1.44}{(R_1 + 2R_2)C} \qquad (6.22)$$

We can understand the operation of the one-shot (explained in detail later in the chapter) by looking at the connections and the corresponding waveforms [47] in Figure 6.28.

6.1.7 Robot Timed Movements

You can use the 555 timer to connect to the relay to produce timed motion, and you can employ the 555 timer as an oscillator so the robot motors can be turned on and off cyclically to have periodic stop-and-go movements by using the circuit shown in Figure 6.29. The schematic of the circuit is shown in Figure 6.30, and the robot movements are shown in Figure 6.31. This circuit is for forward timed movements.

Figure 6.32 shows how the output of the timer chip controls the robot motion.

Table 6.4 presents the parts list for this robot controller.

You can keep the timer part of the circuit, and by simply changing the motor connections to the relay board, you can get different types of timed movements from the robot.

240   *Practical and Experimental Robotics*

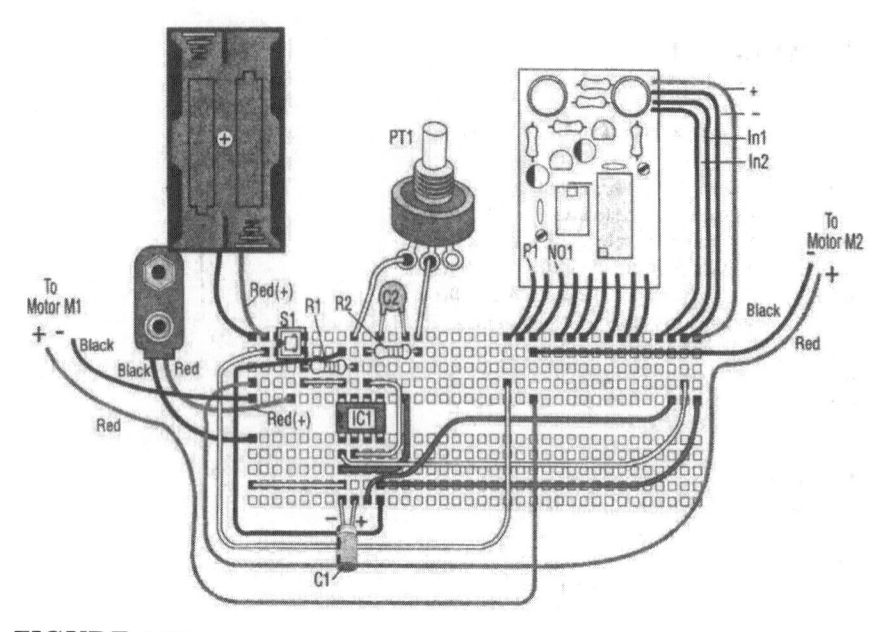

FIGURE 6.29
Stop-and-go circuit layout (Copyright 2006 Chaney Electronics. Reproduced with permission)

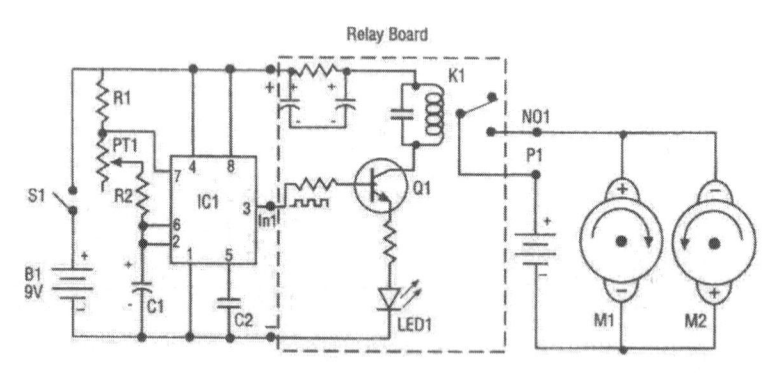

FIGURE 6.30
Stop-and-go circuit schematic (Copyright 2006 Chaney Electronics. Reproduced with permission)

Differential Drive Robot 241

FIGURE 6.31

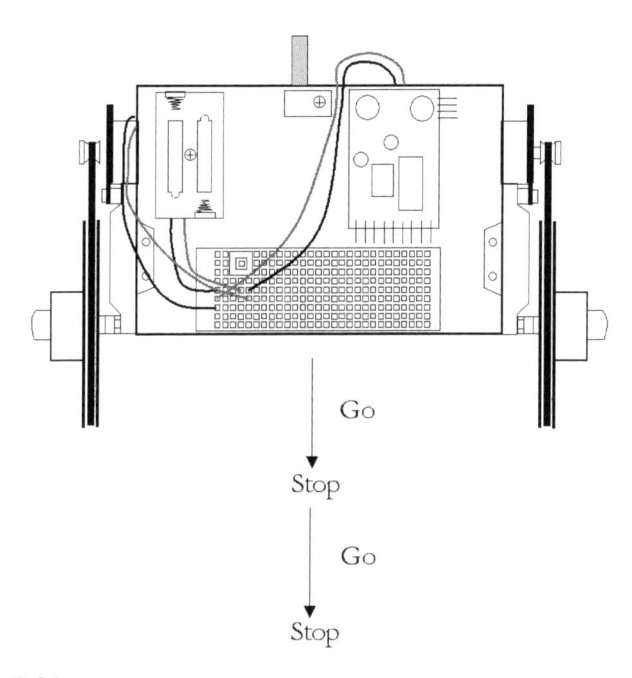

Stop-and-go robot movements

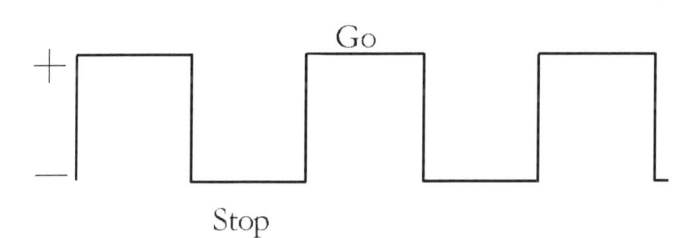

FIGURE 6.32

Timer output

TABLE 6.4
Stop-and-go control parts

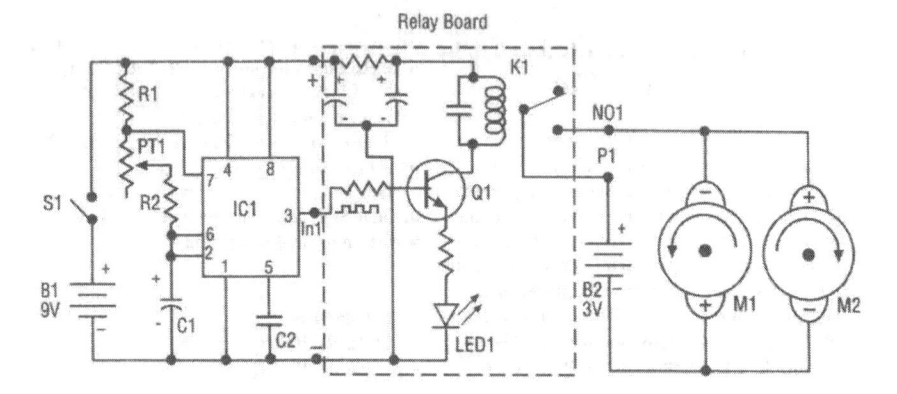

Part	Description
B1	9 V battery
B2	3 V battery (AA batteries)
M1 and M2	Robot motors
R1	Resistor; 10K ohm
R2	Resistor; 22K ohm
PT1	500K potentiometer
C1	Electrolytic capacitor; 10 microF
C2	Disc capacitor; .01 microF
IC1	555 IC
S1	Switch
Misc.	Assembled relay board, wires

FIGURE 6.33

Circuit for backward timed motion (Copyright 2006 Chaney Electronics. Reproduced with permission)

6.1.7.1 Backward Timed Motion

You can make the robot have stop-and-go motion in the backward direction by changing just the connections of the previous circuit to the motors. The circuit with the changed connections looks like the one in Figure 6.33. When the relay is off, the motors are off, and when the relay turns on, the robot moves backward.

6.1.7.2 Time-Controlled Sharp Right Turn

We can make the robot have stop-and-go motion in a sharp right turn by changing just the connections of the previous circuit to the motors. Figure 6.34 shows the circuit with the changed connections. When the relay is off,

Differential Drive Robot 243

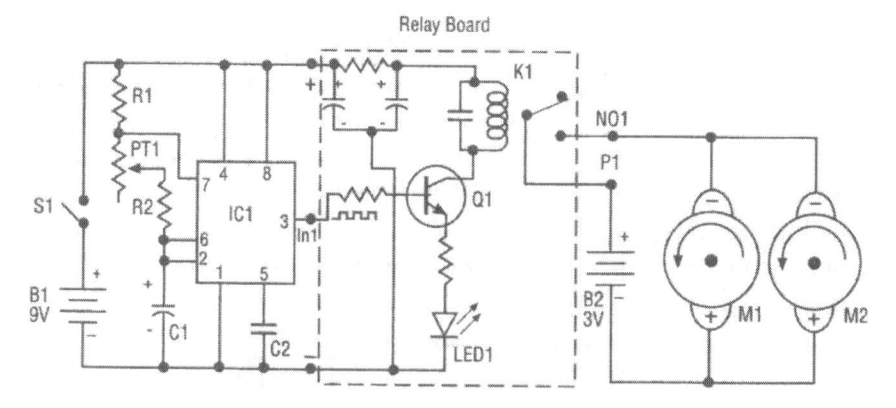

FIGURE 6.34

Circuit for time-controlled sharp right turn (Copyright 2006 Chaney Electronics. Reproduced with permission)

the motors are off, and when the relay turns on, the robot makes a sharp right turn.

6.1.7.3 Time-Controlled Sharp Left Turn

You can make the robot have stop-and-go motion in a sharp left-turn motion by changing just the connections of the previous circuit to the motors. Figure 6.35 illustrates the circuit with the changed connections. When the relay is off, the motors are off, and when the relay turns on, the robot does a sharp left turn.

6.1.7.4 Time-Controlled Combined Forward and Backward Movements

By changing just the connections of the previous circuit to the motors, you can make the robot have timed forward and backward motion. The circuit with the changed connections looks like the one in Figure 6.36. The polarity of the battery connections to both motors changes when the relay turns on, and that changes the direction of each motor.

6.1.7.5 Time-Controlled Combined Forward and Backward Right Turn

To make the robot have a timed forward and backward right turn motion, you simply need to change the connections of the previous circuit to the motors. The circuit with the changed connections looks like the one in Figure 6.37. By changing the relay from off to on, the motion changes from forward to backward right turn. Notice how the motors are connected in this circuit compared to the previous one.

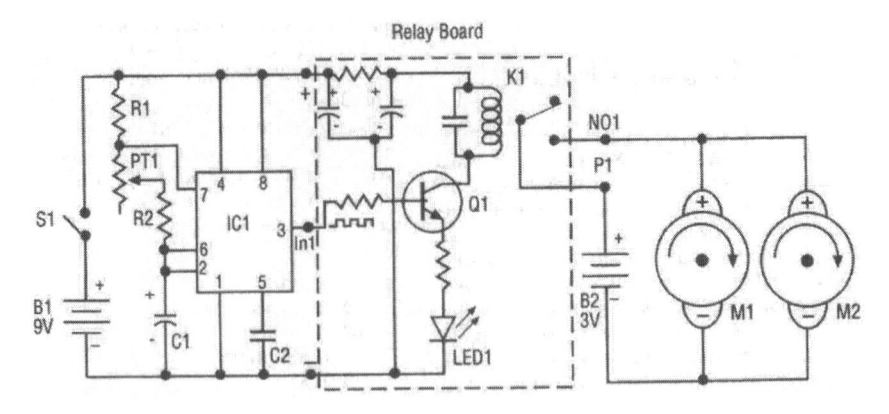

FIGURE 6.35
Circuit for time-controlled sharp left turn (Copyright 2006 Chaney Electronics. Reproduced with permission)

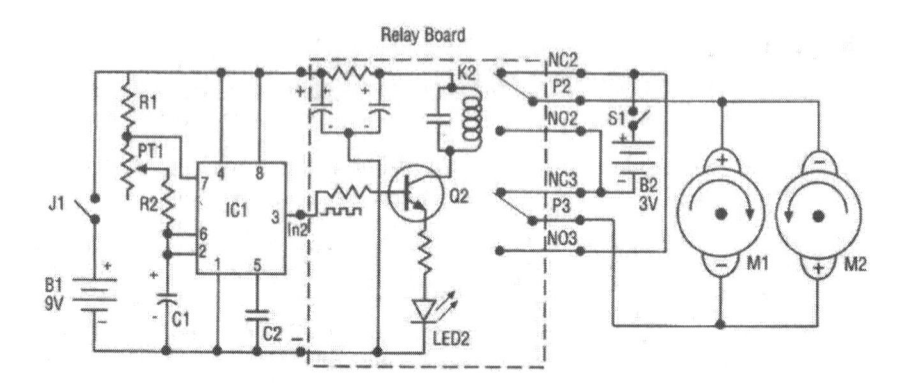

FIGURE 6.36
Circuit for time-controlled combined forward and backward motion (Copyright 2006 Chaney Electronics. Reproduced with permission)

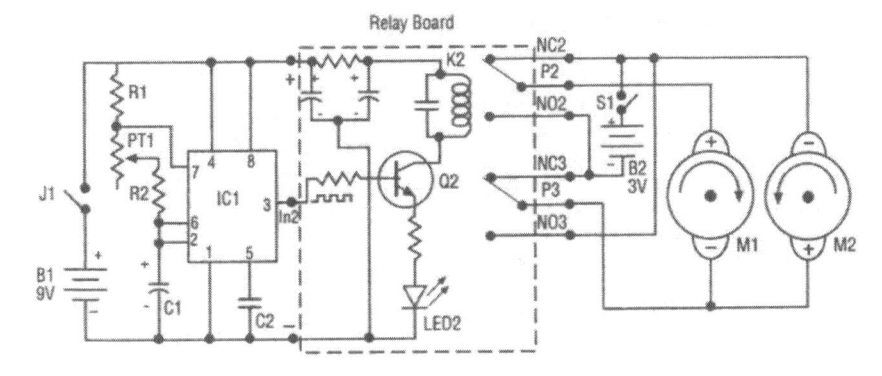

Differential Drive Robot

FIGURE 6.37
  Circuit for time-controlled combined forward and backward right turn (Copyright 2006 Chaney Electronics. Reproduced with permission)

Figure 6.38 shows the top view of the robot's movements.

6.1.7.6 Time-Controlled Combined Forward and Hard Left-Turn Movements

You can give the robot a timed forward and hard left-turn motion by changing just the connections of the previous circuit to the motors. Figure 6.39 shows the circuit with the changed connections.

Figure 6.40 shows the top view of the robot's movements.

6.1.7.7 Double Timer Circuit

You can use two timers to make more complex combined behavior. The circuit in Figure 6.41 combines the backward and sharp left turn on the first timer, and combines on and off control of both motors on the second timer.

6.1.7.8 Night-Runner Robot

The circuit in Figure 6.42 combines the backward and sharp left turn on the first timer, and uses a CDS cell (also called a photocell) to produce on and off control of both motors. A photocell has resistance that depends on light falling on it. When light falls on it, its resistance decreases. In the circuit, when light falls on the cell the resistance is low and therefore the voltage across it is low. Because of the low voltage, transistor Q1 is off. However, when it is dark the voltage across the resistance increases and that turns on the transistor, which turns on the power to the two motors. The motors can now follow the movement commands from the top relay.

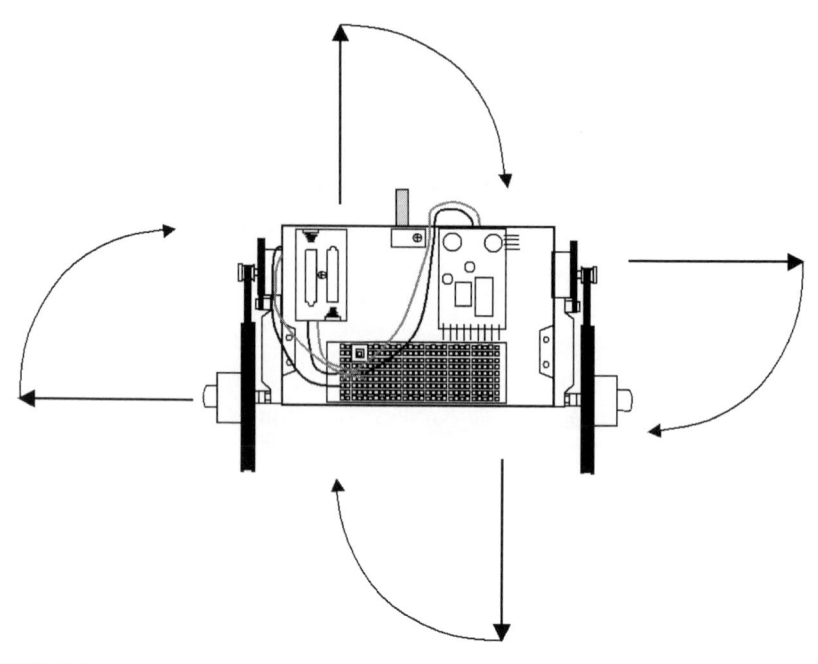

FIGURE 6.38
Top view of time-controlled combined forward and backward right turn

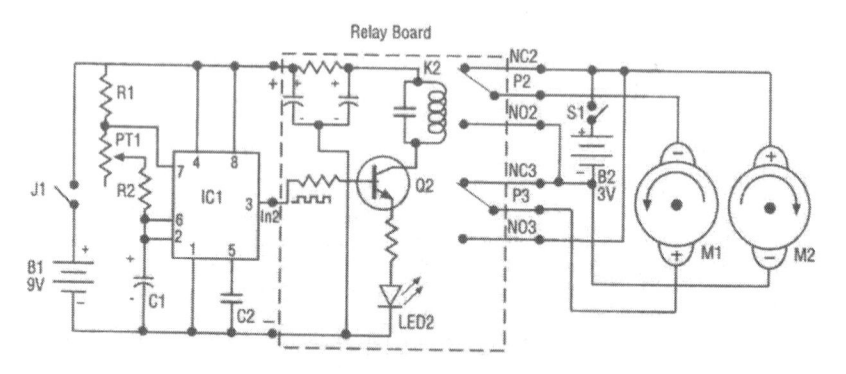

FIGURE 6.39
Circuit for time-controlled combined forward and hard left-turn motion
(Copyright 2006 Chaney Electronics. Reproduced with permission)

Differential Drive Robot 247

FIGURE 6.40
Top view of time-controlled combined forward and hard left-turn motion

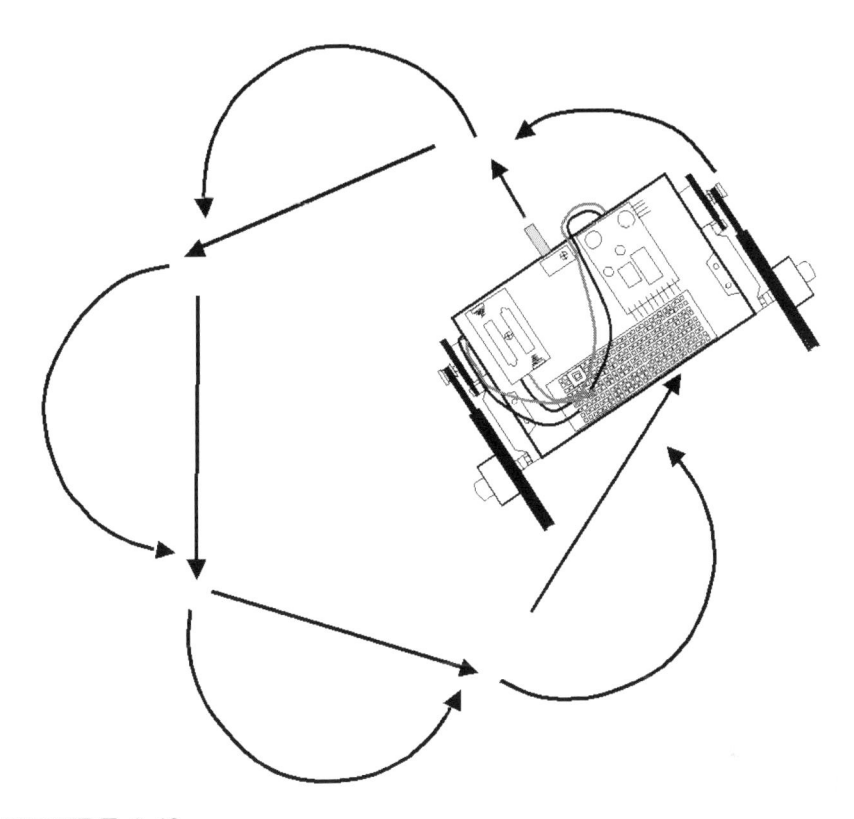

FIGURE 6.41

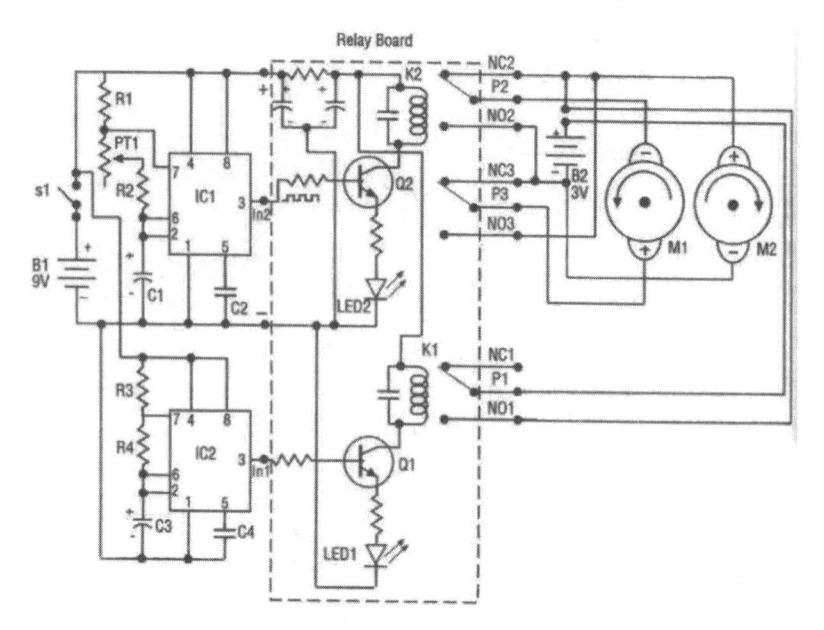

Two-timer circuit (Copyright 2006 Chaney Electronics. Reproduced with permission)

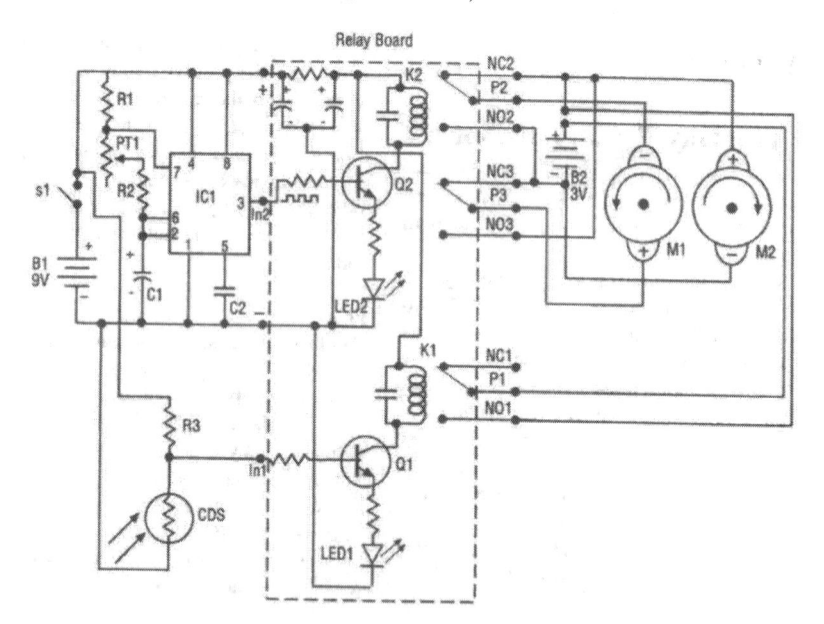

FIGURE 6.42

Night-runner robot circuit (Copyright 2006 Chaney Electronics. Reproduced with permission)

Differential Drive Robot 249

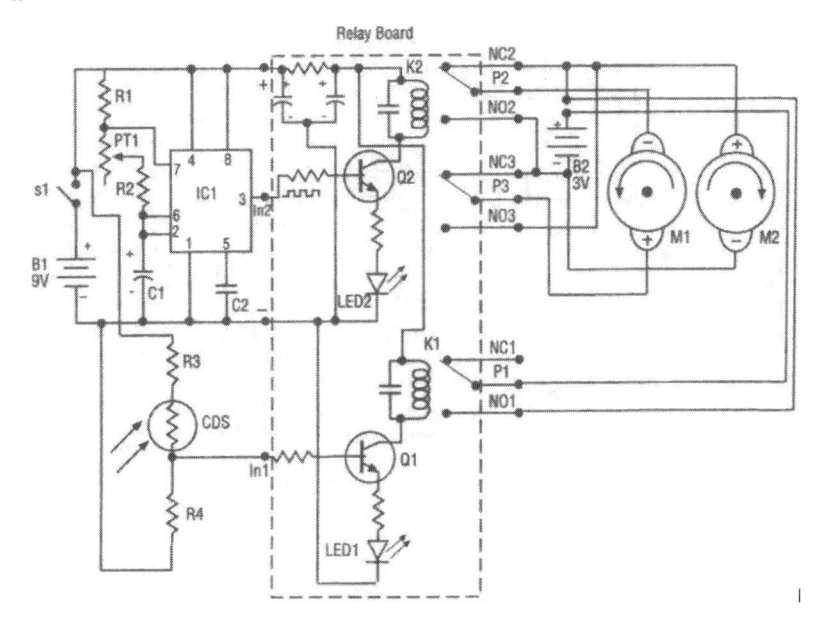

FIGURE 6.43

Day-runner robot circuit (Copyright 2006 Chaney Electronics. Reproduced with permission)

6.1.7.9 Day-Runner Robot

The circuit in Figure 6.43 combines the backward and sharp left turn on the first timer and uses a CDS cell to produce on and off control of both motors. In the circuit, we have simply changed the placement of the photocell to invert the logic of transistor Q1 turning on and off.

6.1.8 Infrared Vision-Based Robot (Robot Eyes)

You can modify the robot to use infrared communication for different tasks. For infrared-based communication you can use the 555 timer to modulate an IR LED to send infrared signals at a specific frequency and have an infrared receiver that responds to that specific frequency. Most remote controls for TVs, VCRs, and so on, use a frequency close to 40Khz for communication. The schematic in Figure 6.44 is derived from the BASIC Stamp programming manual [26] to show an infrared transmitter and receiver using modulated signals. The 555 timer produces pulses whose frequency can be adjusted to match that of the receiver by using the potentiometer attached to the timer. When the receiver gets the infrared light at its desired frequency, its output goes high. You can send digital signals by using the reset input of the 555 timer to transmit serial data and then receive the data from the output pin of the receiver. The reset pin data makes the whole modulation turn on or

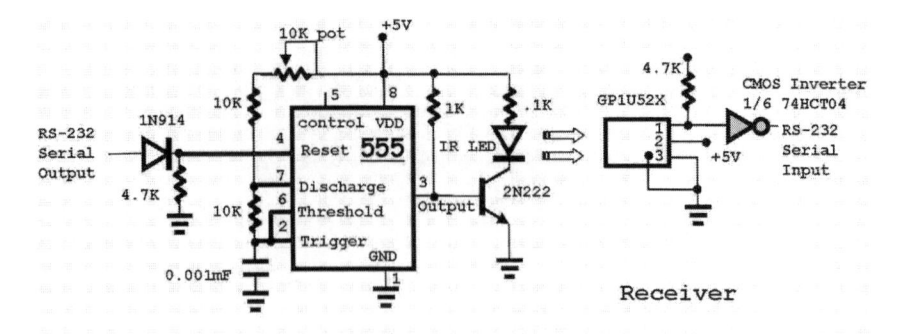

FIGURE 6.44

IR-modulated transmitter and receiver circuits (Copyright 2002 BASIC Stamp Programming Manual 1.9, Parallax Inc. Reproduced with permission)

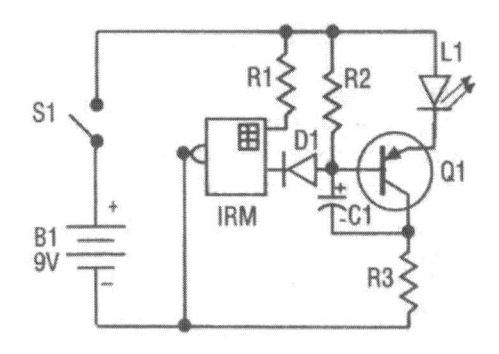

FIGURE 6.45

IR-receiver circuit (Copyright 2006 Chaney Electronics. Reproduced with permission)

off so that the serial data can be transmitted and then received.

The IR receiver that comes with this robotic kit can be connected as shown in Figure 6.45.

Figure 6.46 shows a matching transmitter circuit for the same receiver.

Table 6.5 presents the parts list for this transmitter and receiver circuits.

You can use the IR system to make the robot do various movements. For example, you can make it move backward normally unless you send an IR signal using a TV remote control, and you can make it do a hard left turn until the IR signal is on. Figure 6.47 shows the circuit for this motion.

Differential Drive Robot

FIGURE 6.46

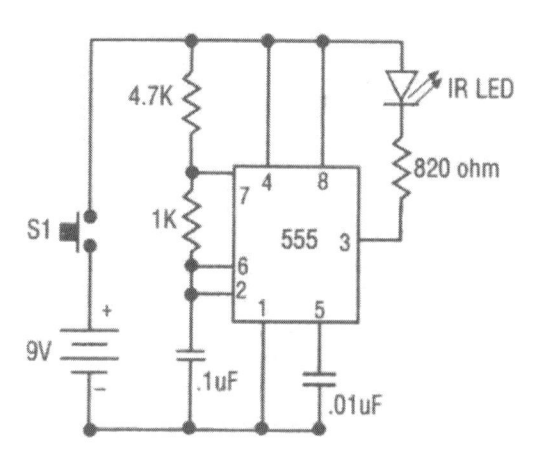

IR-transmitter circuit (Copyright 2006 Chaney Electronics. Reproduced with permission)

TABLE 6.5

IR transmitter and receiver parts list

Part	Description
B1	9 V Battery
IRM	Infrared receiver module
R1	Resistor; 100 ohm
R2	Resistor; 33K ohm
R3	Resistor; 330 ohm
PT1	500K potentiometer
C1	Electrolytic capacitor; 10 microF
D1	Diode; 1N4148
Q1	PNP transistor; 2N3906
S1	Switch
Misc.	Wires

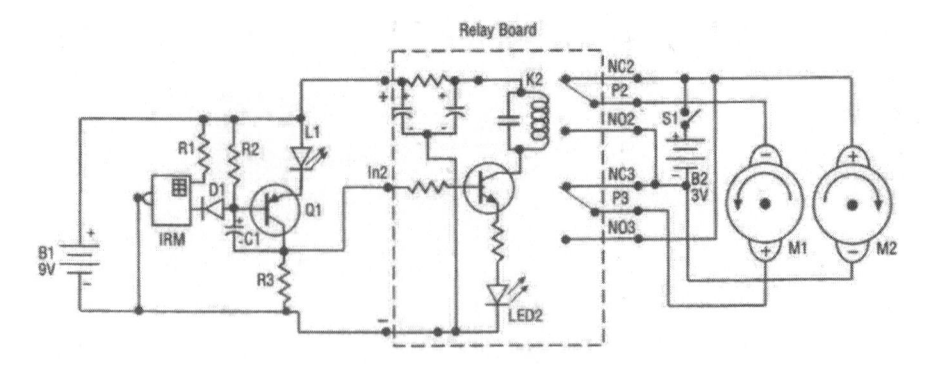

FIGURE 6.47
IR-based backward hard left turn combination (Copyright 2006 Chaney Electronics. Reproduced with permission)

6.1.8.1 Emotional Robot (Robotic Pet)

You can use the IR transmitter and receiver so that the motors are turned off until the IR receiver receives the IR light, and then the robot does a hard left turn. You can place the transmitter and the receiver in a line so that when you bring your hand close to the robot, the IR light reflects from the hand and reaches the receiver. This makes the robot perform a hard left turn. This behavior is similar to petting a dog that is motionless and then moves when you pet it. Figure 6.48 shows the circuit for the robotic pet.

6.1.8.2 Obstacle Avoidance

You can use the IR transmitter and receiver so that the robot keeps moving forward until the IR receiver receives the IR light and then the robot does a hard left turn to avoid the obstacle in its way. You can place the transmitter and the receiver in a line so that when the robot gets close to an obstacle, the IR light reflects from the obstacle and reaches the receiver, making the robot go into a hard left turn. Figure 6.49 shows the circuit for this.

6.1.9 Audio Detection and Response (Robot Ears)

Now we will establish the robot's ability to detect and respond to sounds. The circuit for this uses an electret microphone, an audio amplifier, the 555 timer used as a one-shot, and a flip-flop used as a toggle. We'll cover each in detail before discussing the overall circuit.

6.1.9.1 Electret Microphone

An electret microphone uses the capacitance changes in the microphone caused by the mechanical motions from sound vibrations, which are exploited to give

Differential Drive Robot

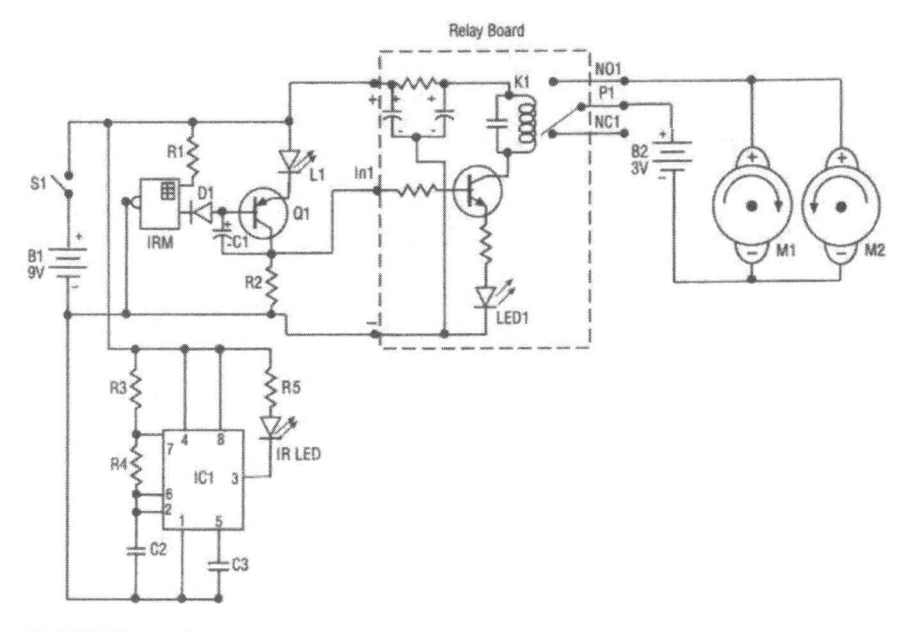

FIGURE 6.48

Robotic pet circuit (Copyright 2006 Chaney Electronics. Reproduced with permission)

out an electric signal that represents the sound signal. These microphones (such as the one in Figure 6.50) contain an integrated preamplifier.

Figure 6.51 shows how to interface the microphone in a circuit.

The potentiometer can control the output level. Table 6.6 provides the parts list for this interface.

6.1.9.2 Audio Amplifier

A popular audio IC (integrated circuit) amplifier is LM386 ([48]), which you can buy from National Semiconductor [33]. Figure 6.52 shows a typical interfacing circuit for this amplifier.

The amplifier's gain is controlled by a capacitor connected between pins 1 and 8; the power is connected to pin 6; the input is connected to pin 3; and capacitive and resistive coupling is done at output pin 5. Figure 6.53 shows the interfacing of the microphone output to the audio amplifier in the robot.

Table 6.7 presents the additional parts list for this interface.

6.1.9.3 555 Timer as a One-Shot

You can use the 555 timer to produce a single output pulse of a fixed duration when the signal at the trigger input transitions from logic 1 (5 V) to logic zero

FIGURE 6.49

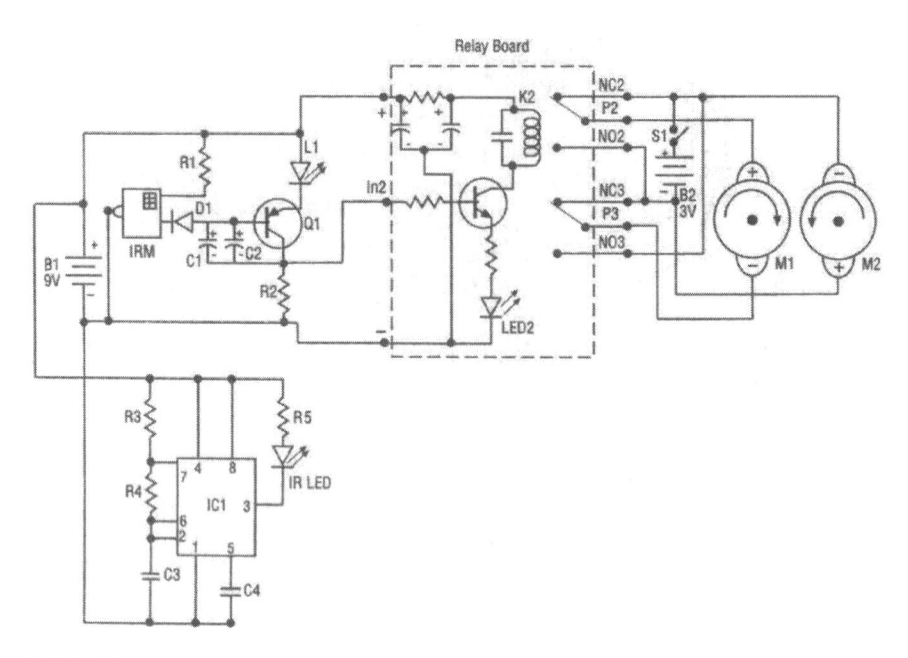

Obstacle avoidance circuit (Copyright 2006 Chaney Electronics. Reproduced with permission)

FIGURE 6.50

An electret microphone (Copyright 2006 All Electronics Corp. Reproduced with permission)

TABLE 6.6
Microphone interfacing parts list

Part	Description
B1	9 V battery
S1	Switch
R1	Resistor; 22K ohm
C1	Electrolytic capacitor; 10 microF
C2	Electrolytic capacitor; 100 microF
Mic	Electret microphone
Misc.	Wires

Differential Drive Robot

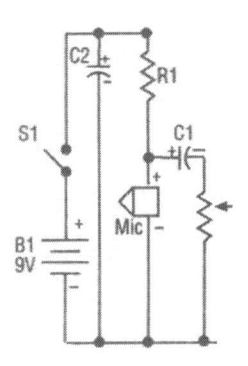

FIGURE 6.51

Electret microphone interfacing (Copyright 2006 Chaney Electronics. Reproduced with permission)

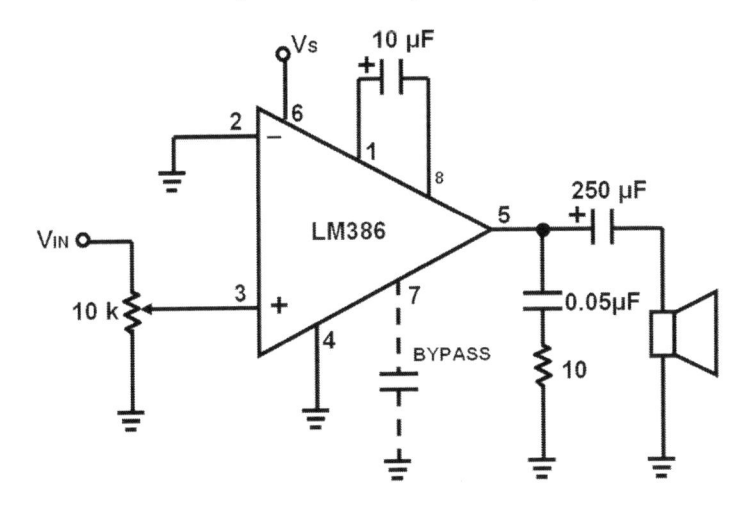

FIGURE 6.52

Audio amplifier interfacing [48].

TABLE 6.7

Microphone amplifier interfacing additional parts list

Parts	Description
PT1	500K potentiometer
R2	Resistor; 100 ohm
R3	Resistor; 3.3K ohm
R4	Resistor; 4.7K ohm
C3	Electrolytic capacitor; 1 microF
C4	Disc capacitor; 0.1 microF
C5	Disc capacitor; 0.1 microF
IC1	Audio amplifier; IC 386
Misc.	Wires

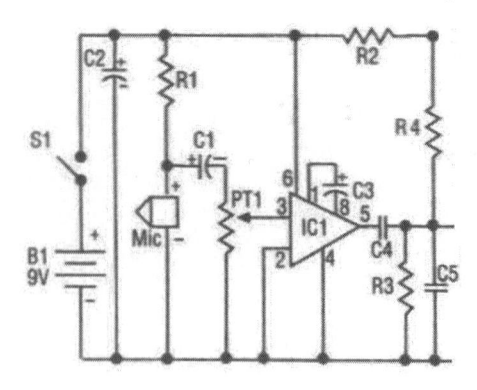

FIGURE 6.53
Audio amplifier interfacing with the microphone (Copyright 2006 Chaney Electronics. Reproduced with permission)

TABLE 6.8
One-shot interfacing with audio amplifier and microphone additional parts list

Parts	Description
R5	Resistor; 3.3M ohm
C6	Disc Capacitor; 0.01 microF
C7	Disc Capacitor; 0.01 microF
IC2	IC 555
Misc.	Wires

(0 V). You can make the timer operate in that *one-shot* mode by connecting it as shown in Figure 6.54. This is called *mono-stable* mode.

To understand its operation, start with the capacitor charge (and therefore its voltage) at 0. When the trigger voltage goes to low, then it is less than $(1/3)V_{CC}$ and it turns on the lower comparator. The R-S flip-flop gets reset so that the output at Q of the flip-flop goes low. This turns off the transistor and allows the capacitor to get charged. When the capacitor's charge is such that the voltage across the capacitor is just above $(2/3)V_{CC}$, the upper comparator turns on and sets the R-S flip-flop output to one again, turning the transistor on. When the transistor turns on, it connects its output to the ground, and that starts discharging the capacitor. The new cycle will not start until the trigger input dips below $(1/3)V_{CC}$ again.

You can understand the operation of the one-shot by looking at the connections and the corresponding waveforms [47] in Figure 6.55, which is sketched based on National Semiconductor's LM555 datasheet.

Figure 6.56 shows the circuit with the microphone, the audio amplifier, and the timer in mono-stable mode.

Table 6.8 provides the additional parts list for this interface.

Differential Drive Robot

FIGURE 6.54

555 timer as a one-shot (mono-stable)

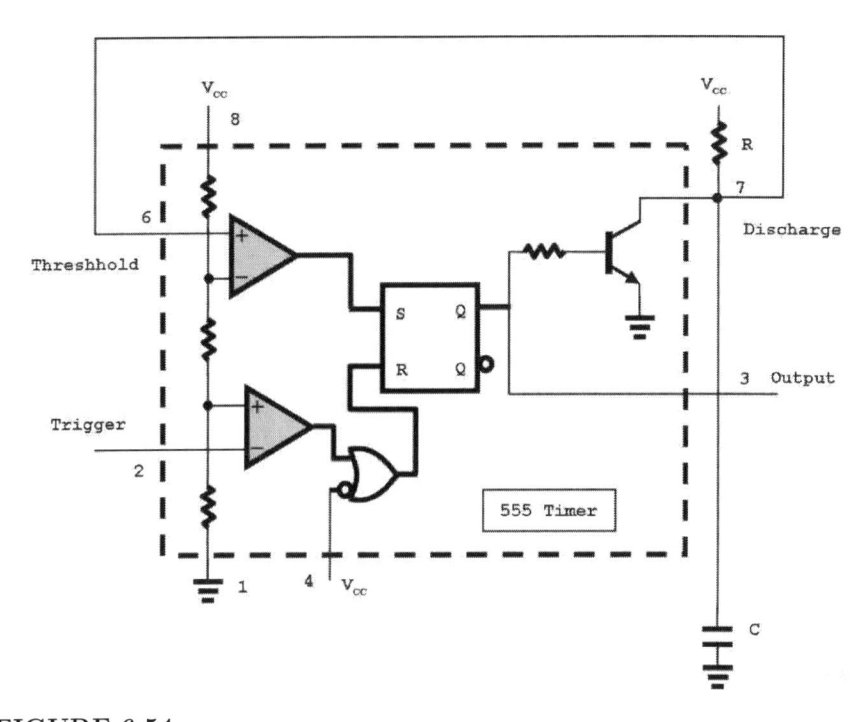

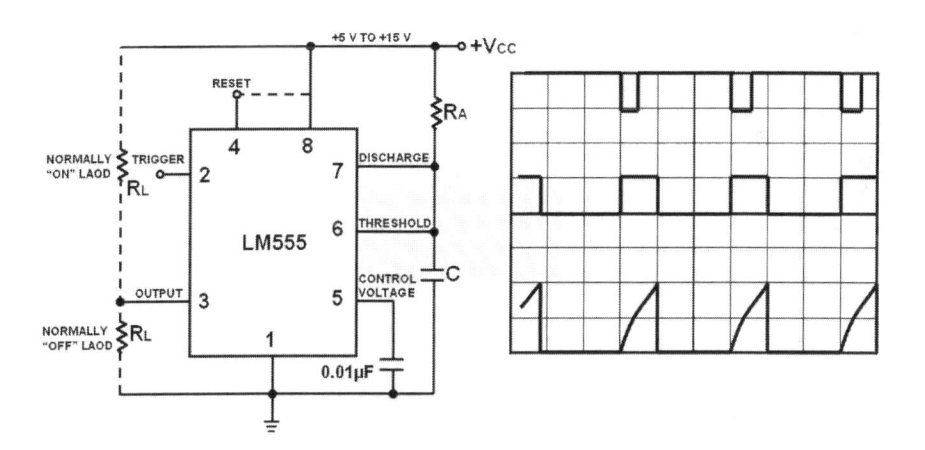

FIGURE 6.55

555 timer as a one-shot [47].

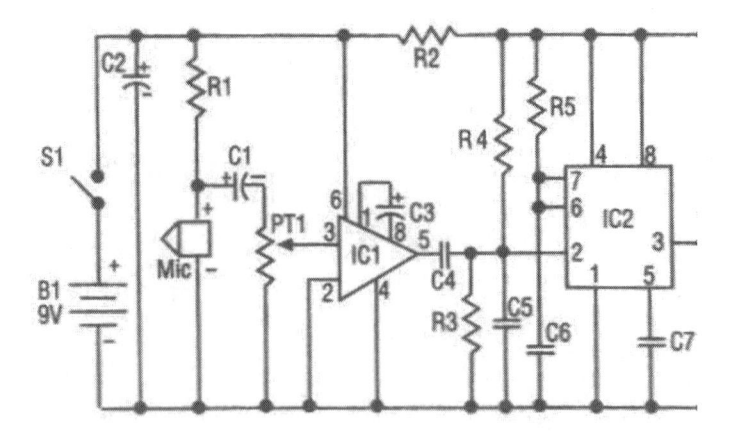

FIGURE 6.56
One-shot interfacing with the audio amplifier and the microphone (Copyright 2006 Chaney Electronics. Reproduced with permission)

TABLE 6.9
Flip-flop interfacing with one-shot, audio amplifier, and microphone additional parts list

Parts	Description
IC3	IC 4013
Misc.	Wires

6.1.9.4 4013 IC Operation (D Flip-Flop)

To understand how the D flip-flop will be used in the robotic circuit, let us take a look at the pin description and the truth table of the flip-flop from Fairchild Semiconductor's CD4013BC datasheet [32]. The figure is reproduced from the datasheet in Figure 6.57.

We will use only F/F 1 from the dual chip. Notice from the truth table that when the R and the S inputs are both at 0, then the output is the complement of the D input. In the robotic circuit, we keep the R and S inputs at 0, and connect the DATA1 input to output. The input to the chip that comes from the one-shot is connected to the CLOCK1 input. Therefore, when the signal at the input from the one-shot goes from low to high, the output changes from 0 to 1 if the previous output was 0; it changes from 1 to 0 if the previous output was 1. Essentially, the output of the flip-flop is toggled whenever the output from the one-shot goes from 0 to 1.

Figure 6.58 shows the circuit with the microphone, the audio amplifier, the timer in mono-stable mode, and the flip-flop.

Table 6.9 lists the additional parts for this interface.

Differential Drive Robot 259

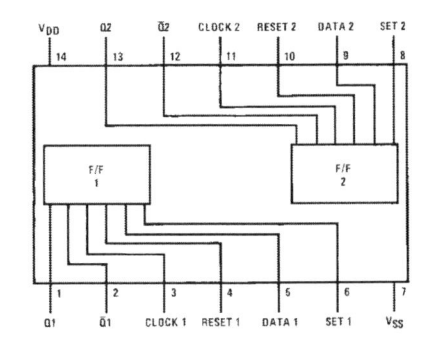

CL (Note 1)	D	R	S	Q	$\overline{Q}$
⤒	0	0	0	0	1
⤒	1	0	0	1	0
⤓	x	0	0	Q	$\overline{Q}$
x	x	1	0	0	1
x	x	0	1	1	0
x	x	1	1	1	1

No Change
x = Don't Care Case
Note 1: Level Change

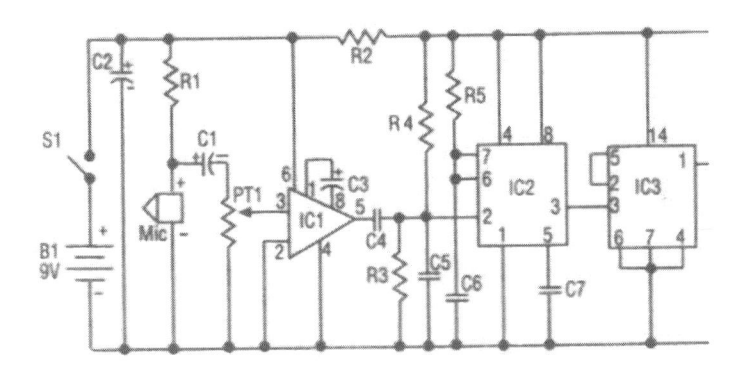

FIGURE 6.57
CD4013BC datasheet from Fairchild Semiconductor (Copyright 2002 Fairchild Semiconductor Datasheet for CD4013BC. Reproduced with permission)

FIGURE 6.58
Flip-flop interfacing with the one-shot, the audio amplifier, and the microphone (Copyright 2006 Chaney Electronics. Reproduced with permission)

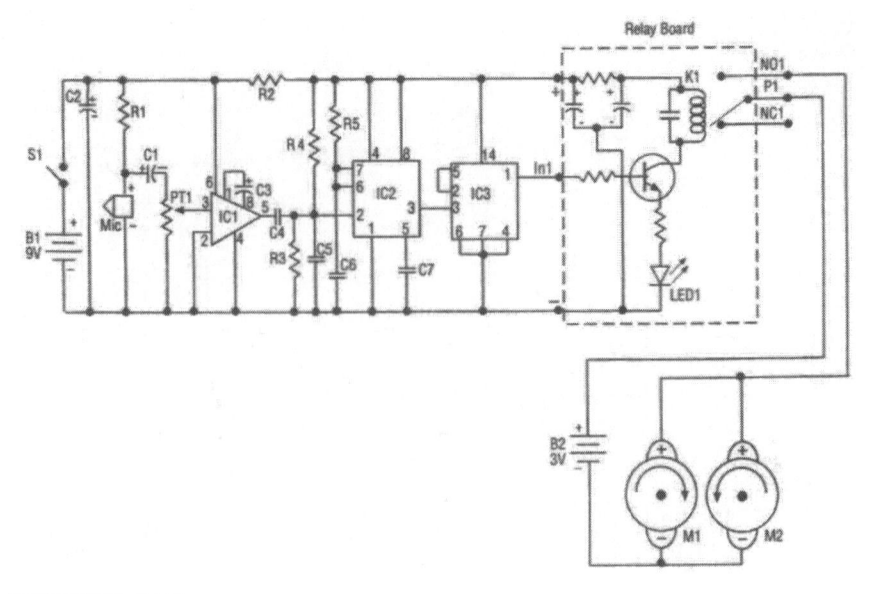

FIGURE 6.59

Sound-activated robot circuit (Copyright 2006 Chaney Electronics. Reproduced with permission)

6.1.10 Sound-Based Robot Movements

You can connect the output from the flip-flop in Figure 6.58 to the relay board and produce different movements depending on how you connect the relay board to the motors.

6.1.10.1 Sound-Activated Robot

The circuit in Figure 6.59 shows a robot that will turn on at the sound of a clap or other loud sound, and will turn off at the next clap or loud sound.

6.1.10.2 Dancer Robot

You can modify the output from the audio amplifier so that we get a DC signal that can charge a capacitor. The output from that can turn on transistors, which can turn the relay on. The circuit in Figure 6.60 uses diode D1 to rectify the audio signal from the audio amplifier. That output charges capacitor C4 which turns on transistors Q1 and Q2. The output of Q2, in turn, turns the relay on. This relay controls the power to both robot motors. When the power to the motors is turned on, the output of the timer makes the robot do sharp right and left turns. This whole scheme makes the robot "dance" when there is music playing (or some continuous sound) and stop moving when the music stops (or when there is no sound).

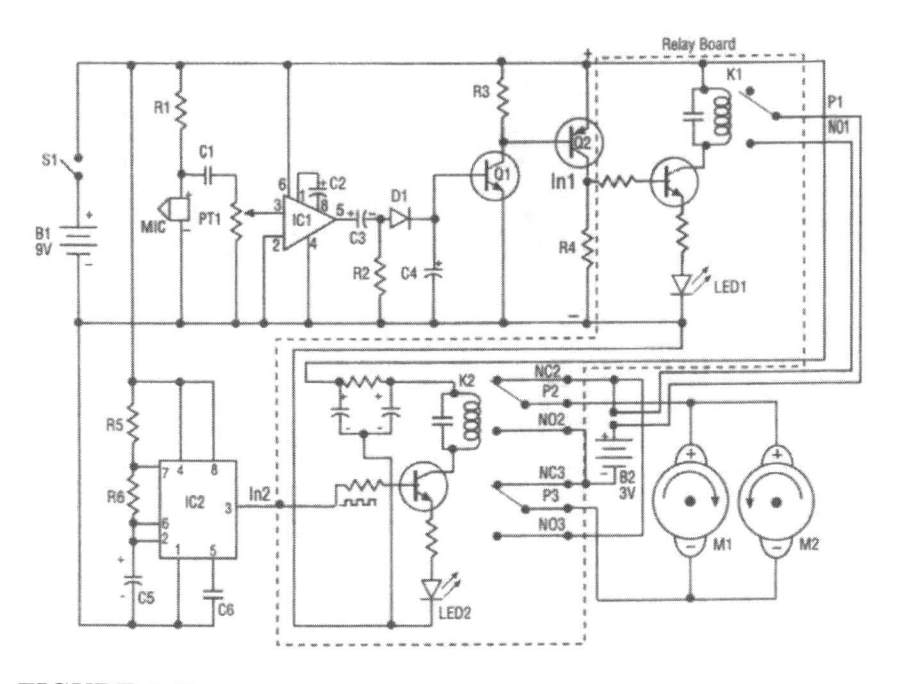

FIGURE 6.60
Dancer robot circuit (Copyright 2006 Chaney Electronics. Reproduced with permission)

TABLE 6.10
Parts list for the dancer robot

Parts	Description
B1	9 V battery
S1	Switch
PT1	500K potentiometer
R1	Resistor; 22K ohm
R2	Resistor; 3.3K ohm
R3	Resistor; 4.7K ohm
R4	Resistor; 1K ohm
R5	Resistor; 10K ohm
R6	Resistor; 100K ohm
C1	Disc capacitor; 0.1 microF (104)
C2	Electrolytic capacitor; 1 microF
C3	Electrolytic capacitor; 10 microF
C4	Electrolytic capacitor; 10 microF
C5	Electrolytic capacitor; 10 microF
C6	Disc capacitor; 0.01 microF (103)
IC1	Audio amplifier; IC 386
IC2	Timer; IC 555
Q1	NPN transistor; 2N3904
Q2	PNP transistor; 2N3906
D1	Diode; 1N4148
Mic	Electret microphone
Misc.	Relay board and wires

Table 6.10 lists the parts for this circuit.

6.2 Robot Speed Control

There are many ways to control the speed of the two DC motors connected to the two wheels of the robot. For details on how to design various speed controllers, please refer to the book Mobile Robotic Car Design [7] mentioned previously. Here you'll learn to use an H-bridge to control the speed of each motor. The H-bridge can be connected to a microcontroller to control the speed of each motor. The H-bridge that we describe here is TPIC0108B [38]. Figure 6.61 shows the chip's functional input-output description.

Figure 6.62 provides the functional block diagram of the chip.

The description of the various pins on the chip appears in Figure 6.63.

Figure 6.64 presents a typical connection for an application of this H-bridge with a DC motor.

Differential Drive Robot

FUNCTION TABLE

IN1	IN2	OUT1	OUT2	MODE
0	0	Z	Z	Quiescent supply current mode
0	1	LS	HS	Motor turns clockwise
1	0	HS	LS	Motor turns counter clockwise
1	1	HS	HS	Brake, both HSDs turned on hard

FIGURE 6.61
Functional input-output representation of TPIC0108B (Courtesy of Texas Instruments)

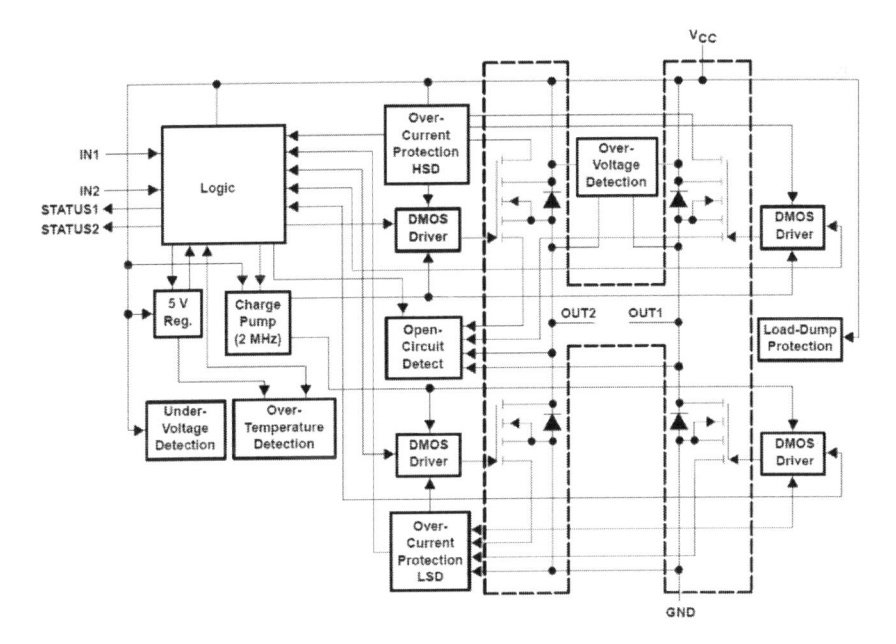

FIGURE 6.62
Functional block diagram of TPIC0108B (Courtesy of Texas Instruments)

Terminal Functions

TERMINAL NAME	NO.	I/O	DESCRIPTION
GND	7, 9, 12, 14	I	Power ground
GNDS	1, 10, 11, 20	I	Substrate ground
IN1	3	I	Control input
IN2	8	I	Control input
OUT1	5, 6	O	Half-H output. DMOS output
OUT2	15, 16	O	Half-H output. DMOS output
STATUS1	13	O	Status output
STATUS2	18	O	Latched status output
V_{CC}	2, 4, 17, 19	I	Supply voltage

NOTE: It is mandatory that all four ground terminals plus at least one substrate terminal are connected to the system ground. Use all V_{CC} and OUT terminals.

FIGURE 6.63

Terminal functions of TPIC0108B (Courtesy of Texas Instruments)

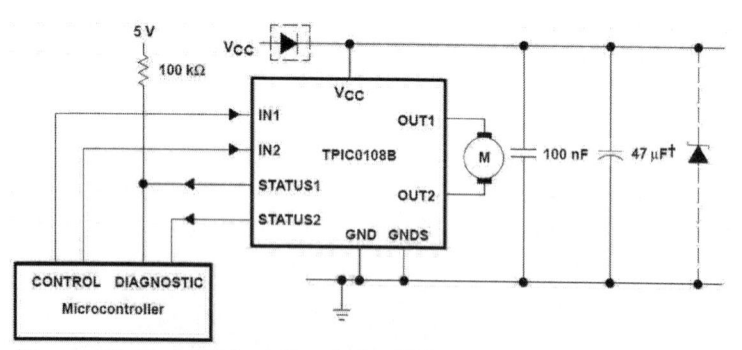

† Necessary for isolating supply voltage or interruption (e.g., 47 µF).
NOTE: If a STATUS output is not connected to the appropriate microcontroller input, it shall remain unconnected.

FIGURE 6.64

Sample application (Courtesy of Texas Instruments)

Differential Drive Robot 265

FIGURE 6.65
Speed control (Courtesy of Texas Instruments)

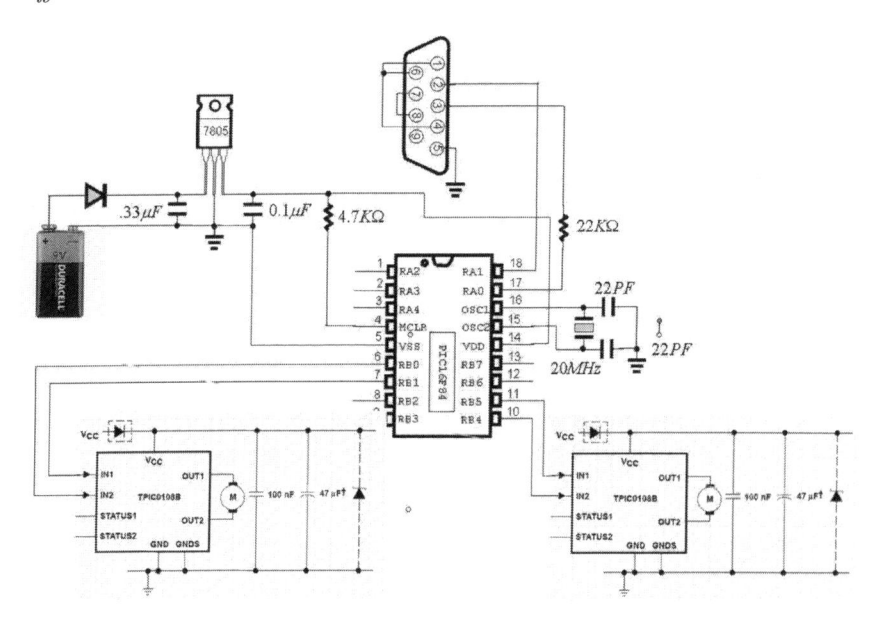

To control the speed of the robot's motors, you can use two H-bridges, one for each motor, and use any microcontroller to give commands to the H-bridges. To control the speed of each motor, you can use PWM signals to each motor. Consult the book called Mobile Robotic Car Design by Kachroo and Mellodge [7] for more details and explanations. To make a wheel move forward at full speed, you can give the signal 0 at IN1, and 1 at IN2 of the H-bridge. To have a low speed, you can keep the IN1 input at 0, and then give a pulsed signal at IN2 that keeps the duty cycle at a percentage corresponding to the desired speed. For example, you can keep the on time and off time at 50% each to give half the voltage to the motor. You can control the motor's speed in the reverse direction by keeping the IN1 input at 1 and pulsing the IN2 input.

Figure 6.65 shows a complete circuit for speed control where PIC16F84A is used as the controller. To learn how to program the PIC controller, a very good reference is PIC Microcontroller Project Book: For PIC Basic and PIC Basic Pro Compliers [14].

6.2.1 PC Control

The circuit also allows for the controller to receive commands from a PC that can be connected using a serial cable. This way, you can do PC control of the robot. To avoid having a cable connected to the PC from the robot, you can

FIGURE 6.66
Maxstream XStream RS232 RF modem (Copyright 2005 MaxStream, Inc. Reproduced with permission)

use an RS-232 wireless modem. You can connect the modem to the RS-232 connector on the robot and the modem on the PC. Figure 6.66 shows one RS-232 RF modem you could use.

6.2.2 Feedback Control

You can add incremental optical encoders to each wheel so that you can connect the outputs from those encoders to the microcontroller inputs. Then you can change the PWM commands to the H-bridges to control the exact speeds of the wheels using feedback control. You can use feedback control theory to design the closed loop speed controllers that use the encoder feedback. One good reference for learning that design is Control Systems Engineering, Fourth Edition, by Norman S. Nise [13]. To design the controller you will need the parameters of the motor and the robot, as well as mathematical models that have been covered in this chapter.

6.3 Robot Kinematics

You can control the forward or backward speed of the robot by controlling the speeds of the two motors, and you can control the robot's rotation using the speeds of the two motors. You can calculate the relationship of the total robot

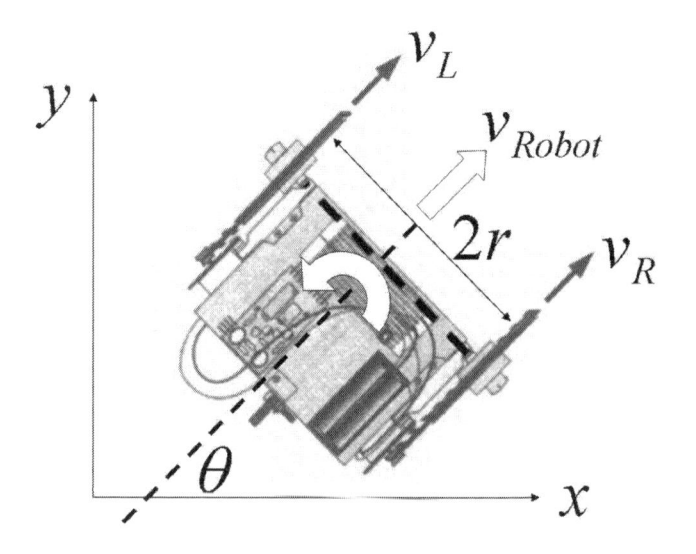

Differential Drive Robot

FIGURE 6.67
Robot kinematic relationships (Copyright 2006 Chaney Electronics. Reproduced with permission)

speed and rotation speed in terms of the angular speed of the two motors (see Figure 6.67).

Figure 6.67 shows the linear speed of the right wheel as v_R and the linear speed of the left wheel as v_L. If the radius of each wheel is R, then these are related to the wheel's angular speeds as follows:

$$v_R = R\omega_R \tag{6.23}$$

$$v_L = R\omega_L \tag{6.24}$$

The angular velocities of the right and left wheels are ω_R and ω_L, respectively. The total linear speed of the robot and its total angular speed can be calculated as

$$v_{Robot} = \frac{(v_R + v_L)}{2} \tag{6.25}$$

$$\omega_{Robot} = \frac{d\theta}{dt} = \frac{(v_R - v_L)}{2r} \tag{6.26}$$

Therefore, by controlling the angular speed of the two wheels, you can control the linear and angular motion of the robot. The combined equations are as follows:

$$v_{Robot} = R\frac{(\omega_R + \omega_L)}{2} \tag{6.27}$$

$$\omega_{Robot} = \frac{d\theta}{dt} = R\frac{(\omega_R - \omega_L)}{2r} \qquad (6.28)$$

Suppose you are given a path that the robot should follow. The path might be given in terms of the x, y coordinates and the angle θ of the robot, rather than the desired wheel speeds (or robot linear and angular speed). In that case, you need a relationship between the world coordinates x, y, and θ, and the control variables of the robot (the linear and angular speeds), which you can control by wheel angular speeds as shown in Figure 6.67. This relationship can be derived as follows:

$$\frac{dx}{dt} = v_{Robot} \cos\theta \qquad (6.29)$$

$$\frac{dy}{dt} = v_{Robot} \sin\theta \qquad (6.30)$$

$$\frac{d\theta}{dt} = R\frac{(\omega_R - \omega_L)}{2r} \qquad (6.31)$$

Using equation 6.27 gives the model in terms of the input angular wheel speeds:

$$\frac{dx}{dt} = \frac{R\cos\theta(\omega_R + \omega_L)}{2} \qquad (6.32)$$

$$\frac{dy}{dt} = \frac{R\sin\theta(\omega_R + \omega_L)}{2} \qquad (6.33)$$

$$\frac{d\theta}{dt} = R\frac{(\omega_R - \omega_L)}{2r} \qquad (6.34)$$

You can make the robot follow trajectories that are given in world coordinates by designing controllers that make use of equations 6.32, 6.33, and 6.34. The trajectories can come from a camera which can observe the robot motions and let a controller decide where the robot should go.

7

Four Wheel Drive Robot

In this chapter we will study the Rigel 4WD robot kit, which is a four-wheeled robot from Budget Robotics. We will look at its construction and mechanics, electrical control, objects and their operations, and sample code to drive the robot. After going through mechanical construction of the robot, we will talk about the electrical components of the robot which will run the robot. Main electrical components are microcontroller (OOPic-R by Savage Innovations [24]) and servomotors. The objects related to Rigel 4WD robot are explained. How servomotors are driven by PWM signals is also briefly explained. Finally, two sample codes are provided to run the robot. The first code drives the robot forward. The second code drives the robot forward for a specified distance, turns the robot 180 degrees about its center, and finally drives it back to the start position.

7.1 Construction and Mechanics

Rigel is a 4WD (four-wheel-drive) differentially steered robot. The robot is 4 1/4 inches high, 6 inches wide, and 6 3/4 inches long. It weighs about 23 ounces without batteries. The robot is shown in Figure 7.1.

The body material is made of 6 millimeter thick PVC rigid expanded plastic. The wheels are 65 millimeters in diameter. Each wheel is controlled independently by a servomotor. The servomotor used with the robot is the GWS S03N servo. This servo is modified by the manufacturer for continuous rotation. The robot can be controlled by any standard microcontroller such as Parallax Board of Education, OOPic-, or BASIC Stamp. Figure 7.2 shows the top view of the robot with the OOPic-R mounted on the robot.

The basic frame of the robot consists of a top deck, a bottom deck, two servo brackets, four servomotors, six angle gussets, and four wheels. Figure 7.3 shows the servo bracket used for mounting the servos.

The servomotor for the front wheel is dropped in through one of the holes and is fastened to the bracket using four 4-40x1/2 inch steel pan head machine screws, #4 size steel washers, and 4-40 size steel hex nuts. Figure 7.4 illustrates one of the servomotors attached to the servo bracket. In addition, Figure 7.5 shows a closer look at the servo and its mounting components.

FIGURE 7.1

Side view of robot

FIGURE 7.2

Rigel 4WD robot with OOPic-R microcontroller

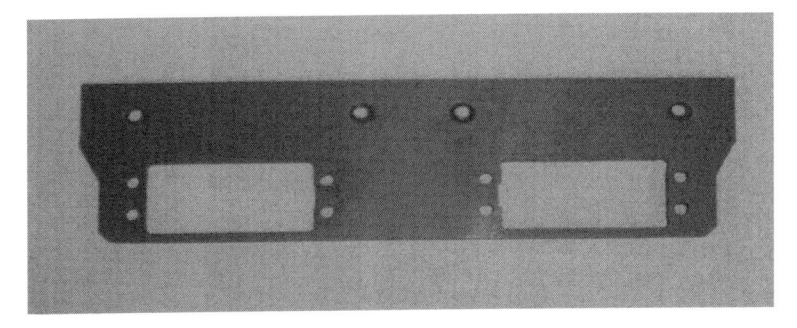

Four Wheel Drive Robot   271

FIGURE 7.3

Servo bracket

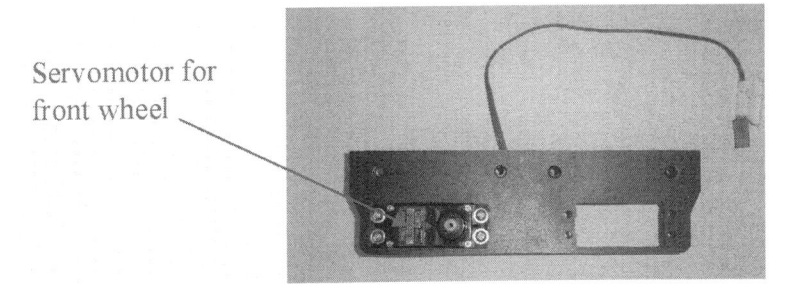

FIGURE 7.4

Servomotor attached to servo bracket

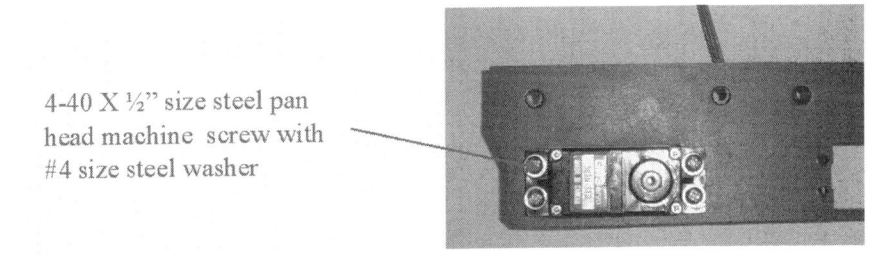

Servomotor for front wheel

4-40 X ½" size steel pan head machine screw with #4 size steel washer

FIGURE 7.5
Zoom-in view of servomotor fastened to servo bracket using steel pan head machine screws and steel washers

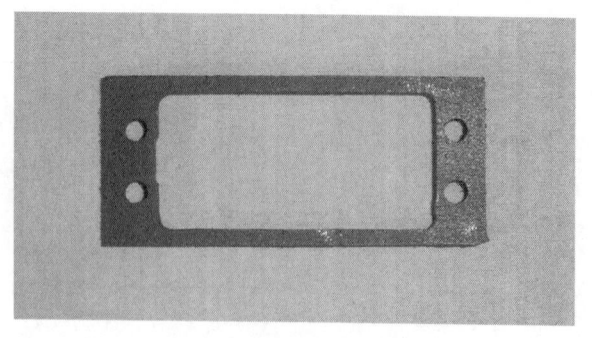

FIGURE 7.6

Servo mounting spacer

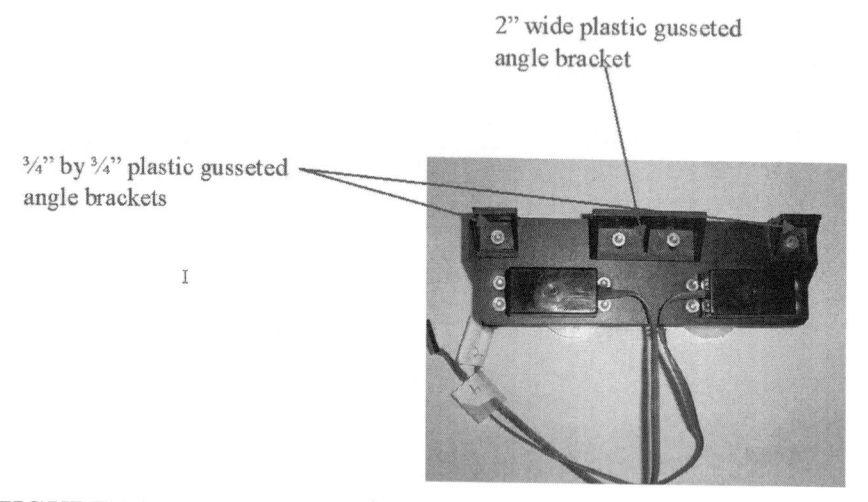

FIGURE 7.7

Angle gussets attached to the servo bracket

The servomotor for the back wheel is fastened to the bracket using a servo mounting spacer, shown in Figure 7.6.

For this purpose four 4-40 x 3/4 inch steel pan head machine screws, #4 size steel washers, and 4-40 size steel hex nuts are used. Two 3/4 x 3/4 inch plastic gusseted angle brackets are mounted at the ends of the servo bracket using one 4-40 x 1/2 inch steel pan head machine screw and one 4-40 size steel hex nut each. Figure 7.7 shows the angle gussets attached to the servo bracket.

One 2-inch-wide plastic gusseted angle bracket is fixed at the center of the servo bracket using two 4-40 x 1/2 inch steel pan head machine screws and 4-40 size steel hex nuts. Note that the screw heads should be facing the outer side of the servo bracket while the hex nuts should be on the inside. The

Four Wheel Drive Robot 273

FIGURE 7.8
Side view of servo bracket showing attached angle gussets

angle gussets are used to secure the servo brackets to the robot base. Figure 7.8 shows the side view of the servo bracket showing attached angle gussets.

The front wheel is attached to the servo horn through the holes, as shown in Figure 7.9. Four 1 1/16 inch screws, #4 size steel washers, and 4-40 size steel hex nuts are used for this purpose. Figure 7.10 shows the servo horn attached to the wheel. Figure 7.11 gives the back view of the wheel.

Note: The washers are fixed from the back just before threading in the nut.

The servo horn is attached to the servomotor using a horn mounting screw. The same procedure is repeated for the back wheel. Figure 7.12 shows the front and back wheels attached to the completed servo bracket.

The second servo bracket is assembled in the same way as explained above. Figure 7.13 shows the second completed servo bracket.

After both the servo brackets have been assembled completely, the bottom deck (shown in Figure 7.14) is attached to the servo brackets through the 2-inch-wide angle gussets using four 4-40 x 1/2 inch size steel pan head machine screws and 4-40 size steel hex nuts. Figure 7.15 shows the servo bracket attached to the bottom deck through 2-inch-wide angle gussets.

Note: The heads of the screws should be facing the top of the deck, while the nuts should be on the inside.

To further secure the bottom deck firmly to the servo brackets, four 4-40 x 1/2 inch size steel pan head machine screws and 4-40 size steel hex nuts are used to fasten the 3/4 x 3/4 inch angle gussets to the bottom deck as shown in Figure 7.16 and Figure 7.17 (top view).

The top deck is now mounted on the bottom deck using four 1 1/4 inch aluminum risers as shown in Figure 7.18 and Figure 7.19 (side view). The risers are screwed onto the bottom deck using 3/8 inch black oxide screws

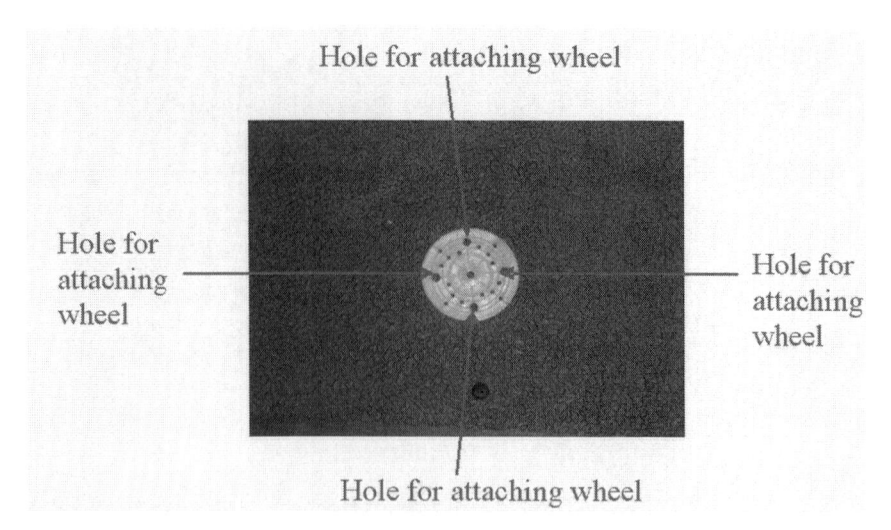

FIGURE 7.9

FIGURE 7.10

Wheel attached to servo horn

FIGURE 7.11

Back view of wheel attached to servo horn

FIGURE 7.12
Front and back wheels attached to completed servo bracket

FIGURE 7.13
Second completed servo bracket

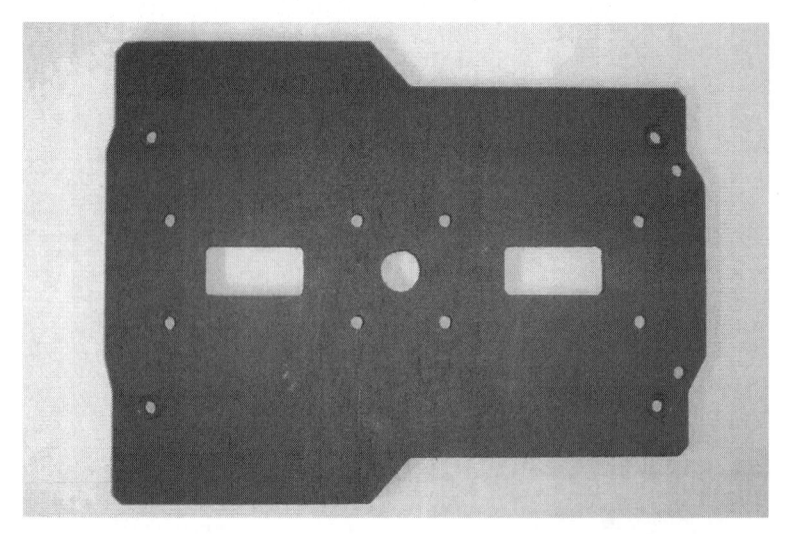

FIGURE 7.14

Bottom deck

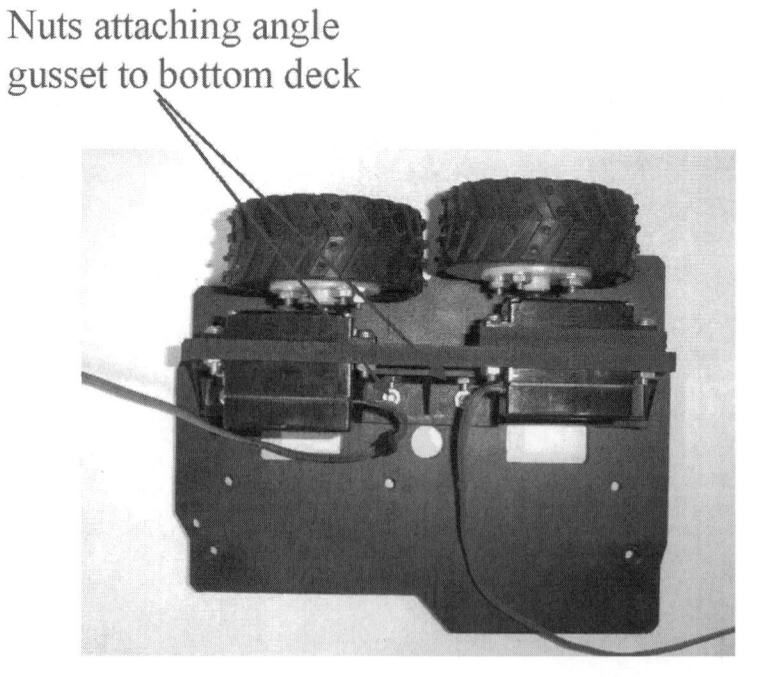

FIGURE 7.15
Servo bracket attached to bottom deck through 2-inch-wide angle gussets

Four Wheel Drive Robot 277

FIGURE 7.16
Both servo brackets attached to bottom deck through 2-inch-wide and 3/4 x 3/4 inch angle gussets

FIGURE 7.17
Top view of bottom deck

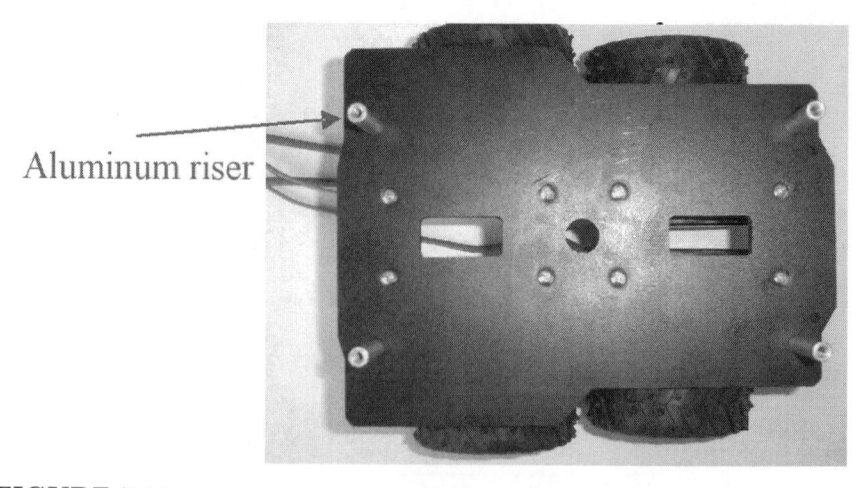

FIGURE 7.18
Aluminum risers attached to bottom deck

and then attached to the top deck in a similar fashion.

Figure 7.20 shows the top deck attached to the aluminum risers.

This completes the mechanical construction of the robot. The completed robot is illustrated in Figure 7.21.

7.2 Electrical Control

In this section we will explore how to connect the OOPic-R to the Rigel 4WD robot as well as the necessary software components to write a sample program to operate the robot. Then we will provide an example program which moves the robot forward.

7.2.1 Connecting the OOPic-R with the Rigel 4WD Robot

Figure 7.22 illustrates the connection of the GWS S03N RC servos (orange/brown wires shown in Figure 7.22) to the I/O bank I of the OOPic-R. For more information on the various I/O banks available with the OOPic-R refer to [24]. The connection of the main battery (the red and black wires show the connection to the battery) is shown in Figure 7.23. Also, the setting of the power selection jumper S1 is shown in Figure 7.24. The battery is a 7.2 volt NiMh battery which supplies power for the motors.

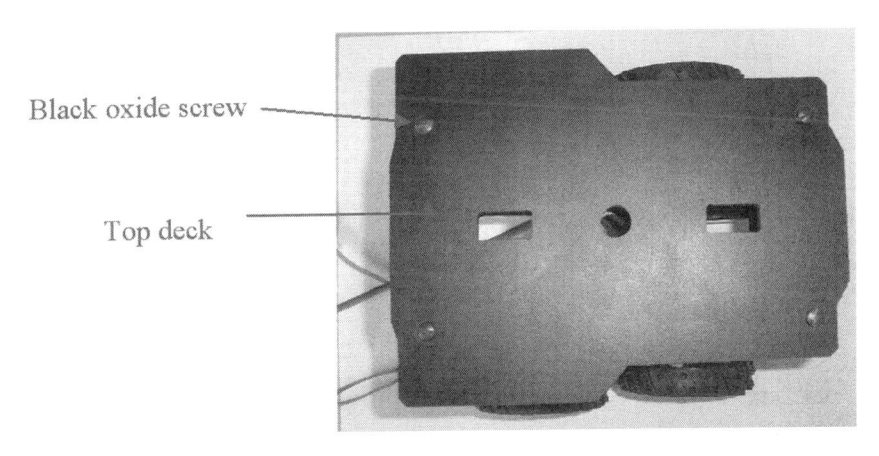

Four Wheel Drive Robot

FIGURE 7.19
 Side view of bottom deck with attached aluminum risers

FIGURE 7.20
 Top deck attached to aluminum risers

FIGURE 7.21

Completed four-wheeled robot

FIGURE 7.22

Connections of the I/O lines of the RC servos (servos not shown) to the I/O bank I

FIGURE 7.23
Connection of the battery to the main I/O

FIGURE 7.24
Setting of jumper S1 when RC servos are connected to I/O bank I

7.2.2 A Review of OOP in Reference to the OOPic-R

In this section, some of the tables and explanations are prepared based on the material from OOPic website [24]. The detailed tables for all the objects are provided in the website as well [24]. OOP is an acronym for *object-oriented programming*. It is a term used to identify languages that are oriented toward the use of objects. OOP is an effort to make the task of writing computer programs easier by allowing the programmer to write bits of code that interact with what appears to be physical objects.

The OOPic has been specially designed for robotics. It is an acronym for object-oriented PIC. OOPic provides an object-oriented language model designed to interact with the electrical hardware components that are attached to the OOPic. The concept behind OOPic involves the use of preprogrammed objects from a library to do all the work of interacting with the hardware and then writing small scripts in BASIC, C, or Java syntax styles to control the objects.

Note: In this chapter the OOPic-R is used to control the Rigel four-wheeled robot. The OOPic-R can be programmed using BASIC, C, or Java. However, in this chapter all programming is done in BASIC.

Table 7.2 gives a list of object types used in OOPic language.

7.2.2.1 Creating and Working with Objects

The object declaration statement syntax is given below:

```
Dim <id>[(<sub>)] as New <type>
```

The term *id* is the name of the identifier for the new object. The name given should follow the standard naming conventions as given below:

1. Identifier names *must* begin with a letter.

2. Identifier names *cannot* contain a period.

3. Identifier names *must not* exceed 32 characters.

4. Identifier names *must* be unique within the application. (Identifier names are case-insensitive.) This includes local variable and object names within subprograms and functions.

The term *sub* is the optional dimension of an array. It may be a positive numeric value but no expressions are allowed. The term *type* is the type of the object of which an instance is to be created. An example is:

```
Dim myvar as New oByte
```

In this example, a new instance of the object oByte is created with the name *myvar*. The term *oByte* is a variable object which can store an 8-bit value in its value property. This means that myvar is a variable that can have any value between 0 and 255.

TABLE 7.1
Important terms used in OOP

Term	Description
Application	A computer program written for the purpose of controlling electronic equipment connected to an OOPic microcontroller.
Keyword	Any word whose meaning has already been defined in the OOPic language. Keywords may be a command, a function, a statement, or any other word that OOPic uses for any other purpose. All keywords in OOPic Basic language are case insensitive.
Identifier	The names given to the components of an application program. Objects, variables, subprocedures, and functions are given identifying names.
Constant	A predefined keyword that literally represents a value. Constants are used in place of numbers so that code is easier to read.
Variable	An identifier that represents a value, but unlike a constant it can be changed by assigning it a value. The value represented by a variable will remain the same until another value is assigned.
Command	A keyword that instructs the OOPic to take a particular action.
Function	A keyword that instructs the OOPic to perform a calculation and return a result.
Statement	A "sentence" of commands, functions, identifiers, and constants that instructs the OOPic to take a particular action and describes how to do it.
Operator	Operators are special commands that perform calculations, evaluations, and assignments.
Event	An action recognized by an object to which code can be written to respond to. Events may occur as a result of a hardware condition or software manipulation of an object.
Event-driven	A term used to describe a programming model. An event-driven application consists of code that remains idle until called upon by an object to respond to an event.
Method	A keyword (similar to a function or a statement) that is part of the logical unit of an object and whose operation acts directly upon that object.
Object	A term used to describe a set of variables and code that acts as one logical unit.
Object-oriented	A term used to describe a programming model where certain variables and code act as a single logical unit.
Property	A variable that is part of the logical unit of an object.
Instance	An instance of an object that is defined by a class. An object in an OOP is a class that defines certain characteristics.

TABLE 7.2
Object types used in OOPic language

Type	Description
Hardware	An object that represents and/or encapsulates a physically implemented piece of hardware in the OOPic. In an application, multiple instances of these objects can be declared, but only one per piece of hardware present in the OOPic can be operational at any given time. The hardware list is different for the different types of OOPic.
Processing	An object that retrieves values from other objects, performs a specified calculation, and then stores the resulting value in another object. A processing object can be declared for use in as many multiple instances as memory can hold. The processing object list is different for the different types of OOPic.
Variable	An object that stores a value and provides evaluation properties about that value. A variable object can be declared for use in as many multiple instances as memory can hold.
System	An object that controls one of several system functions. A system object is intrinsic and is present at the time the OOPic powers up. You cannot declare new instances of these objects.

7.2.2.2 Controlling Objects with Their Properties

An object's properties are the values that it holds. The value of a property is set when the behavior of an object is to be changed. The following is the syntax for setting an object's property to a value:

`Object.property=expression`

An example code regarding objects is given below:

```
Dim myswitch as New oButton
Sub Main()
   myswitch.IOLine = 5
   myswitch.mode = 0
   myswitch.style = 2
End Sub
```

In the previous example an instance of the object oButton is created having the name *myswitch*. The term *oButton* is a hardware object that reads the state of a switch and controls the state of an LED using a single I/O line. The line *myswitch.IOLine*=5 means that the switch is connected to I/O line 5. The line with *myswitch.mode*=0 sets the switch as a push-button switch.

Style is a value that specifies how the LED looks when it is on. The line with *myswitch.style*=2 causes the LED to blink at 4Hz.

7.2.2.3 Performing Actions with Methods

A method is an action that an object can perform. When invoked, methods instruct the object to do a specified function. The following shows the syntax of invoking an object's method.

```
Object.method
```

Note: Unlike the property of an object, the method does not have a value. The method of an object performs an action, while the property of an object sets a value.

An example is

```
Dim num as New oByte
Sub Main()
  num.value=10
  num.inc
  num.invert
End Sub
```

In the previous example, **num** is an instance of the variable object oByte. First, the property, value of num, is set to 10. The line with *num.inc* invokes the inc method, which increments the value of num by 1. The value property of oByte can be any number between 0 and 255. Here, it is set to 10 and then incremented to 11. The line with *num.invert* inverts the bits in the value property (one's complement).

7.2.3 Important Objects Used While Interfacing the OOPic-R with the Rigel 4WD Robot

This section explores the necessary objects for using OOPic-R (by Savage Innovations [24]) on the Rigel 4WD robot.

7.2.3.1 oButton Object

The oButton object is a three byte hardware object that reads the state of a switch and controls the state of an LED using a single I/O line. The switch can be configured as a push button or a toggle button and the LED can be set to on, off, or three different patterns of blinking. An oButton object continuously cycles between digital input (i.e., reading a push-button style switch) and digital output (i.e., controlling an LED). In order for both the LED and the push button to be connected to the same I/O line, they must be connected to the I/O line through a resister, as indicated by the wiring diagram shown in Figure 7.25.

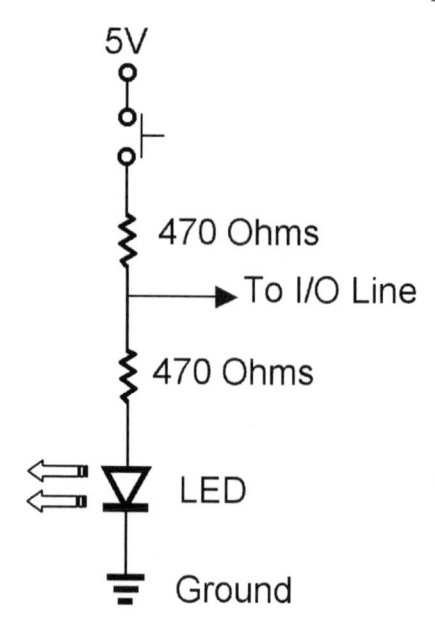

FIGURE 7.25
Wiring diagram for connecting LED and push button to the same I/O line.

The LED can optionally be disconnected from the push button state and be controlled by a different value.

Table 7.3 lists the properties of the oButton object.

The table 7.4 describes the behaviour of a button based on the properties Mode and InvrtIn.

7.2.3.2 oServo Object

The oServo object is a hardware object that uses one digital I/O line to set the position of an RC servomotor. A commercially available Hitec HS-422 servo is shown in Figure 7.26.

The size of an oServo object is 4 bytes. Table 7.5 presents the properties of the oServo object in detail.

7.2.3.3 Basic Introduction to Servo Operation

An RC servo is commonly found in radio frequency-controlled devices such as airplanes. A servo is a motor in which the shaft does not continuously rotate. The output of a servo is the movement of its shaft. The shaft assumes a position based upon a pulse width modulation (PWM) input shown in Figure 7.27.

The PWM input is a positive-going pulse that has a width between 1 and 2 milliseconds. The rate at which the pulses are sent is called a *refresh rate*.

TABLE 7.3
Properties of the oButton object

Property	Description
Address	Returns a pointer to the address of the oButton object instance. Data-Type: Address, Read-Only; Data-Range: 0 - 127
InvrtIn	A value that specifies if the state of the button is inverted when it is read. Data-Type: Bit; Data-Range: 0 - 1.
IOLine	The I/O Line for the button. Data-Type: Nibble; Data-Range: 0 - 31
Mode	Selects whether or not the connected switch is treated as a push button or a toggle switch. Data-Type: Bit; Data-Range: 0 - 1
Option	A value that specifies which value property controls the LED. Data-Type: Bit; Data-Range: 0 - 1. When Option is 0, the LED is controlled by the value property. When it is 1, it is controlled by the valueL property.
Style	A value that specifies how the LED looks when it is on. Data-Type: Nibble; Data-Range: 0 - 3. When Style is 0, the LED will be solid ON. When it is 1 and 2, the LED will blink at 1 Hz and 4 Hz respectively. When it is 3, the LED will blink twice at 4 Hz, stay off for 1/2 seconds, and repeat.
value	A value that indicates the current state of the button. Data-Type: Bit, Flag, Default; Data-Range: 0 - 1. When value is 0 and Constant is cvFalse, the button state is OFF. When value is 1 and Constant is cvTrue, the button state is ON.
valueL	A value that optionally controls the LED. Used when the Option property is set to 1. Data-Type: Bit, Flag; Data-Range: 0 - 1. When value is 0 and Constant is cvFalse, the the LED state is OFF. When value is 1 and Constant is cvTrue, the LED state is ON.

TABLE 7.4
Behavior of a button based on Mode and InvrtIn properties

Mode	InvrtIn	Description
0	0	The button is treated as a push button. Pressed is ON, Released is OFF.
1	0	The button is treated as a toggle switch, Pressed once is ON, Pressed second time is OFF.
0	1	The button is treated as a push button. Pressed is OFF, Released is ON.
1	1	The button is treated as a toggle switch, Released once is ON, Released a second time is OFF.

FIGURE 7.26
Commercially available Hitec HS-422 servo. (Copyright 2006 Lynxmotion Inc.)

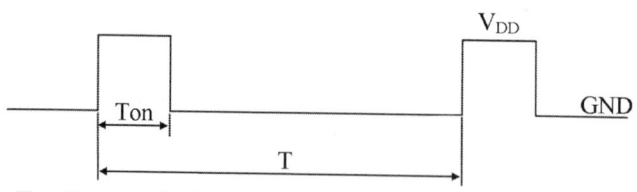

Ton: Between 1 to 2 ms
T: Period about 20 ms

FIGURE 7.27
Pulse Width Modulation (PWM) input

TABLE 7.5
Properties of the oServo object

Property	Description
Address	Returns a pointer to the address of the oServo Object instance. Data-Type: Number-Pointer, Read-Only; Data-Range: 0 - 127.
Centre	Adjusts the value for the servo's mechanical center. Data-Type: Byte; Data-Range: 0 - 63.
InvrtOut	Value that specifies if the output pulse is reversed. Useful when using servos as drive motors on opposite sides of a robot. Data-Type: Bit, Flag; Data-Range: 0 - 1. When InvrtOut property is 0, Constant is cvFalse and the servo control signal is normal. When it is 1, Constant is cvTrue and the servo control is reversed.
IOLine	The physical I/O Line to use. Data-Type: Byte; Data-Range: 0 - 31.
Operate	A value that specifies whether or not the pulse is outputted. Data-Type: Bit, Flag; Data-Range: 0 - 1. When Operate is 0, the Constant is cvFalse and the servo control signal is not outputted. When it is 1, the Constant is cvTrue and the servo control signal is outputted.
Refresh	When = 1 the servo refresh rate is doubled. When = 0 the servo refresh rate is normal. Data-Type: Bit, Flag; Data-Range: 0 - 1. When Refresh is 0, the Constant is cvFalse and the Servo is normal. When it is 1, the Constant is cvTrue and the Servo control signal is doubled.
String	The value property represented as a string. Data-Type: String.
value	The value specifying the position of the servomotor. Data-Type: Byte Default; Data-Range: 0 - 127.

The default refresh rate while using the oServo object is 35 pulses per second. When the pulse width is 1 millisecond, the servo shaft is at 0 degrees. With a pulse width of 2 milliseconds, the servo moves to 180 degrees. As the pulse width varies between 1 and 2 milliseconds the servo's shaft will move between the 0 degree and 180 degree positions.

An RC servo requires only one I/O line connection to the OOPic. It is absolutely necessary that the servo be driven by a different power source than the one that the OOPic is using and that the ground of each power supply be connected together. The oServo object is capable of positioning an RC servo anywhere within its 180 degree rotational span. The position within the 180 degree span is specified by a value with a range of 0 to 63. The maximum number of oServo objects that can be dimensioned in a single application program is 21; however, in general, more than 12 is not a good idea. An oServo object positions an RC servomotor connected to the I/O line specified by the IOLine property to a position specified by the value, the Center, and the InvertOut property. The Operate property specifies whether the servo is active. When set to 1 it is active, and when set to 0 the oServo object is inactive, and the specified I/O line is set to 0 volts, which releases the servo. The Center property must be adjusted for the mechanical differences of each different servomotor connected to the OOPic.

The oServo object is tailored to control a standard RC servo by generating a PWM servo control pulse with the logical high-going pulse ranging from 0 to 3 milliseconds in duration in 1/36 millisecond increments. A typical servo with a rotational range of 180 degrees is positioned by the duration of a logical high-going pulse in the range of 0.61 to 2.36 milliseconds. If the InvertOut property is set to 0, the PWM's logical high-going pulse time is calculated by the following formula in the line of code:

```
Pulse duration =   ((value + Center) * (1/36)) ms
```

If the InvertOut property is set to 1, the pulse time is calculated by the following formula.

```
Pulse duration =   ((107 - (value + Center)) * (1/36)) ms
```

7.2.3.4 PWM Servo Control Pulse Duration Examples

To position a servo at 0 degrees, the value property is set to 0, and the Center property is set to 22:

```
pulse duration = ((0 + 22) * (1/36)) = .61ms
```

To position a servo at 180 degrees (full swing) the value property is set to 63 and the Center property is set to 22:

```
pulse duration = ((63 + 22) * (1/36)) = 2.36ms
```

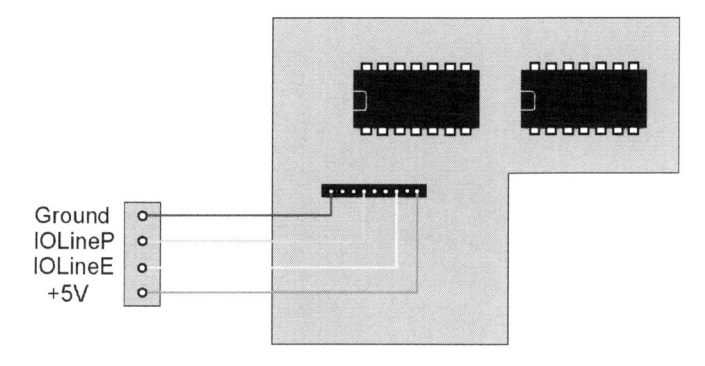

Four Wheel Drive Robot   291

FIGURE 7.28
   Polaroid sonar ranging unit connections

TABLE 7.6
Pin descriptions of polaroid sonar ranging unit

Pin No.	Pin Name	Description	I/O Name	OOPic I/O Line
1	Gnd	Ground	–	G
2	BLNK	Blanking	–	–
3	–	–	–	–
4	INIT	Intialization	IOLineP	Any
5	–	–	–	–
6	OSC	Oscillator	–	–
7	ECHO	Echo	IOLineE	Any
8	BINH	Blanking Inhibit	–	–
9	V+	–	–	+5

To position a servo at 180 degrees (full swing) while turning in reverse, the value property is set to 0, the InvertOut property is set to 1, and the Center property is set to 22:

```
pulse duration = ((107  -(0 + 22)) * (1/36)) = 2.36ms
```

7.2.3.5 oSonarPL Object

The oSonarPL is an object that controls a Polaroid ultrasonic rangefinder and measures the distance between the sonar transducer and its target in 64 steps-per-foot increments. The oSonarPL uses two I/O lines and two power lines as shown in Figure 7.28. The I/O lines can be any two of the OOPic's 31 I/O lines. The Polaroid sonar requires a 5 to 6 volt power supply that is capable of handling roughly 2A during the transmit period and 100mA after the transmit period. Figure 7.29 shows a Polaroid sonar ranging unit. The pin descriptions of the unit are presented in Table 7.6.

   The oSonarPL object handles all the necessary I/O timing required to com-

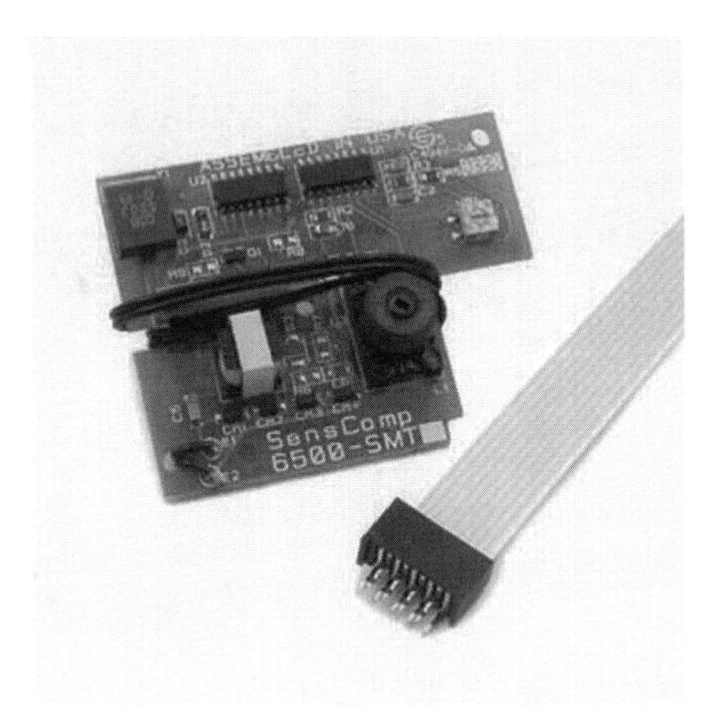

FIGURE 7.29
  Commercially available Polaroid sonar ranging unit from Acroname.
    (Copyright 2002-2007 SensComp, Inc.)

Four Wheel Drive Robot

TABLE 7.7
Properties of the oSonarPL object

Property	Description
Address	Returns a pointer to the address of the oSonarPL object instance. Data type: Address, Read Only; Data range: 0-127.
IOLineE	A value that specifies the physical I/O line to use for the echo signal. Data type: Byte; Data range: 0-31.
IOLineP	A value that specifies the physical I/O line to use for the ping signal. Data type: Byte; Data range: 0-31.
Operate	A value that specifies whether the data is updated. Data type: Bit, Flag; Data range: 0-1.
Received	A value that indicates that an echo was received. Data type: Bit, Flag; Data range: 0-1.
TimeOut	A value that indicates that the sonar ping is out of range. Data type: Bit, Flag; Data range: 0-1.
Transmitting	A value that is set to 1 when the sonar pings. Data type: Bit, Flag; Data range: 0-1.
value	A value that indicates the last sonar reading. Data type: Word, Default; Data range: 0-32768.

municate with the Polaroid ultrasonic rangefinder and measure the distance between the sonar transducer and its target. The size of an oSonarPL object is 6 bytes. When the oSonarPL's Operate property transitions from 0 to 1, a ping signal is sent out to the I/O line specified by the IOLineP property, the Transmitting property is set to 1, and the Received property is cleared to 0. Then the oSonarPL monitors the IOLineE property, waiting for an echo. Once the echo is received, the time of flight is stored in the value property, the Transmitting property is cleared to 0, and the Received property is set to 1.

If no echo is received after a short period of time, the Transmitting property is cleared to 0 and the TimeOut property is set to 1. The properties of the oSonarPL object are listed in Table 7.7.

7.2.3.6 oIRPD1 Object

This is a hardware object that uses one I/O line to read the reflection state of an IR LED contained within an IR Proximity Detector (IRPD). A commercially available Lynxmotion IRPD is shown in Figure 7.30. The connections of the device are sketched in Figure 7.31. The pin descriptions of the device are listed in Table 7.8.

The oIRPD1 object monitors the I/O line specified by the IOLine property and updates the value property with the current state. The I/O line is expected to be connected to the output of an IRPD. The size of an oIRPD1

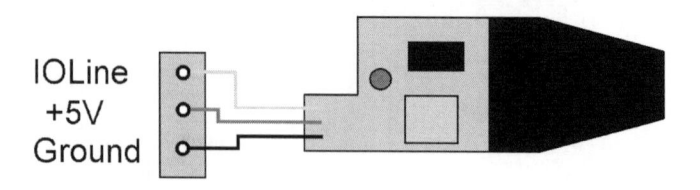

FIGURE 7.30

Pins of the Lynxmotion IRPD

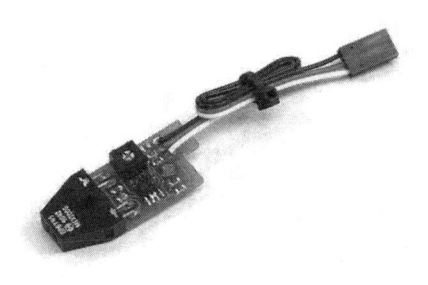

FIGURE 7.31

Commercially available Lynxmotion IRPD (Copyright 2006 Lynxmotion Inc.)

TABLE 7.8
Pin descriptions of the Lynxmotion IRPD and OOPic I/O lines

Pin No.	Pin Name	Description	I/O Name	OOPic I/O Line
1	Signal	Sensor output	I/O line	Any
2	5V	Power voltage	–	5V
3	GND	Ground	–	Gnd

Four Wheel Drive Robot   295

TABLE 7.9
Properties of the oIRPD1 object

Property	Description
Address	Returns a pointer to the address of the oIRPD1 Object instance. Data-Type: Address, Read-Only; Data-Range: 0 - 127.
IOLine	A value that specifies the physical I/O Line to connect to the IR Sensor. Data-Type: Nibble; Data-Range: 0 - 31.
Value	A value that indicates the value of the reflection state. Data-Type: Word, Default; Data-Range: 0 to 3. When value is 0 the IR LED is not reflecting back to the IR sensor. When value is 1 the IR LED is reflecting back to the IR sensor.

object is 1 byte. The properties of the oIRPD1 are listed in Table 7.9.

7.2.3.7 OOPic Object

This is a System object. The OOPic object maintains and controls the internal operations of the OOPic chip. Several of the internal operations can be controlled and watched from the properties provided by the OOPic object. The OOPic object is intrinsic. There is no class definition for it, and new instances of it are not allowed. The size of an OOPic object is five bytes. The properties of the OOPic object are listed in Table 7.10.

7.2.4 Sample Code to Drive the Rigel 4WD Robot Using the OOPic-R

This section presents two sample codes. The first code drives the robot forward. The second code drives the robot forward for a specified distance, turns the robot 180 degrees about its center, and finally drives it back to the start position.

7.2.4.1 Sample Code 1

The following program makes the Rigel 4WD robot move forward. It uses the oServo Hardware object to control the R/C servos of the robot, and the OOPic System object to provide the necessary delay.

```
Dim frontR As New oServo       'this makes an oServo object
Dim frontL As New oServo
Dim backR  As New oServo
Dim backL  As New oServo

Sub main()
```

TABLE 7.10
Properties of the OOPic object

Property	Description
Delay	A value that specifies how long, in 1/100ths of a second, the OOPic will delay the next instruction. The maximum the OOPic will delay the next instruction is 655.35 seconds or 10.9225 minutes. Data-Type: Word; Data-Range: 0 - 65535.
ExtVRef	A value that specifies the source of the voltage reference for the OOPic's analog to digital module. Data-Type: Bit; Data-Range: 0 - 1. When the ExtVRef is 0 and the Constant is cvOff, oA22D objects use a 5 volt range. When the ExtVRef is 1 and the Constant is cvOn, oA2D objects use a voltage range specified by the voltage present on I/O line 4.
Hz1	A 1-bit value that cycles once every second. The actual cycle time is .99957Hz. Data-Type: Bit, Flag; Data-Range: 0 - 1.
Hz60	A 1-bit value that cycles once every 1/60 second. Data-Type: Bit, Flag; Data-Range: 0 - 1.
Node	A value used when two or more OOPics "talk" to each other via the I2C network. A Node value of more than 0 is the OOPic's I2C network address. Data-Type: Byte; Data-Range: 0 - 127. When the value of Node is 0 and the Constant is cvFalse, the Servo control signal is not outputted. When the Node is 1 and the Constant is cvTrue, the Servo control signal is outputted.
Operate	A value that specifies the power mode of the OOPic Chip. Data-Type: Bit, Flag, Default; Data-Range: 0 - 1. When the Operate is 0 and the Constant is cvfalse, the OOPic is powered down. When the Operate is 1 and the Constant is cvTrue, the OOPic is on.
Pause	A value that specifies if the program flow is suspended. Data-Type: Bit, Flag; Data-Range: 0 - 1. When the value of Pause is 0 and the Constant is cvFalse, the pull-up resistors are not connected. When the Pause is 1, the pull-up resistors are connected.
Reset	A value that resets the OOPic when set. Data-Type: Bit, Flag; Data-Range: 0 - 1. When the Reset is 0 and the Constant is cvFalse, the OOPic operates normally. When the Reset is 1 and the Constant is cvTrue, the OOPic will reset by setting the StartStat property to 3.
StartStat	A value that indicates the cause of the last OOPic reset. Data-Type: Nibble, Read Only; Data-Range: 0 - 3. When the StartStat is 0, 1, or 2, the last reset was caused by power-on, by the reset line, or by power-brownout, respectively. When it is 3, the last reset was caused by Watch-Dog-Time.

Four Wheel Drive Robot

```
Call setup            'this calls the subroutine Setup

frontR.value = -3.5   'sets the front right wheel into motion
frontL.value = 3      'sets the front left wheel into motion
backR.value = -3.5    'sets the back right wheel into motion
backL.value = 3       'sets the back left wheel into motion

OOPic.delay = 1000    'delay needed to keep the robot in motion

frontR.value = 0      'this "stops" the front right wheel
frontL.value = 0      'this "stops" the front left wheel
backR.value = 0       'this "stops" the back right wheel
backL.value = 0       'this "stops" the back left wheel

end Sub

Sub Setup()
' ----------this will setup the right front wheel ------------
frontR.IOLine = 1     'set the servo to use I/O line 1
frontR.center = 52    'set the servo center to 52
frontR.operate = 1    'turn the servo on

'------------this will setup the left front wheel-------------
frontL.IOLine = 2     'set the servo to use I/O line 2
frontL.center = 53    'set the servo center to 53
frontL.operate = 1    'turn the servo on

'------------this will setup the right back wheel-------------
backR.IOLine = 3      'set the servo to use I/O line 3
backR.center = 52     'set the servo center to 52
backR.operate = 1     'turn the servo on

'-------------this will setup the left back wheel-------------
backL.IOLine = 4      'set the servo to use I/O line 4
backL.center = 53     'set the servo center to 53
backL.operate = 1     'turn the servo on

end Sub
```

In the above code, frontR, frontL, backR, and backL are four different instances of the hardware object oServo. These refer to the servos connected to the front right, front left, back right, and back left wheels of the robot, respectively. The subroutine setup connects each servo to a particular I/O line of the OOPic, sets up the center of the servo, and finally turns it on. In this program, I/O group1 of the OOPic is used for connecting the servos.

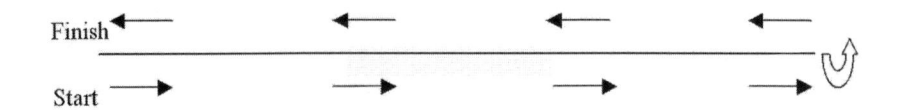

FIGURE 7.32
Path to demonstrate maneuverability

The wheels are set into motion by using the value property of the oServo object. A delay of ten seconds is provided by using the OOPic System object to maintain the robot in motion for ten seconds. Finally, at the end of ten seconds, the motion is stopped by setting the value property of each instance of the oServo object to zero.

7.2.4.2 Sample Code 2

This program makes the robot follow the path given in Figure 7.32.

Since all the wheels are driven independently, an extra caution needs to be taken to drive the robot straight. The sample code is given below.

```
Dim frontR As New oServo      'this makes an oServo object
Dim frontL As New oServo
Dim backR  As New oServo
Dim backL  As New oServo

Sub main()

Call setup              'this calls the subroutine Setup

FrontR.Value = -3.5     'sets the front right wheel into motion
frontL.Value = 3        'sets the front left wheel into motion
backR.Value = -3.5      'sets the back right wheel into motion
backL.Value = 3         'sets the back left wheel into motion

OOPic.delay = 1000      'delay needed to keep the robot in motion

frontR.Value = -3       'this section turns the robot 180 degrees
frontL.Value = -3
backR.Value = -3
backL.Value = -3
OOPic.delay = 390

frontR.Value = -3.5     'this section moves the robot forward
frontL.Value = 3        'three feet again
backR.Value = -3.5
```

```
backL.Value = 3

OOPic.delay = 1000

frontR.Value = 0     'this "stops" the front right wheel
frontL.Value = 0     'this "stops" the front left wheel
backR.Value = 0      'this "stops" the back right wheel
backL.Value = 0      'this "stops" the back left wheel

end Sub

Sub Setup()
' ----------this will setup the right front wheel ------------
frontR.Ioline = 1    'set the servo to use I/O line 1
frontR.center = 52   'set the servo center to 52
frontR.operate = 1   'turn the servo on

'------------this will setup the left front wheel-------------
frontL.Ioline = 2    'set the servo to use I/O line 2
frontL.center = 53   'set the servo center to 53
frontL.operate = 1   'turn the servo on

'------------this will setup the right back wheel-------------
backR.Ioline = 3     'set the servo to use I/O line 3
backR.center = 52    'set the servo center to 52
backR.operate = 1    'turn the servo on

'------------this will setup the left back wheel-------------
backL.Ioline = 4     'set the servo to use I/O line 4
backL.center = 53    'set the servo center to 53
backL.operate = 1    'turn the servo on
end Sub
```

The four-wheel differential drive robot is much more elegant than the two-wheel differential drive robot. The main advantages of this robot are maneuverability and the ability to traverse terrain. The four-wheel differential drive robot can drive in a straight line, turn around 180 degrees, and return along the same line. This shows a four-wheel vehicle's ability over a two-wheel vehicle in regard to turning radius. The two-wheel vehicle could not turn around in a confined area; the four-wheel robot has no trouble.

8
Hexapod Robot

In this chapter we will study a six-legged robot: the Extreme Hexapod II and III by Lynxmotion Inc. The figures and some of the technical content of the chapter is recreated from the Lynxmotion webpage [25] with permission. The hexapod can move more than 12 inches per second. It is 5.5 inches high, 17 inches wide and 13.5 inches long. It weighs about 44 ounces without batteries. The front and side view of the robot are shown in Figure 8.1 and 8.2.

In the figure you can see that the robot has six legs with two degrees of freedom (DOF) per leg. That means each leg can move in two independent ways. The two independent movements are created in two joints: a shoulder joint and an elbow joint. The shoulder joint allows 150 degrees rotational motion. The elbow joint allows 1.5 inches of linear motion to raise and lower the leg.

The legs and the body of the robot are made from ultra-tough laser-cut Lexan, a type of break-resistant plastic. As described previously, each leg has two joints controlled by two servomotors. The servomotor used in the robot is Hitec HS-422. The servomotors need to be adjusted so that they are in the mid position (center of rotation). The calibration of the servomotors will be discussed in the "Electrical Control" section.

The hexapod can be controlled by a standard microcontroller such as OOPic [24], Basic Stamp 2 [26], and Basic Atom Pro [25]. In addition to a microcontroller, a servomotor driver, such as the Mini SSC II or SSC-12 [25], is used to control the servomotors. A top view of the robot with a Basic Atom Pro microcontroller is shown in Figure 8.3.

The Mini SSC II is an old standard. It uses serial input and it can control up to eight servomotors. There is no speed control. The SSC-12 driver is a variable-speed version of the Mini SSC II and can control 12 servomotors. There are other servo drivers that use USB. These drivers can be used to control the robot with a PC through a serial port or a USB cable.

8.1 Construction and Mechanics

First we will study the servomotors and then the mechanical construction of the hexapod: the legs and the body.

FIGURE 8.1
Extreme Hexapod II front view. (Copyright 2006 by Lynxmotion Inc.)

FIGURE 8.2
Extreme Hexapod II side view. (Copyright 2006 by Lynxmotion Inc.)

FIGURE 8.3
Extreme Hexapod with a Basic Atom Pro microcontroller (Copyright 2006 by Lynxmotion Inc.)

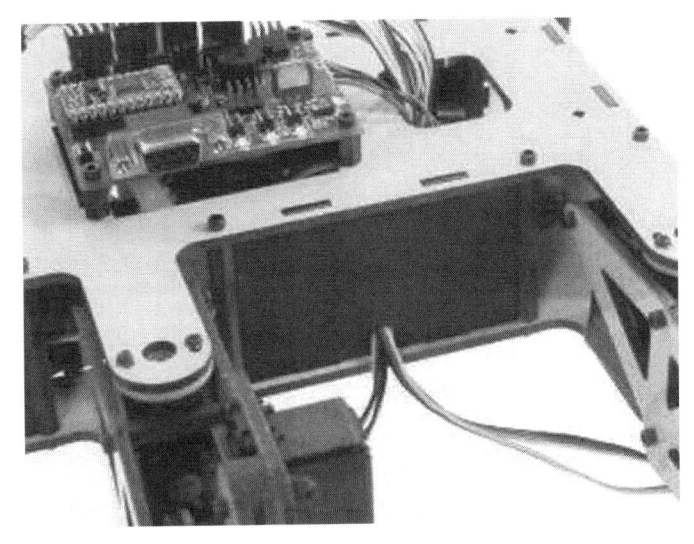

8.1.1 Servomotors

The servomotor used in the robot is the Hitec HS-422, shown in Figure 8.4.

The servomotor is controlled by 4.9 to 6 volts. It can create a torque of 57 oz-in. The speed of the servomotor is 0.16 seconds per 60 degrees. It weighs about 1.66 ounces.

8.1.2 Mechanical Construction of Extreme Hexapod II

The hexapod has two main mechanical parts: the legs and the body. As stated previously, each leg has two joints: shoulder and elbow. The leg structure is shown in Figure 8.5.

8.1.2.1 Shoulder Joint

The shoulder has two panels attached to both sides of the servomotor. The back plate is attached to the servomotor with L brackets and nylon rivet fasteners, shown in Figure 8.6.

Attach the L brackets to the back plate by pushing the rivet fasteners in from the back of the panel as shown in Figure 8.7.

Next, after cleaning the bottom of the servomotor with alcohol and peeling the green plaid cover from the tape on the hinge, attach the hinge, shown in Figure 8.8, to the bottom of the servomotor near the wire. The hinge has to be pressed on to the servomotor very firmly for a full ten seconds to ensure a

FIGURE 8.4
The Hitec HS-422 servomotor. (Copyright 2006 by Lynxmotion Inc.)

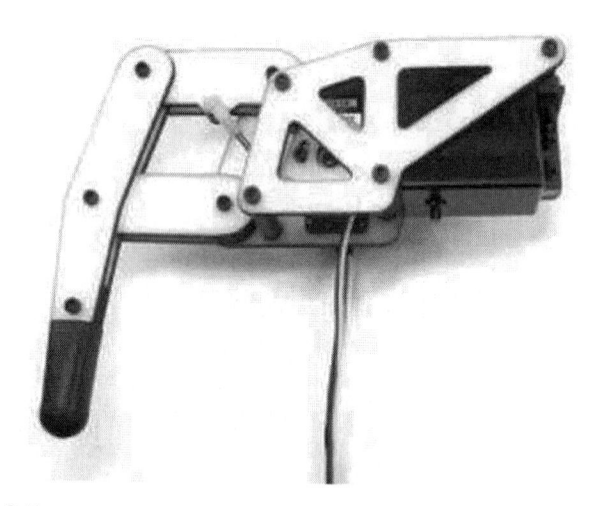

FIGURE 8.5
Hexapod leg structure with servomotors. (Copyright 2006 by Lynxmotion Inc.)

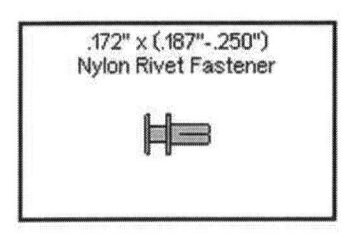

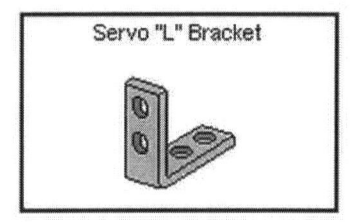

FIGURE 8.6
Nylon rivet fastener and servo L bracket. (Copyright 2006 by Lynxmotion Inc.)

FIGURE 8.7
The back panel with L brackets attached. (Copyright 2006 by Lynxmotion Inc.)

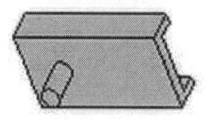

FIGURE 8.8

HD hinge for standard-size servomotor. (Copyright 2006 by Lynxmotion Inc.)

FIGURE 8.9

The servomotor with an attached hinge. (Copyright 2006 by Lynxmotion Inc.)

good bond.

The servomotor with the hinge attached is shown in Figure 8.9. The hinge will allow you to attach the leg to the body of the robot.

Attach the servomotor to the back panel by pushing the four fasteners in from the top. The servomotor's mounting tabs should be on top of the L brackets, as shown in Figure 8.10.

The 1-inch nylon spacer bars are attached to the back panel using 4-40x1/4 hex socket head cap screws, shown in Figure 8.11.

The spacer bar to the left of the servomotor hole should be aligned so the vertical servomotor will fit properly, as shown in Figure 8.12.

Next, drop the second (vertical) servomotor through the hole and attach it to the back panel with four rivet fasteners as shown in Figure 8.13.

Before we go to the next step, the vertical servomotor needs to be adjusted to the mid position. Centering a servomotor requires generating a 1.5-millisecond positive pulse that repeats every 20 milliseconds. This signal

Hexapod Robot

FIGURE 8.10
The servomotor attached to the back panel. (Copyright 2006 by Lynxmotion Inc.)

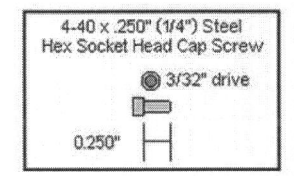

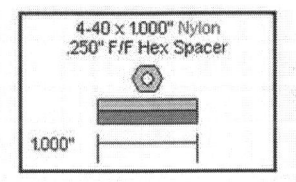

FIGURE 8.11
4-40x1/4 hex socket head cap screws and F/F hex spacer. (Copyright 2006 by Lynxmotion Inc.)

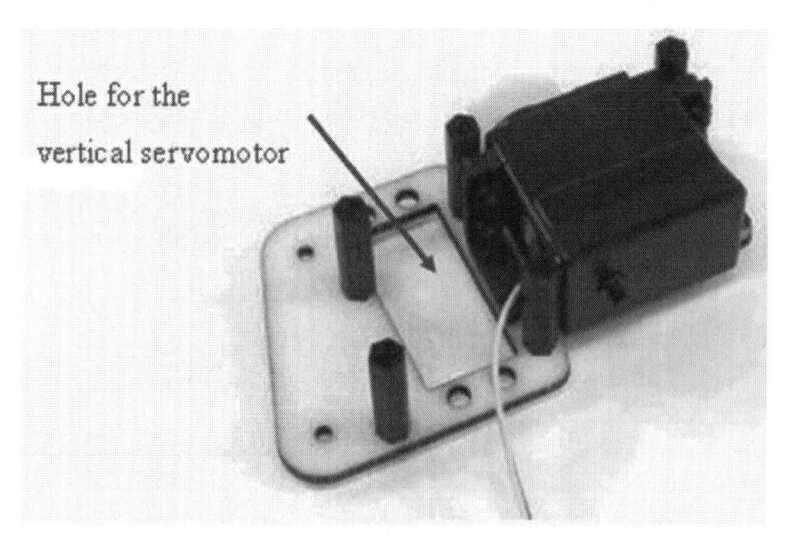

FIGURE 8.12
The back plate with servomotor and spacer. (Copyright 2006 by Lynxmotion Inc.)

FIGURE 8.13
The back panel with both servomotors. (Copyright 2006 by Lynxmotion Inc.)

Hexapod Robot 309

FIGURE 8.14
Phillips head tapping screws, external tooth lock washer, and the servo horn. (Copyright 2006 by Lynxmotion Inc.)

FIGURE 8.15
Shoulder with servo leg lever. (Copyright 2006 by Lynxmotion Inc.)

can be generated by using Bot Board, SSC drivers, or other servomotor controllers or microcontrollers. Adjust the servomotors using the procedure laid out in the "Electrical Control" section later in the chapter.

After the servomotor is centered, the servo leg lever should be installed onto the servo horn using two #2x1/4 Phillips head tapping screws and two washers as shown in Figure 8.14.

The installed servo leg lever is shown in Figure 8.15.

8.1.2.2 Elbow (Knee) Joint

To assemble the elbow, or knee, joint, the upper driven link is attached as shown in Figure 8.16. It consists of the top link (with mounting tab), a round Lexan spacer, and the lower link.

As shown in Figure 8.17, a 4-40x5/8 hex socket head cap screw and a 4-40x1/4 nylon acorn locking nut are used for this purpose. The screw goes in from the top, and the nut is on the bottom.

Attach two ball links and nuts, as shown in Figure 8.18 to the upper link

FIGURE 8.16
Connection of the upper driven link to the shoulder. (Copyright 2006 by Lynxmotion Inc.)

FIGURE 8.17
4-40x0.625 hex socket head cap screw and 4-40x0.25 nylon acorn locking nut. (Copyright 2006 by Lynxmotion Inc.)

Hexapod Robot

2-56 Ball Link
2-56 Nut

FIGURE 8.18
　　Ball link and nut. (Copyright 2006 by Lynxmotion Inc.)

Ball link attached to upper link

Ball link attached to servo lever

FIGURE 8.19
　　Ball links attached to upper link and servo lever. (Copyright 2006 by Lynxmotion Inc.)

and the servo lever. The ball links and the nuts will hold the pieces of the leg together as shown in Figure 8.19.

The ball links should be facing in opposite directions, i.e., the ball on the servo lever should be facing down, and the ball on the upper link should be facing up.

In order to connect the two ball links together, a ball socket assembly (called a dog bone) is required. The dog bone is shown in Figure 8.20. The length of the dog bone should be exactly 1.75 inches. The ball sockets should face in opposite directions to match the ball links.

The ball link socket and threaded rod required for this connection is shown in Figure 8.21.

The dog bone is snapped onto the ball joints firmly, as shown in Figure 8.22.

The front panel is then attached using five of the 4-40x1/4 screws, as shown in Figure 8.23.

The hex socket head cap screw to be used is shown in Figure 8.24.

The lower link is attached the same way as the upper link, as shown in Figure 8.25. A 4-40x5/8 hex socket head cap screw and a 4-40x1/4 nylon acorn locking nut is used.

To assemble the leg, align the two leg pieces with the leg spacer, which looks like a shortened leg piece, as shown in Figure 8.26. A 4-40x1/2 screw

FIGURE 8.20
    Dog bone. (Copyright 2006 by Lynxmotion Inc.)

FIGURE 8.21
    The ball link socket and threaded rod required to make the dog bone. (Copyright 2006 by Lynxmotion Inc.)

FIGURE 8.22
    Dog bone connecting upper link and servo lever. (Copyright 2006 by Lynxmotion Inc.)

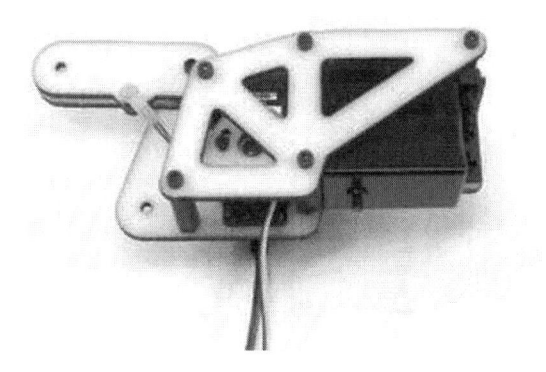

FIGURE 8.23
  Front panel connection. (Copyright 2006 by Lynxmotion Inc.)

FIGURE 8.24
The 4-40x0.25 hex socket head cap screw. (Copyright 2006 by Lynxmotion Inc.)

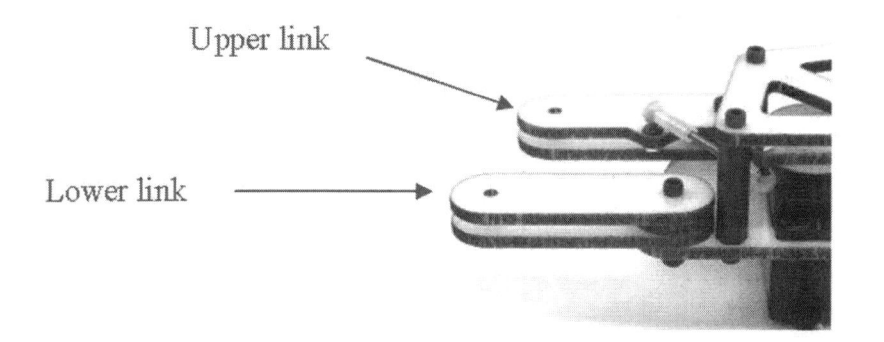

FIGURE 8.25
  Connection of the lower link. (Copyright 2006 by Lynxmotion Inc.)

FIGURE 8.26
Assembly of the leg. (Copyright 2006 by Lynxmotion Inc.)

FIGURE 8.27
Leg fitted with rubber foot end cap. (Copyright 2006 by Lynxmotion Inc.)

and acorn nut (shown in Figure 8.17) is used to hold the leg together. The screw should go in through the front, and the nut should be in the back.

Slide the rubber end cap of the robot foot onto the end of the leg, as shown in Figure 8.27.

Attach the leg assembly to the linkage using two 4-40x5/8 nylon screws and acorn nuts to hold it together, as shown in Figure 8.28. The screws should go in through the front, and the nuts should be in the back.

Adjust the hardware associated with the leg pivots. If it is too loose, the leg will be sloppy. If it is too tight, it will cause unnecessary friction.

This completes the construction of one leg of the hexapod. Out of the remaining five legs, two have to be constructed in the same way since they will be on the same side of the robot. The remaining three should be mirror images of the leg described in the procedure above since they will be on the other side of the robot.

FIGURE 8.28
   Connection of leg to linkage. (Copyright 2006 by Lynxmotion Inc.)

FIGURE 8.29
3/8-inch M/F hex spacers and 1 1/2-inch F/F hex spacers. (Copyright 2006 by Lynxmotion Inc.)

8.1.2.3 Body Construction

The robot kit is available with either 3/8-inch M/F hex spacers and 1 1/2-inch F/F hex spacers, or the longer 1 7/8-inch F/F aluminum spacers, as shown in Figure 8.29.

The 4-40x3/8-inch hex socket screws (shown in Figure 8.30) are used to attach the spacers to the bottom of the robot's top, as shown in Figure 8.31.

In order to attach the microcontroller (Bot Board or OOPic-R), you use four 1/4-inch hex screws and four 3/8-inch nylon hex spacers, as shown in Figure 8.32.

The screws should go through the Lexan from bottom to top. Four additional 1/4-inch hex screws are provided for attaching the board to the spacers, as shown in Figure 8.33.

You can connect an optional IRPD (infrared proximity detector by Lynxmotion Inc.) to the front of the robot. For this purpose the mounts should be made using two 3/8-inch spacers and two 1/4-inch hex screws (shown in Figure 8.32). Two additional 1/4-inch hex screws are provided for attaching

FIGURE 8.30
4-40x3/8 hex socket head cap screw. (Copyright 2006 by Lynxmotion Inc.)

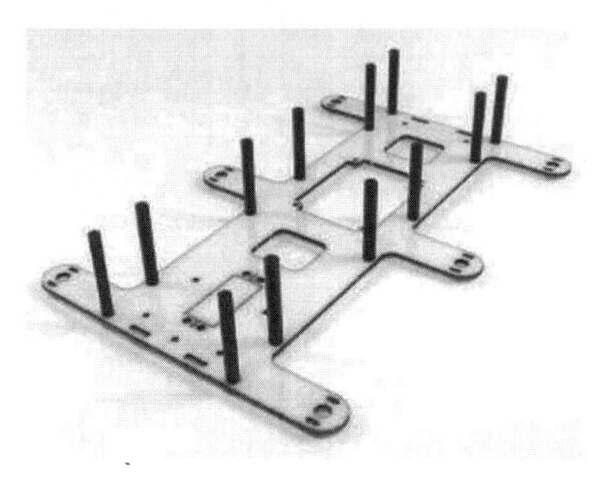

FIGURE 8.31
Spacers attached to the inner side of the top of the robot. (Copyright 2006 by Lynxmotion Inc.)

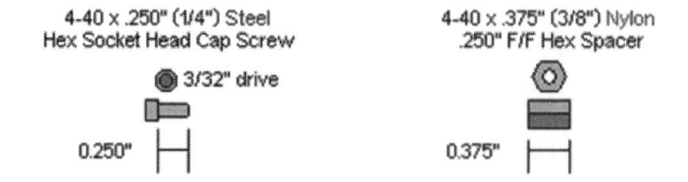

FIGURE 8.32
4-40x0.25 hex socket head cap screw and 4-40x0.375 F/F hex spacer. (Copyright 2006 by Lynxmotion Inc.)

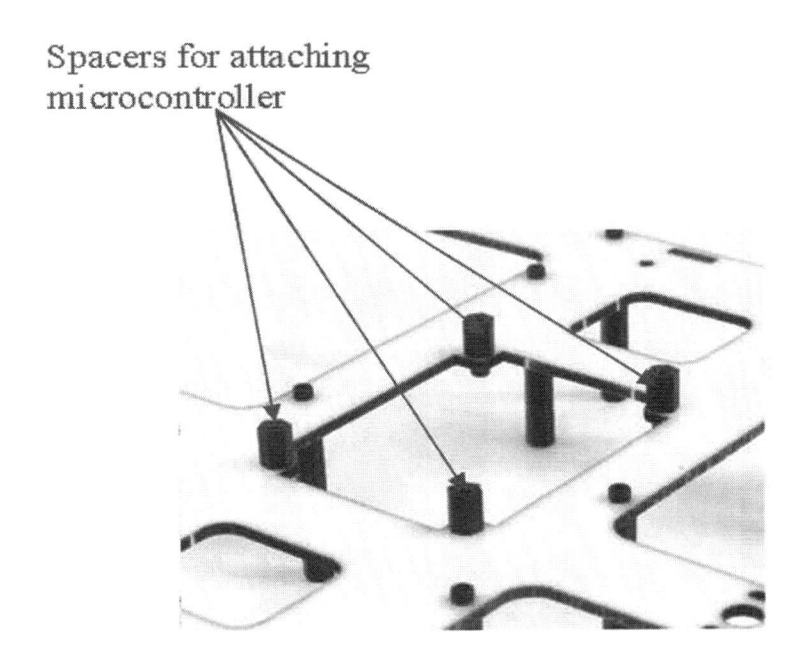

FIGURE 8.33
Mounts prepared for attaching microcontroller board. (Copyright 2006 by Lynxmotion Inc.)

FIGURE 8.34
Mounts prepared for attaching IRPD. (Copyright 2006 by Lynxmotion Inc.)

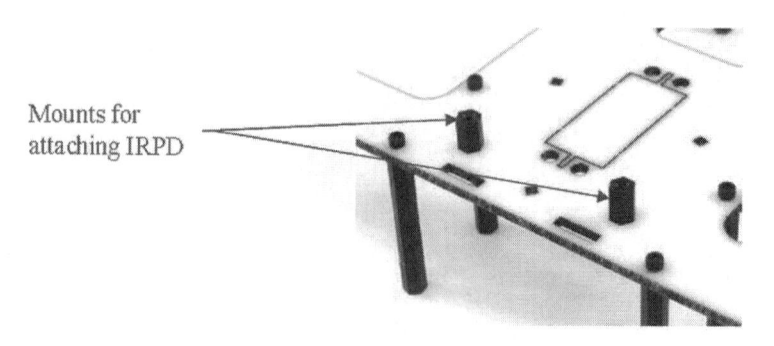

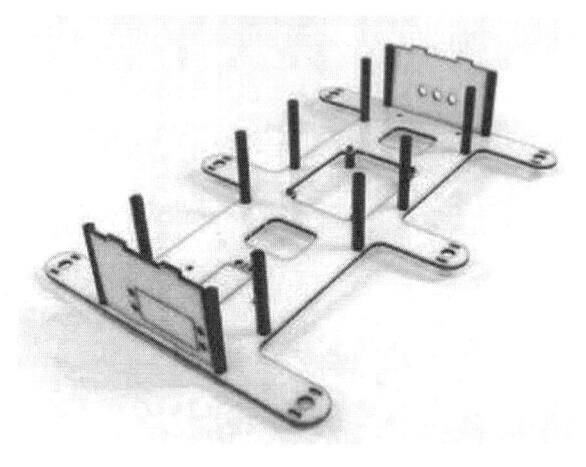

FIGURE 8.35
Robot frame with end panels. (Copyright 2006 by Lynxmotion Inc.)

IRPD to the spacers as shown in Figure 8.34.

Slide in the end panels as shown in Figure 8.35. Mount the panel with the servo hole in the front and the panel with the switch holes in the back.

If the kit is provided with 3/8-inch M/F hex spacers and 1 1/2-inch F/F hex spacers, you need to mount the bottom of the robot using 12 1/4-inch hex screws, shown in Figure 8.36.

Note: The bottom panel is symmetrical, meaning there is no front or back.

Figure 8.37 shows the bottom and top panels attached.

If the kit provides the longer 1 7/8-inch F/F aluminum spacer, mount the bottom of the robot with 12 3/8-inch hex screws, shown in Figure 8.38.

Note: The bottom panel is symmetrical, meaning there is no front or back.

To maintain compatibility with the program which will move the robot, the robot legs have to be mechanically aligned. In order to do this, move the servo

Hexapod Robot

FIGURE 8.36
4-40x1/4 hex socket head cap screw. (Copyright 2006 by Lynxmotion Inc.)

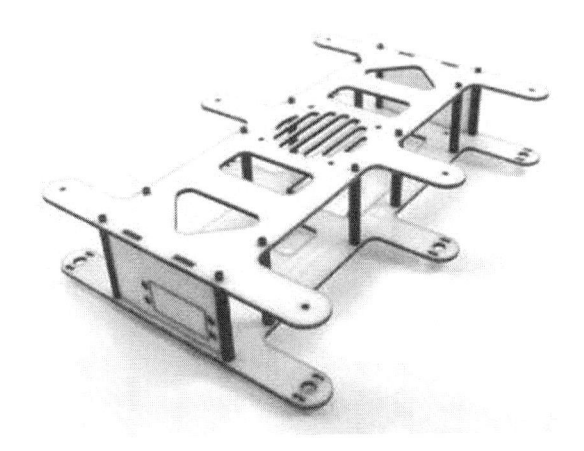

FIGURE 8.37
Bottom panel attached to robot frame. (Copyright 2006 by Lynxmotion Inc.)

FIGURE 8.38
4-40x3/8 hex socket head cap screw. (Copyright 2006 by Lynxmotion Inc.)

FIGURE 8.39

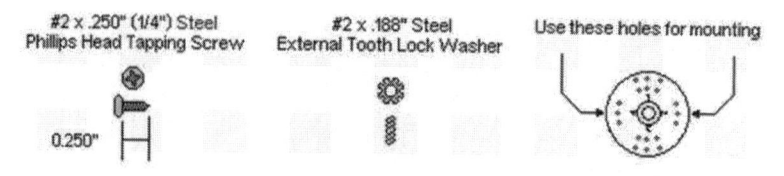

Phillips head tapping screw, external tooth lock washer, and servo horn. (Copyright 2006 by Lynxmotion Inc.)

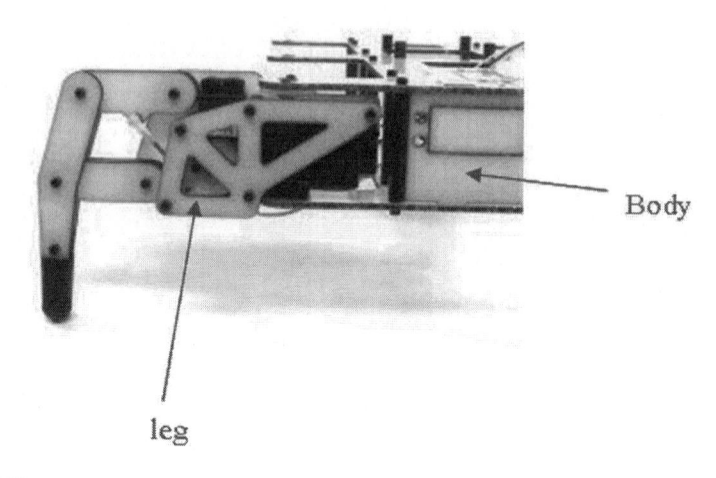

FIGURE 8.40

Leg attached to robot body. (Copyright 2006 by Lynxmotion Inc.)

to its center of rotation. Centering a servo simply requires generating a 1.5mS positive-going pulse that repeats every 20 milliseconds. This signal can be generated by using Bot Board, SSC drivers, or other servomotor controllers or microcontrollers. Before going onto the next step, adjust all the servomotors using the procedure laid out in the "Electrical Control" section later in this chapter.

While the servo is centered, install the leg onto the robot base. This step requires drilling holes with a 1/16-inch drill bit. Use two #2x1/4 Phillips head tapping screws with external tooth lock washers to mount to the servo horn, shown in Figure 8.39. The result is shown in Figure 8.40.

The leg should line up with the body, as shown in Figure 8.41. This will help the robot walk in a straight line.

Install all the legs the same way, making sure that they all line up. This completes the mechanical assembly of the robot, as shown in Figure 8.42.

FIGURE 8.41
Robot leg lined up correctly with body. (Copyright 2006 by Lynxmotion Inc.)

FIGURE 8.42
Completed hexapod. (Copyright 2006 by Lynxmotion Inc.)

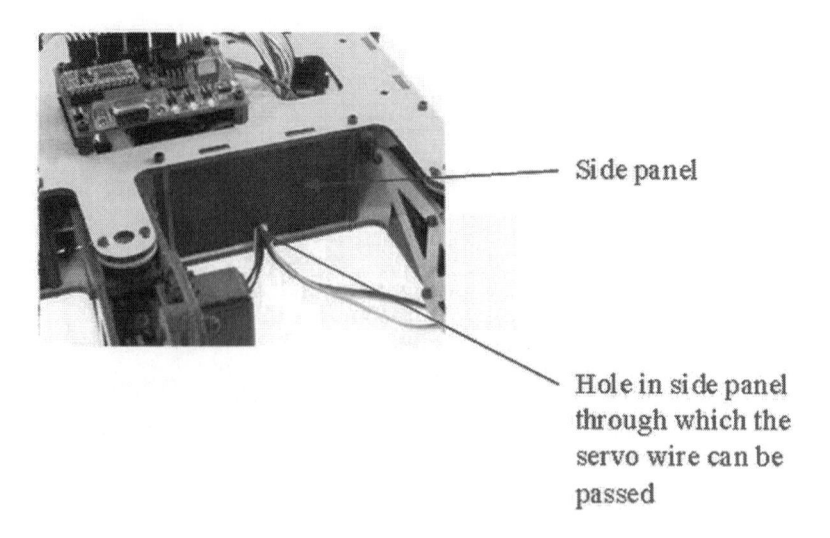

Side panel

Hole in side panel through which the servo wire can be passed

FIGURE 8.43
Side panel attached to the hexapod. (Copyright 2006 by Lynxmotion Inc.)

8.1.2.4 Addendum

Newer kits include four black plastic pieces, and the body panels have eight extra holes, shown in Figure 8.43. These pieces are side panels, and the holes are used to mount them. To attach these side panels, align the bottom of the plastic pieces with the holes. Make sure the servo leg wires are directed through the hole in the panel. Then press the top of the panel into place. To remove the side panels, press outward at the top of the panels and they will pop out.

8.1.3 Mechanical Construction of Extreme Hexapod III

The Extreme Hexapod III has two main mechanical parts: the legs and the body. This hexapod has a three degrees of freedom leg design that enables the robot to walk side to side (crab walk) as well as keep the feet moving in a straight line when walking forward. The robot uses 18 Hitec HS-475 servos for the legs.

To construct the leg, a servo is dropped in through the hole of the leg panel, and four nylon rivet fasteners, shown in Figure 8.44, are used to hold it in place.

The servo tabs should be on top of the Lexan. The rivet fasteners should go in from the top, through the servo tabs and the Lexan, as shown in Figure 8.45.

Three 4-40x1/2 screws and three acorn nuts, shown in Figure 8.46, are used to hold the spacers and second leg panel together with the first.

Hexapod Robot

FIGURE 8.44
Nylon rivet fastener. (Copyright 2006 by Lynxmotion Inc.)

FIGURE 8.45
The servo connected to a leg panel. (Copyright 2006 by Lynxmotion Inc.)

FIGURE 8.46
4-40x1/2 hex socket head cap screw and 4-40x1/4 acorn locking nut. (Copyright 2006 by Lynxmotion Inc.)

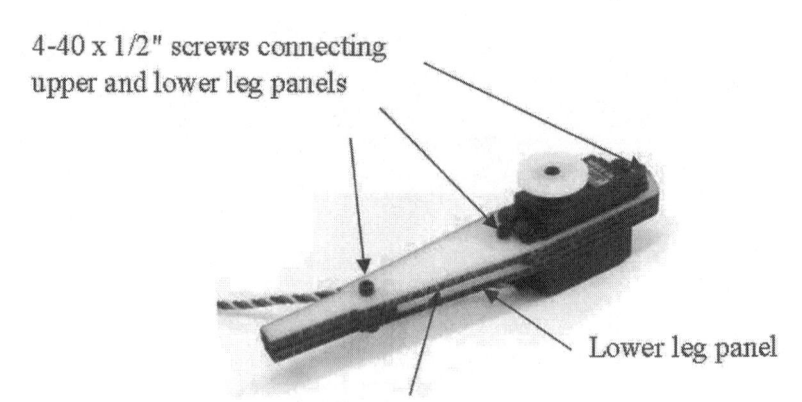

4-40 x 1/2" screws connecting upper and lower leg panels

Lower leg panel

Upper leg panel

FIGURE 8.47

Upper and lower leg panels attached together. (Copyright 2006 by Lynxmotion Inc.)

The screws should go in from the top, with the nuts on the bottom, as shown in Figure 8.47.

Slide the robot's foot rubber end cap onto the end of the leg as shown in Figure 8.48.

Clean the bottom of the servo with alcohol and allow it to dry. Peel the green plaid cover off of the tape on the hinge, shown in Figure 8.49.

Line up the hinge with the edge of the servo and press it onto the servo very firmly for a full ten seconds to ensure a good bond (see Figure 8.50).

Attach two of the L brackets using four of the nylon rivet fasteners, shown in Figure 8.51.

Push the rivet fasteners in from the back (the bottom of panel) as shown in Figure 8.52.

Use four rivet fasteners, shown in Figure 8.53, to attach the servo.

Push the fasteners in from the top. Make sure the servo's mounting tabs are on top of the L bracket. Figure 8.54 shows the horizontal servo attached to the panel.

Put the second servo through the servo hole. Use four rivet fasteners to hold it in place as shown in Figure 8.55.

Clean the bottom of the servo with alcohol and allow it to dry. Peel off the green plaid cover from the tape on the hinge (shown in Figure 8.51). Line the hinge up with the edge of the servo and press it on to the servo very firmly for a full ten seconds to ensure a good bond (see Figure 8.56).

Use four 4-40x3/8 hex socket head cap screws, shown in Figure 8.57, to attach the 1-inch nylon spacer bars, shown in Figure 8.58.

Figure 8.59 shows the resulting structure with the hex spacers attached.

Hexapod Robot

FIGURE 8.48
Rubber end cap attached to robot leg. (Copyright 2006 by Lynxmotion Inc.)

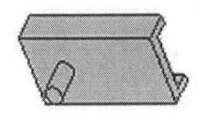

FIGURE 8.49
Servo hinge. (Copyright 2006 by Lynxmotion Inc.)

Servo hinge attached to servo

FIGURE 8.50
Servo hinge attached to servo. (Copyright 2006 by Lynxmotion Inc.)

FIGURE 8.51
Nylon rivet fastener and servo L bracket. (Copyright 2006 by Lynxmotion Inc.)

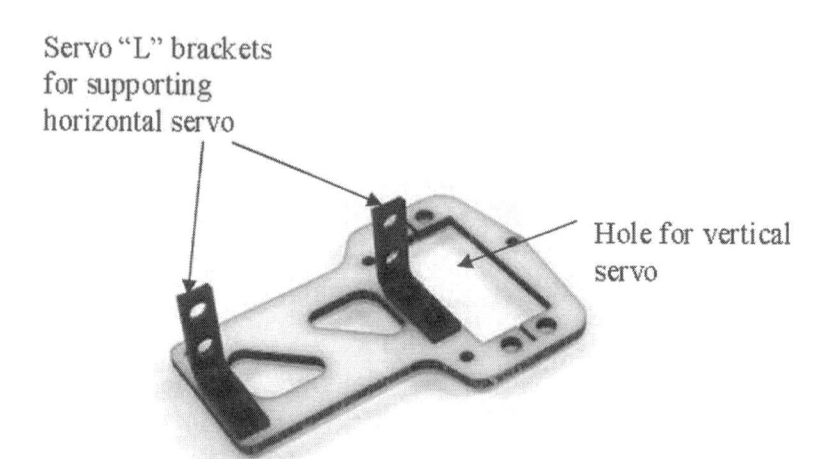

FIGURE 8.52
The L brackets attached to bottom panel. (Copyright 2006 by Lynxmotion Inc.)

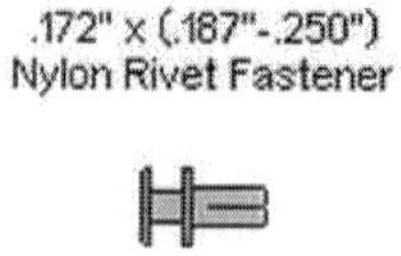

FIGURE 8.53
Nylon rivet fastener. (Copyright 2006 by Lynxmotion Inc.)

Hexapod Robot 327

FIGURE 8.54
Horizontal servo attached to panel. (Copyright 2006 by Lynxmotion Inc.)

FIGURE 8.55
Vertical servo attached to panel. (Copyright 2006 by Lynxmotion Inc.)

FIGURE 8.56
Servo hinges attached to horizontal and vertical servos. (Copyright 2006 by Lynxmotion Inc.)

FIGURE 8.57
4-40x3/8 hex socket head cap screw. (Copyright 2006 by Lynxmotion Inc.)

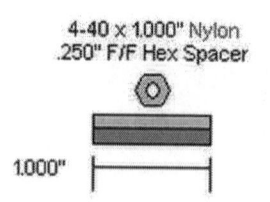

FIGURE 8.58
4-40x 1inch hex spacer. (Copyright 2006 by Lynxmotion Inc.)

Hexapod Robot 329

FIGURE 8.59
Nylon hex spacers attached to bottom panel. (Copyright 2006 by Lynxmotion Inc.)

Attach the front panel with four of the 4-40x3/8 screws, as shown in Figure 8.60.

If the kit has 3/8-inch M/F hex spacers and 1 1/2-inch F/F hex spacers, shown in Figure 8.61, assemble the leg cross member as shown in Figure 8.62.

Use two 1/4-inch and two 3/8-inch hex socket head cap screws, shown in Figure 8.63, to assemble the leg cross member.

Assemble the cross member lever as shown in Figure 8.64 if the kit has the longer 1 7/8-inch F/F aluminum spacer, shown in Figure 8.65.

Use four 3/8-inch hex socket head cap screws, shown in Figure 8.65, to assemble the cross member lever.

To maintain compatibility with the program which moves the robot, the robot legs have to be mechanically aligned. In order to do this, move the servo to its center of rotation. While the servo is centered, install the servo leg lever on to the servo horn as shown in Figure 8.66.

This step requires the indicated holes to be drilled with a 1/16-inch drill bit. Use two #2x1/4-inch Phillips head tapping screws and two washers. Use the servo horn holes as illustrated in Figure 8.67. Make sure that the bottom edge of the lever and the bottom edge of the main panel are parallel.

While the servo is centered, install the lower leg onto the servo horn, as shown in Figure 8.68.

Installing the lower leg onto the servo horn requires the indicated holes to be drilled with a 1/16-inch drill bit. Use two #2x1/4 Phillips head tapping screws and two washers. Make sure that the lower leg is perpendicular to the leg lever.

This completes the construction of the Hexapod III legs, shown in Figure 8.69.

FIGURE 8.60
Connection of front panel. (Copyright 2006 by Lynxmotion Inc.)

FIGURE 8.61
4-40x3/8 M/F hex spacer and 4-40x1.5 F/F hex spacer. (Copyright 2006 by Lynxmotion Inc.)

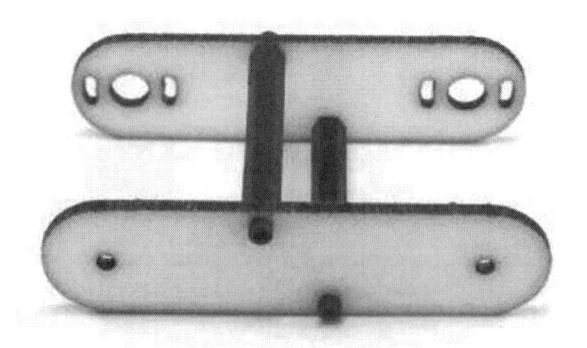

FIGURE 8.62
Assembly of the leg cross member using 3/8-inch M/F hex spacer and 1 1/2-inch F/F hex spacer. (Copyright 2006 by Lynxmotion Inc.)

Hexapod Robot

FIGURE 8.63
4-40x1/4 and 4-40x3/8 hex socket head cap screws. (Copyright 2006 by Lynxmotion Inc.)

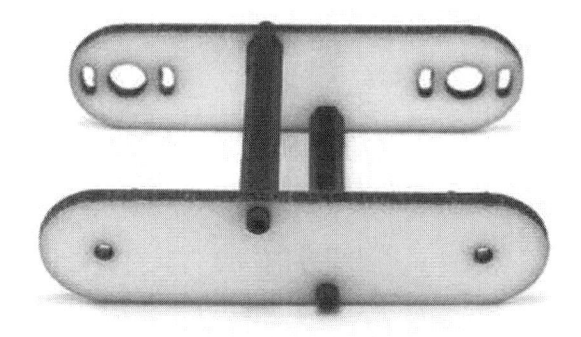

FIGURE 8.64
Assembly of the leg cross member using 1 7/8-inch F/F aluminum spacer. (Copyright 2006 by Lynxmotion Inc.)

FIGURE 8.65
4-40x3/8 hex socket head cap screw and 4-40x1 7/8 F/F hex spacer. (Copyright 2006 by Lynxmotion Inc.)

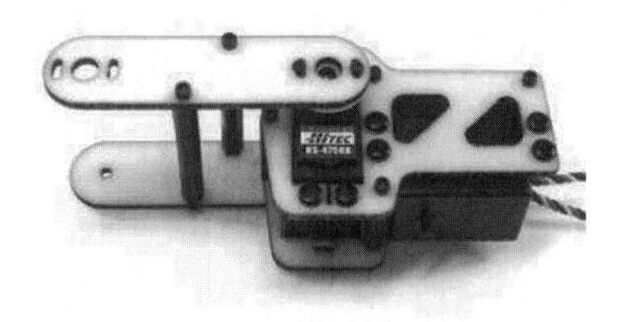

FIGURE 8.66
Servo leg lever attached to servo horn. (Copyright 2006 by Lynxmotion Inc.)

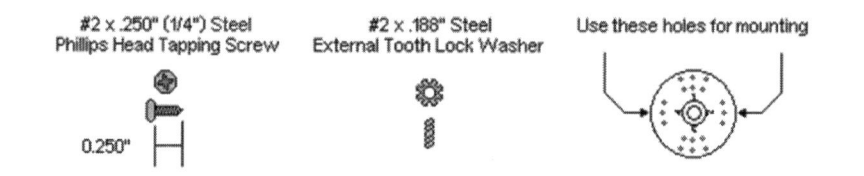

FIGURE 8.67
Phillips head tapping screw, external tooth lock washer, and servo horn. (Copyright 2006 by Lynxmotion Inc.)

FIGURE 8.68
Lower leg attached to servo horn. (Copyright 2006 by Lynxmotion Inc.)

Hexapod Robot

FIGURE 8.69
Completed leg. (Copyright 2006 by Lynxmotion Inc.)

FIGURE 8.70
3/8-inch M/F hex spacers and 1 1/2 F/F hex spacers. (Copyright 2006 by Lynxmotion Inc.)

Out of the remaining five legs, two have to be constructed in the same way while the remaining three should be mirror images of the leg described in the procedure above.

8.1.3.1 Body Construction

The kit is available with either 3/8-inch M/F hex spacers and 1 1/2-inch F/F hex spacers, or the longer 1 7/8-inch F/F aluminum spacers, shown in Figure 8.70.

Use the 4-40x3/8 hex socket screws (shown in Figure 8.71) to attach the spacers to the bottom of the robot's top, as shown in Figure 8.72.

Use four 3/8-inch nylon hex spacers and four 1/4-inch hex screws, shown in Figure 8.73, to attach the microcontroller (Bot Board or OOPic-R).

The screws should go through the Lexan from bottom to top. Four additional 1/4-inch hex screws are provided for attaching the board to the spacers, as shown in Figure 8.74.

FIGURE 8.71
4-40x3/8 hex socket head cap screw. (Copyright 2006 by Lynxmotion Inc.)

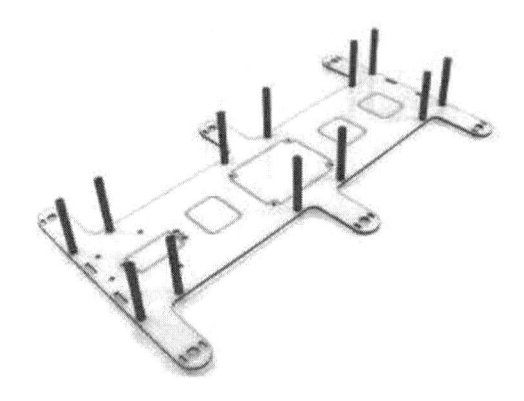

FIGURE 8.72
Spacers attached to the inner side of the top of the robot. (Copyright 2006 by Lynxmotion Inc.)

FIGURE 8.73
4-40x1/4 hex socket head cap screw and 4-40x3/8 F/F hex spacer. (Copyright 2006 by Lynxmotion Inc.)

Hexapod Robot 335

FIGURE 8.74
Mounts prepared for attaching the microcontroller board. (Copyright 2006 by Lynxmotion Inc.)

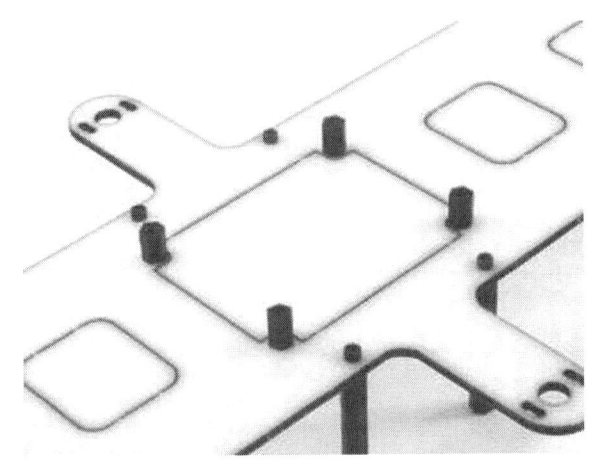

An optional IRPD can be connected at the front of the robot. For this purpose, make the mounts using two 3/8-inch spacers and two 1/4-inch hex screws (shown in Figure 8.73). Two additional 1/4-inch hex screws are provided for attaching IRPD to the spacers as shown in Figure 8.75.

Slide the end panels in as shown in Figure 8.76. Mount the panel with the servo hole in the front and the panel with the switch holes in the back.

If the kit has 3/8-inch M/F hex spacers and 1 1/2-inch F/F hex spacers, shown in Figure 8.77, mount the bottom of the robot using twelve 1/4-inch hex screws, as shown in Figure 8.78.

Note: The bottom panel is symmetrical, meaning there is no front or back.

The kit has the longer 1 7/8-inch F/F aluminum spacer, shown in Figure 8.79. Mount the bottom of the robot using 12 3/8-inch hex screws and these spacers, as shown in Figure 8.80.

Note: The bottom panel is symmetrical, meaning there is no front or back.

To maintain compatibility with the program which moves the robot, the robot legs have to be mechanically aligned. In order to do this, move the servo to its center of rotation using the Bot Board, SSC, or any other servo controller or microcontroller. Centering a servo simply requires generating a 1.5 millisecond positive-going pulse that repeats every 20 milliseconds.

While the servo is centered, install the leg onto the robot base. This step requires the indicated holes to be drilled using a 1/16-inch drill bit. Use two #2x1/4 Phillips head tapping screws with external tooth lock washers. Use the servo horn holes as illustrated in Figure 8.81. See the installed leg in Figure 8.82.

Make sure that the leg lines up with the body, as shown in Figure 8.83. This will help the robot walk in a straight line.

FIGURE 8.75
Mounts prepared for attaching IRPD. (Copyright 2006 by Lynxmotion Inc.)

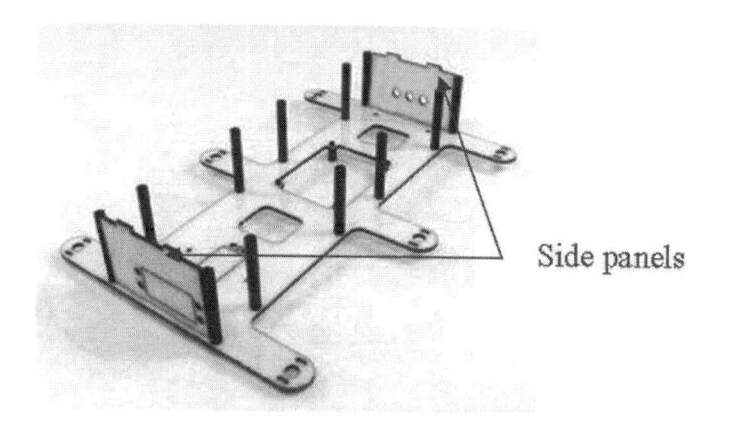

FIGURE 8.76
Side panels attached to robot frame. (Copyright 2006 by Lynxmotion Inc.)

FIGURE 8.77
4-40x1/4 hex socket head cap screw. (Copyright 2006 by Lynxmotion Inc.)

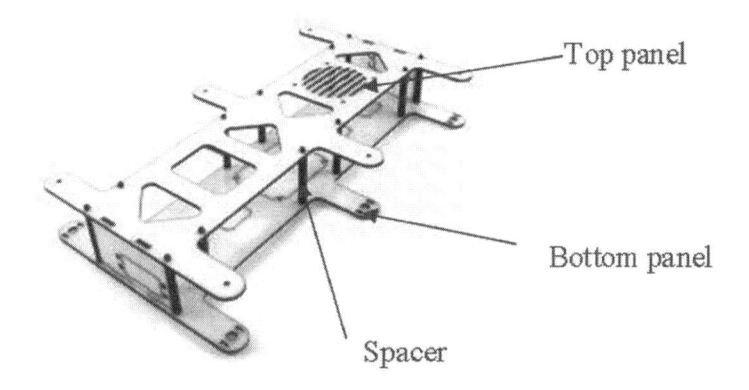

Hexapod Robot

FIGURE 8.78
Bottom panel attached to robot frame. (Copyright 2006 by Lynxmotion Inc.)

FIGURE 8.79
4-40x3/8 hex socket head cap screw.

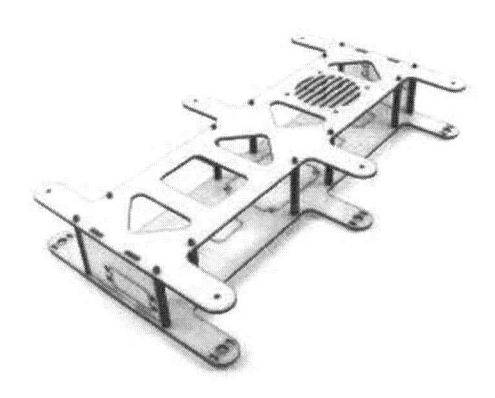

FIGURE 8.80
Bottom panel attached to robot frame. (Copyright 2006 by Lynxmotion Inc.)

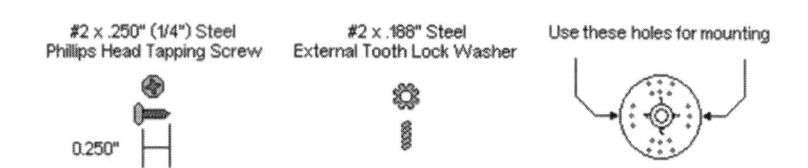

FIGURE 8.81
Phillips head tapping screw, external tooth lock washer, and servo horn. (Copyright 2006 by Lynxmotion Inc.)

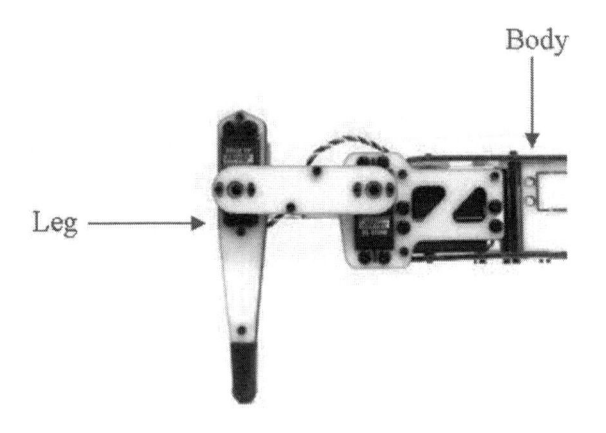

FIGURE 8.82
Leg attached to robot body. (Copyright 2006 by Lynxmotion Inc.)

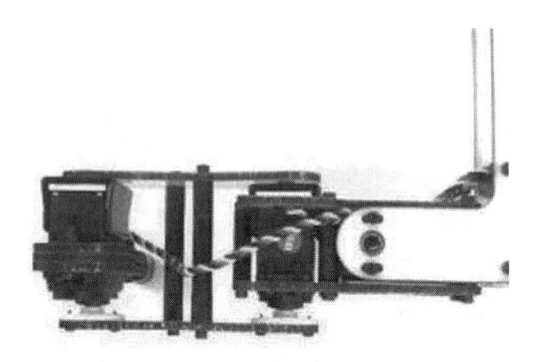

FIGURE 8.83
Robot leg lined up correctly with body. (Copyright 2006 by Lynxmotion Inc.)

Hexapod Robot

FIGURE 8.84
Completed Hexapod III. (Copyright 2006 by Lynxmotion Inc.)

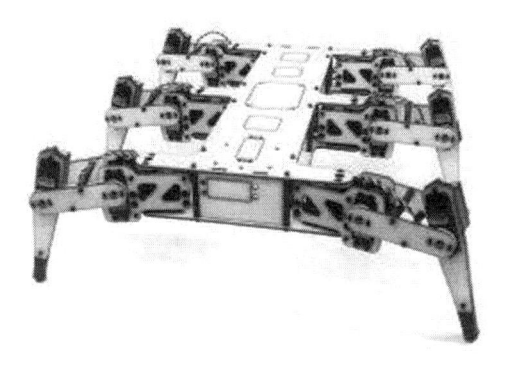

All the legs are installed in the same way, making sure that they all line up. This completes the mechanical assembly of the robot. Figure 8.84 illustrates the completed Hexapod III.

8.1.3.2 Addendum

In newer kits, four black plastic pieces are included, and the body panels have eight extra holes. These pieces are side panels, and the holes are used to mount them. In order to attach these side panels, align the bottom of the plastic pieces with the holes. Make sure that the servo leg wires are directed through the hole in the panel. Then press the top of the panel into place. In order to remove the side panels, press outward at the top of the panels and they will pop out. See the attached side panels in Figure 8.85.

8.2 Electrical Control

This section explains how Hexapod robots can be controlled and moved. We will present a controller with the robot and provide walking schemes in addition to necessary programs to move the robot.

8.2.1 Lynxmotion 12-Servo Hexapod with BS2e

The 12-servo hexapod, shown in Figure 8.86, has two degrees of freedom in each leg. Forward, backward, right turn, and left turn can be accomplished

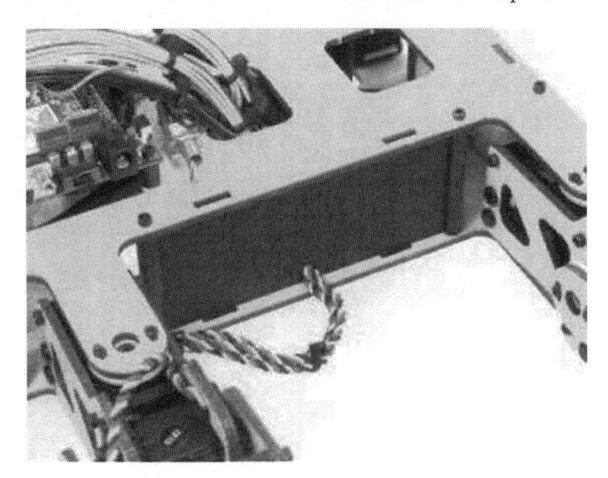

FIGURE 8.85
Side panel attached to hexapod. (Copyright 2006 by Lynxmotion Inc.)

with variable speed. Gradual turns can even be accomplished by the Hexapod. There are two servomotors per leg-one to control striding and the other to control climbing. The hexapod has a three-way switch for Servo On, Off, and Download.

8.2.1.1 Next Step Carrier Board

The Next Step Carrier board, shown in Figure 8.87, is a Basic Stamp 2 - based carrier board that has been specifically designed for robot control. By programming the controller on the Next Step Carrier board, you can program the robot for forward, backward, left turn, right turn, and gradual turn moves.

The list of necessary Next Step features for the robot builders:

- A 5V 500mA regulator is available for the peripherals.

- Up to four hobby servos can be plugged right into the first four I/O positions.

- Provisions for up to four bumper switches are available.

- An external reset switch is included.

- Powering options allow the Basic Stamp to share the servo power supply or be powered separately.

- Two on-board push buttons and LEDs can be used for debugging, as a simple user interface, or to allow certain parameters to be altered while the program is running.

Hexapod Robot

FIGURE 8.86
Lynxmotion 12-servo hexapod interfaced with Next Step carrier Board (for BS2e), SSC-12 servo controller, and serial LCD display

FIGURE 8.87
Next Step Carrier board v2.0 for BS2e-IC. (Copyright 2006 by Lynxmotion Inc.)

FIGURE 8.88
SSC-12 servo controller. (Copyright 2006 by Lynxmotion Inc.)

To program the Next Step, plug the cable from the PC's serial port into the micro. Run the Windows editor and type in, or load an example Basic program. Click Run. The program is downloaded to the micro and automatically begins running. The programmable cable can be disconnected for wire-free operation. The code is stored in an EEPROM, and therefore protected from power loss. If the program needs to be changed, new code can be downloaded at any time. The micro will stop, accept the new code, and then immediately begin running the new code.

8.2.1.2 SSC-12 Servo Controller

The servos are controlled by an SSC-12 servo controller, shown in Figure 8.88.

When a program is downloaded into the BS2e, it sends out serial data to the serial servo controller. This servo controller then sends out pulse signals to control the corresponding servos. You do not need the SSC-12 servo controller to control the servos; they can be directly controlled by the BS2e microcontroller. However, as the number of servos increases, the microcontroller can become bogged down with servo controls, leaving no time for monitoring the sensors. The advantage of the SSC-12 is that it takes care of the timing for the servos and leaves the microcontroller open to handle other issues. The positioning data sent to the servo controller consists of 3 bytes:

- Byte1: Sync data (i.e., the number 255)

- Byte2: Servo number (i.e., any number between 0 and 11) + (speed)

- Byte3: Position number (i.e., any number between 0 and 254)

Hexapod Robot   343

For example, if the position information sent to the servo controller is 255, 3, 180, it will cause servo number 3 to move to position number 180. Byte 2 is actually 3 + 0. Here, 0 is the servo speed value. A speed value of 0 will cause the servo to move to a particular position very fast. Now suppose it is necessary to move servo number 8 to position number 90 very slowly. Byte 1 is 255, Byte 2 is 8 + 16, and Byte 3 is 90. So the position information sent to the servo controller will be 255, 24, 90. Sending a new destination while the servo is in motion will result in the servo immediately responding to the new command.

Here are servo speed values listed in increasing order:

- 16 = 0.5 unit/frame (this means 10.16 seconds are required to move a full 180 degrees)

- 32 = 1.0 unit/frame (this means 5.08 seconds are required to move a full 180 degrees)

- 48 = 1.5 unit/frame (this means 3.39 seconds are required to move a full 180 degrees)

- 64 = 2.0 unit/frame (this means 2.55 seconds are required to move a full 180 degrees)

- 80 = 2.5 unit/frame (this means 2.03 seconds are required to move a full 180 degrees)

- 96 = 3.0 unit/frame (this means 1.69 seconds are required to move a full 180 degrees)

- 112 = 3.5 unit/frame (this means 1.45 seconds are required to move a full 180 degrees)

- 128 = 4.0 unit/frame (this means 1.27 seconds are required to move a full 180 degrees)

- 144 = 4.5 unit/frame (this means 1.13 seconds are required to move a full 180 degrees)

- 160 = 5.0 unit/frame (this means 1.02 seconds are required to move a full 180 degrees)

- 176 = 5.5 unit/frame (this means 0.92 seconds are required to move a full 180 degrees)

- 192 = 6.0 unit/frame (this means 0.85 seconds are required to move a full 180 degrees)

- 208 = 6.5 unit/frame (this means 0.78 seconds are required to move a full 180 degrees)

FIGURE 8.89

Pulse width modulated input signal

- 224 = 7.0 unit/frame (this means 0.73 seconds are required to move a full 180 degrees)

- 240 = 7.5 unit/frame (this means 0.68 seconds are required to move a full 180 degrees)

- 0 = as fast as possible (depends on servo)

8.2.1.3 Basic Introduction to Servo Operation

A servo is a motor in which the shaft does not continuously rotate. The output of a servo is the movement of its shaft. The shaft assumes a position based upon a pulse width modulated (PWM) input, shown in Figure 8.89.

The PWM input is a positive-going pulse that has a width between 1 millisecond and 2 milliseconds. The rate at which the pulses are sent is called the refresh rate. When the pulse width is 1 millisecond, the servo shaft is at 0 degrees. With a pulse width of 2 milliseconds, the servo moves to 180 degrees. As the pulse width varies between 1 millisecond and 2 milliseconds, the servo's shaft will move between the 0 and 180 degree positions. The servo used with the hexapod is the HS-422 from Hitec, shown in Figure 8.90.

The 180-degree rotation span of the servo is divided into 255 divisions. The distance a servo should move is not specified by the number of degrees, but by the position number (any number between 0 and 254). Hence, if it is required to move the servo to 90 degrees, it has to be moved to position number 63.

8.2.1.4 Connecting the BS2e to the Hexapod

The BS2e controller is plugged into a Next Step v2.0 carrier board (behind the serial port), shown in Figure 8.91.

The SSC-12 servo controller is interfaced with the carrier board by connecting the S/in pin of the SSC-12 to the I/O pin 8 of the carrier board. Each of the 12 servos of the hexapod has an I/O line made up of three wires of three different colors. The red wire is the connection to power, the black wire is the connection to ground, and the yellow wire is the PWM signal. These I/O

Hexapod Robot

FIGURE 8.90
Commercially available Hitec H-422 servo. (Copyright 2006 by Lynxmotion Inc.)

FIGURE 8.91
Zoomed in view of Next Step carrier board connected to SSC-12

lines of the servos are connected to I/O pins 0 through 11 of the SSC-12 servo controller.

An LCD display is interfaced with the carrier board by connecting it to I/O pin 13 of the Next Step carrier board. The +9V and the Servo Power connection on the carrier board are connected to the + connection on the SSC-12 servo controller, and this connection is given to the positive of a 7.2V NiMH battery. The two GND connections (in between the +9V and the Servo Power connection) on the carrier board are connected to the - sign connection (next to +) on the SSC-12, and this connection is given to the negative of the battery.

8.2.2 Programming the Hexapod

The Windows version of the Basic compiler software is used here to program the hexapod. The source code is written in the Windows text editor. The BS2e on board the hexapod is interfaced with the PC through a DB-9 cable as explained earlier. Set the three-way switch to Download. After entering the code into the editor, check the syntax by pressing Ctrl+T (or selecting the Check Syntax option in the Run menu). If the program is free of syntactical errors and all the connections to the BS2e are proper, a tab at the bottom right of the editor goes green and displays the message "Tokenize successful." Download the code into the BS2e by pressing Ctrl+ R (or selecting the Run option in the Run menu). Disconnect the DB-9 cable for wire-free (untethered) operation. Then with the three-way switch set to Servo On, the hexapod executes the downloaded code.

8.2.2.1 Walking Scheme for the Hexapod

Figure 8.92 shows the configuration of the robot at home position with the servo numbers and legs matched. Home position is when all the limbs of the robot are straight and touching the ground. S stands for stride, and L stands for lift. Notice that the S and L on leg six are different than what you would assume. Since servos are numbered starting from 0, the 11th and 12th servos are numbered as 10 and 11. During programming extra caution should be given to this numbering scheme.

Table 8.1 and 8.2 show two walking schemes with the initial position of the hexapod as shown in Figure 8.92.

8.2.2.2 Sample Code to Execute Walking Scheme 1

To create the walking program for the robot, certain features of the robot servos must be known. To make the right side of the robot stride forward/lift (legs 1, 2, and 3), a distance must be added to the current position. To make the left side of the robot stride forward/lift (legs 4, 5, and 6), a distance must be subtracted from the current position. The following code shows how to execute walking scheme 1.

Hexapod Robot

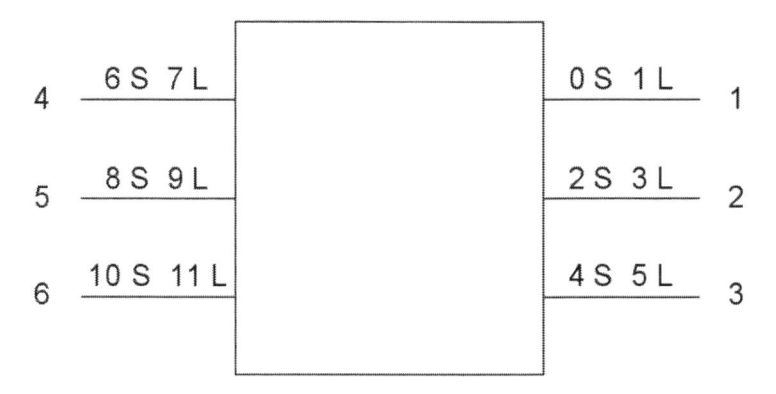

```
         6 S 7 L              0 S 1 L
   4  ───────────┌────────┐ ───────────  1
                 │        │
         8 S 9 L │        │   2 S 3 L
   5  ───────────│        │ ───────────  2
                 │        │
       10 S 11 L │        │   4 S 5 L
   6  ───────────└────────┘ ───────────  3
```

FIGURE 8.92
                Configuration of robot at home position

TABLE 8.1
Hexapod walking scheme 1

Step	Action
Step 1	Lift legs 1, 3, and 5.
Step 2	Move legs 1, 3, and 5 forward by distance x, and at the same time move legs 2, 4, and 6 backward by the same distance x as shown in Figure 8.93.
Step 3	Put legs 1, 3, and 5 down.
Step 4	Lift legs 2, 4, and 6.
Step 5	Move legs 2, 4, and 6 forward by a distance of $2x$, and at the same time move legs 1, 3, and 5 backward by the same distance $2x$ as shown in Figure 8.94.
Step 6	Put legs 2, 4, and 6 down.
Step 7	Lift legs 1, 3, and 5.
Step 8	Move forward legs 1, 3, and 5 by $2x$ and at the same time move legs 2, 4, and 6 backward by the same distance $2x$.
Step 9	Put legs 1, 3, and 5 down.
Step 10	Repeat from step 4 onward.

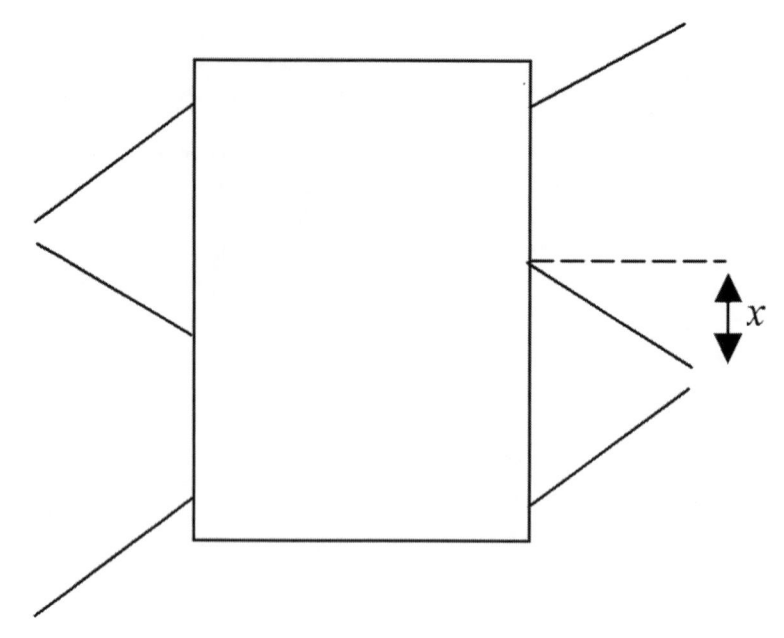

FIGURE 8.93

First move of scheme 1

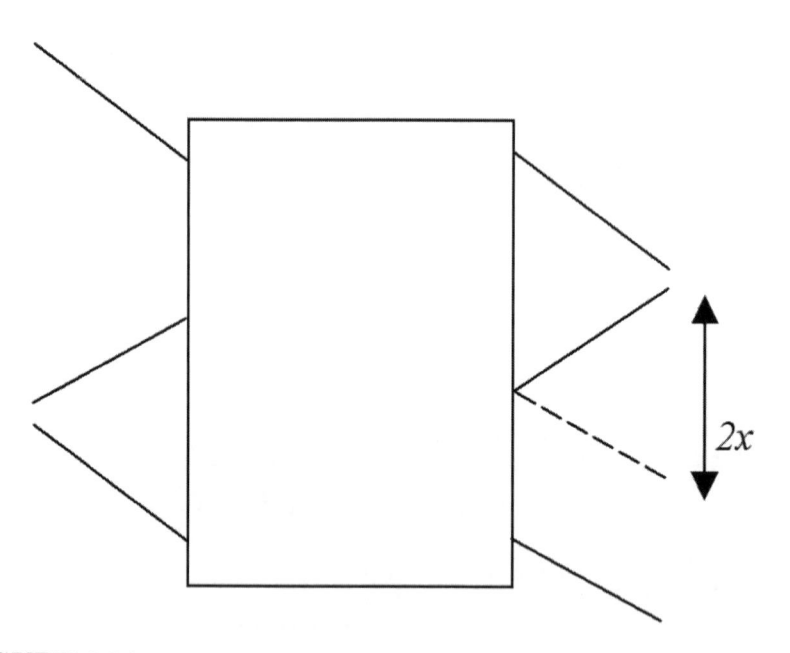

FIGURE 8.94

Second move of scheme 1

TABLE 8.2
Hexapod walking scheme 2

Step	Action
Step 1	Lift legs 1, 3, and 5.
Step 2	Move forward legs 1, 3, and 5 by some distance x as shown in Figure 8.95.
Step 3	Put legs 1, 3, and 5 down, and at the same time move legs 2, 4, and 6 backward by the same distance x as shown in Figure 8.93.
Step 4	Lift legs 2, 4, and 6.
Step 5	Move forward legs 2, 4, and 6 by a distance of $2x$ as shown in Figure 8.96.
Step 6	Put legs 2, 4, and 6 down, and at the same time move backward legs 1, 3, and 5 by a distance of $2x$ as shown in Figure 8.94.
Step 7	Lift legs 1, 3, and 5.
Step 8	Move forward legs 1, 3, and 5 by $2x$.
Step 9	Put legs 1, 3, and 5 down, and at the same time move backward legs 2, 4, and 6 by $2x$.
Step 10	Repeat from step 4 onward.

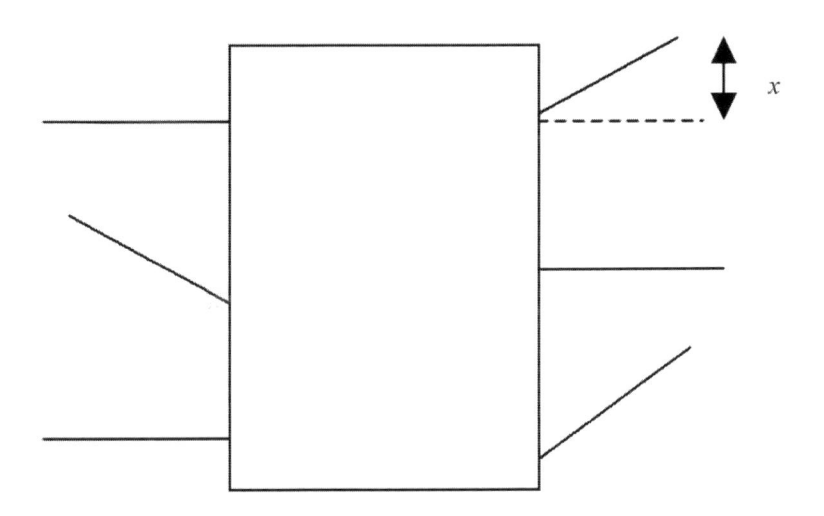

FIGURE 8.95
First move of scheme 2

FIGURE 8.96

Second move of scheme 2

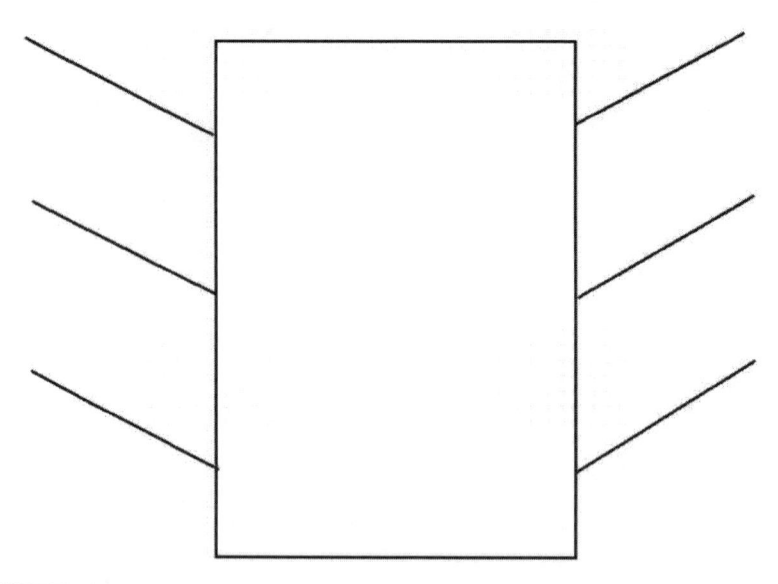

```
'{$STAMP BS2E}
'Jan 02, 2005

serpos     VAR   Byte(12)
x          VAR   Byte
lift       VAR   Byte
stride     VAR   Byte
y          VAR   Byte
i          VAR   Byte

lift = 120
y = 30

PAUSE 1000

SEROUT 13,$418C,["Lynxmotion Inc. H2 Setup Program"]
GOSUB homepos
stride = y
GOSUB limbup_1
GOSUB limbforward_1
GOSUB limbdown_1
GOSUB limbup_2
stride = 2*y
```

Hexapod Robot

```
FOR i=1 TO 4
    GOSUB limbforward_2
    GOSUB limbdown_2
    GOSUB limbup_1
    GOSUB limbforward_1
    GOSUB limbdown_1
    GOSUB limbup_2
NEXT
PAUSE 200
GOSUB homepos
END

homepos:
  serpos(0)  = 107  '127 right front H (+ forward)
  serpos(1)  = 55   '55  right front V (+ lift)
  serpos(2)  = 122  '127 right middle H (+ forward)
  serpos(3)  = 55 '55  right middle V (+ lift)
  serpos(4)  = 132  '127 right rear H (+ forward)
  serpos(5)  = 200  '200 right rear V (+ lift)
  serpos(6)  = 137 '127 left front H (- forward)
  serpos(7)  = 200  '200 left front V (- lift)
  serpos(8)  = 117  '127 left middle H (- forward)
  serpos(9)  = 200  '200 left middle V (- lift)
  serpos(10) = 117   '127 left rear H (- forward)
  serpos(11) = 55   '55 left rear V (- lift)
GOSUB sout
RETURN

sout:
  FOR x = 0 TO 11
    SEROUT 8,$4054,[255,x+32,serpos(x)]
  NEXT
  PAUSE 20
RETURN

limbup_1:
  serpos(1) = serpos(1) + lift
  serpos(5) = serpos(5) - lift
  serpos(9) = serpos(9) - lift
GOSUB sout
PAUSE 1000
RETURN

limbdown_1:
```

```
  serpos(1) = serpos(1) - lift
  serpos(5) = serpos(5) + lift
  serpos(9) = serpos(9) + lift
GOSUB sout
PAUSE 1000
RETURN

limbup_2:
  serpos(3) = serpos(3) + lift
  serpos(7) = serpos(7) - lift
  serpos(11) = serpos(11) + lift
GOSUB sout
PAUSE 1000
RETURN

limbdown_2:
  serpos(3) = serpos(3) - lift
  serpos(7) = serpos(7) + lift
  serpos(11) = serpos(11) - lift
GOSUB sout
PAUSE 1000
RETURN

limbforward_1:
  serpos(0) = serpos(0) + stride
  serpos(2) = serpos(2) - stride
  serpos(4) = serpos(4) + stride
  serpos(6) = serpos(6) + stride
  serpos(8) = serpos(8) - stride
  serpos(10) = serpos(10) + stride
GOSUB sout
PAUSE 1000
RETURN

limbforward_2:
  serpos(0) = serpos(0) - stride
  serpos(2) = serpos(2) + stride
  serpos(4) = serpos(4) - stride
  serpos(6) = serpos(6) - stride
  serpos(8) = serpos(8) + stride
  serpos(10) = serpos(10) - stride
GOSUB sout
PAUSE 1000
RETURN
```

Hexapod Robot 353

Analysis: In the above code, first the variables are declared. Serpos is an array of 12 bytes and others are byte variables. The first SEROUT instruction, i.e., SEROUT 13,\$418C,["Lynxmotion Inc. H2 Setup Program"], indicates that an LCD display has been connected to I/O pin 13 of the Next Step carrier board. The output argument list causes "Lynxmotion Inc. H2 Setup Program" to be displayed on the LCD. Then the command GOSUB calls the homepos subroutine which causes the hexapod to go to home position.

The GOSUB sout instruction after the homepos subroutine sens serial data to the SSC-12 servocontroller. The SEROUT 8,\$4054,[255,x+32,serpos(x)] instruction indicates that the SSC-12 is connected to I/O pin 8 of the Next Step carrier board. $4054 specifies the Baudmode. Since x is the index variable of the array serpos, [255,x+32,serpos(x)] indicates the 3 bytes of information (to the SSC-12) sent to control each of the 12 servos one at a time (x goes from 0 to 11). 32 is the chosen servo speed value.

The homepos subroutine instructs each of the servos to go to a particular position (as specified by the position number). However, this information is actually sent to the servos by means of the SSC-12 servo controller by executing the sout subroutine. For each servo, 3 bytes of information is required in order to send it to a particular servo to a particular position. The sout subroutine sends these 3 serial bytes to the SSC-12 by means of the SEROUT instruction. Depending on the servo number specified in the 3 bytes, the SSC-12 causes movement of that particular servo. For example, consider the first line of the homepos subroutine: serpos(0) = 107. This information is sent to the SSC-12 as SEROUT 8,$4054,[255,0+32,serpos(0)] (here index $x = 0$). By setting servo(0) to 107 (serpos(0) = 107), you move the 0th servo to position number 107 which corresponds to a specific angle. Other subroutines, such as sout, limbup_1, limbforward_1, and so on, work pretty much the same way and are used to make the hexapod make a stride or lift a leg.

8.2.3 Adjusting Servomotors to Mid Position

In this section we will study how to adjust HS-422 to its mid position by Mini SSC II, SSC-12, Basic Atom, Basic Atom Pro, and Basic Stamp 2 (BS2e) [25] [26].

8.2.3.1 Mini SSC II Servomotor Driver

A 9-volt (DC) battery is connected to the SSC power input. In addition, a 4.8 to 7.2 volt (DC) battery is connected to the SVO power input. When the board is powered correctly, the green LED should illuminate. The SSC board generates the pulses automatically. HS-422 servomotors should be connected to any port, with the black wire closest to the outside of the board, as shown in Figure 8.97.

FIGURE 8.97
Adjusting a servomotor to mid position with the Mini SSC II. (Copyright 2006 by Lynxmotion Inc.)

Hexapod Robot 355

FIGURE 8.98
Adjusting the servomotor to mid position with the SSC-12. (Copyright 2006 by Lynxmotion Inc.)

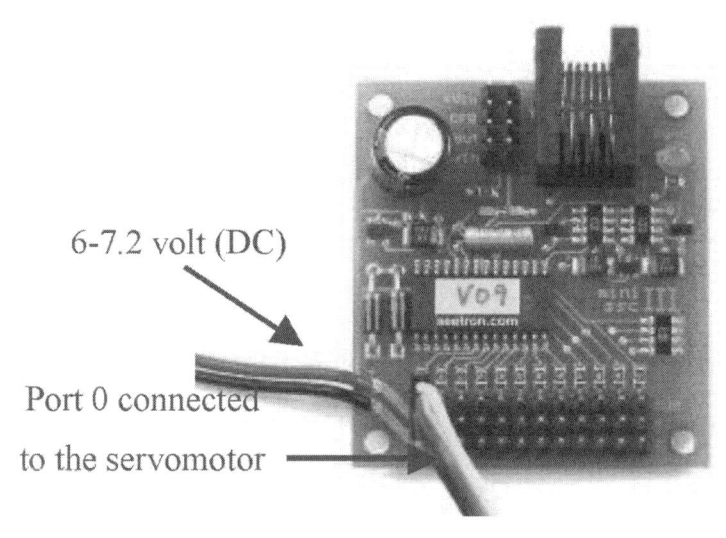

8.2.3.2 SSC-12 Servomotor Driver

In the SSC-12, the whole board is powered by a 6 to 7.2 volt (DC) battery. It has a green LED that will light if the power is connected correctly, as shown in Figure 8.98. The servomotor can be plugged into any of the ports, since the board generates the pulses automatically.

8.2.3.3 Basic Atom Microcontroller

The Basic Atom board is powered by a 6 to 7.2 volt (DC) battery through the VS connector. VS jumpers for the first four I/O pins should be installed as shown in Figure 8.99. The servomotor is connected to port 0 with the black wire closest to the edge of the board.

The program for generating the positioning pulse (1.5 millisecond positive pulse with 20 millisecond period) on I/O pin 0 is shown in the code below.

```
low p0
start:
        pulsout p0,1500
        pause 20
```

8.2.3.4 Basic Atom Pro Microcontroller

The power, the VS jumpers for the first four ports, and the servomotor connections are connected the same way as the Basic Atom, shown in Figure 8.100.

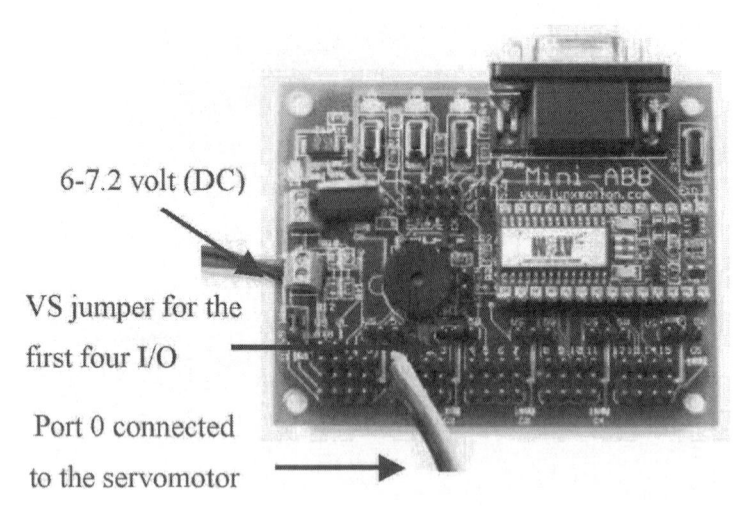

FIGURE 8.99
Basic Atom connections for adjusting a servomotor. (Copyright 2006 by Lynxmotion Inc.)

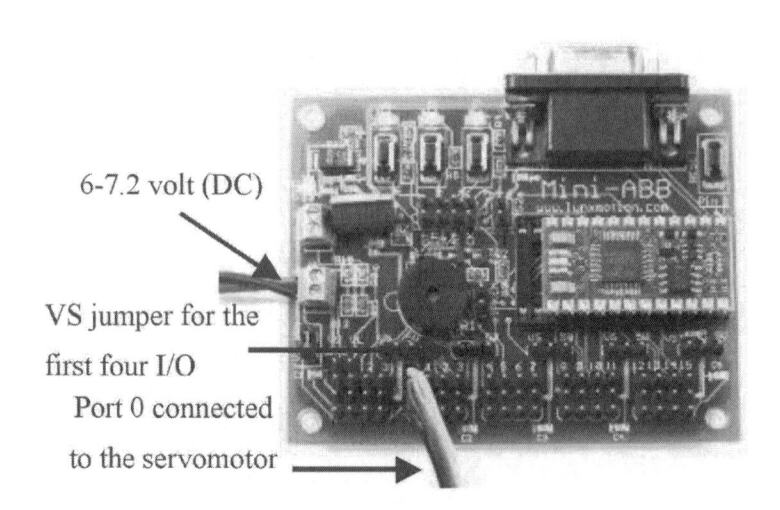

FIGURE 8.100
Basic Atom Pro connections for adjusting a servomotor. (Copyright 2006 by Lynxmotion Inc.)

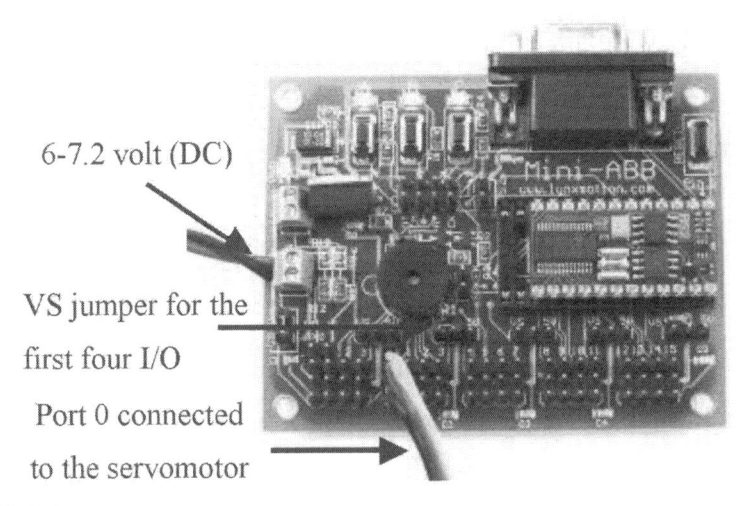

Hexapod Robot

6-7.2 volt (DC)

VS jumper for the first four I/O

Port 0 connected to the servomotor

FIGURE 8.101
Basic Stamp 2 connections for adjusting a servomotor. (Copyright 2006 by Lynxmotion Inc.)

The program for generating the position-adjusting pulse is given in the code below.

```
enablehservo %0000000000000001,1000,2000,20
start:
      hservo [0/128/255]
end
```

8.2.3.5 Basic Stamp 2/BS2e Microcontroller

For the Basic Stamp 2, the same wiring scheme as in the Basic Atom Pro is followed as shown in Figure 8.101. The only difference is the microcontroller chip used on the board.

The program for generating position-adjusting pulses for Basic Stamp is given below.

```
start:
  pulsout   0,750
  pause     20
```

9

Biped Robots

In this chapter we will study biped (two-legged walking) robots: their construction and control.

9.1 Bigfoot: The Walker

The picture of the assembled Bigfoot robot is shown in Figure 9.1. This robotics kit belongs to Milford Instruments Limited, England.

The Bigfoot robot is a two-legged walking robot, which uses two servomotors to accomplish its walking motion. It uses a static gait as compared to a dynamic gait. In a static gait, robots keep their center of gravity on the base of the robot so that the robot does not fall if kept in that position. The dynamic gait is a walking or a running pattern where the stability of the robot is accomplished by the momentum of the robot. For example, we can design single-legged hopping robots that would fall if stopped but do not fall because of their motion [12].

Bigfoot uses the Basic Stamp microcontroller. The Bigfoot uses one servomotor to transfer the center of gravity to the standing foot so that the other foot can be moved by the other servomotor. It uses toe switches so that it can know when it has hit something and it can then back up.

9.1.1 Construction

The construction steps for various parts of the robot are presented here.

9.1.1.1 Servomotor Mounting

The two servomotors are glued to a board during construction. This is shown in Figure 9.2.

9.1.1.2 Ankle Construction

In ankle construction, you need to take 1 mm diameter brass wire and bend it into the shape shown in Figure 9.3.

The brass wire needs to be glued to the ankle piece as shown in Figure 9.4.

FIGURE 9.1
   Bigfoot robot (Copyright Milford Instruments Ltd)

FIGURE 9.2
   Servomotors on the board (Copyright Milford Instruments Ltd)

Biped Robots

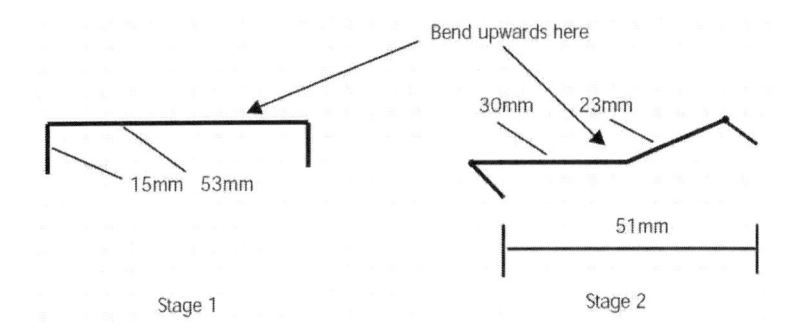

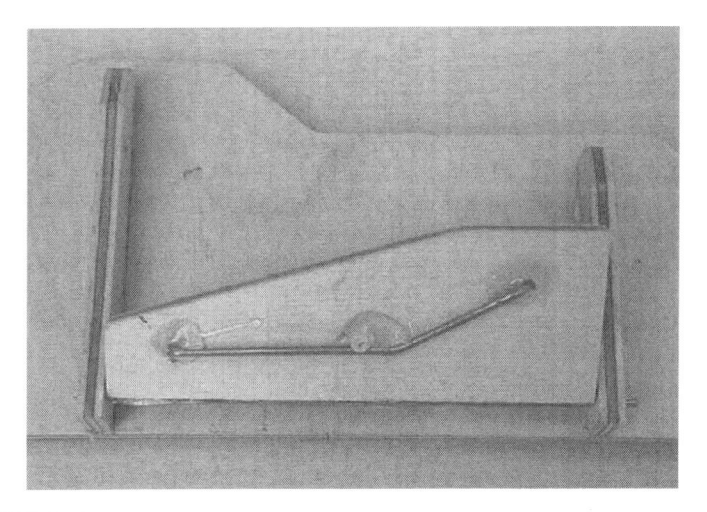

FIGURE 9.3
Brass wire bending (Copyright Milford Instruments Ltd)

FIGURE 9.4
Brass wire glued to the ankle piece (Copyright Milford Instruments Ltd)

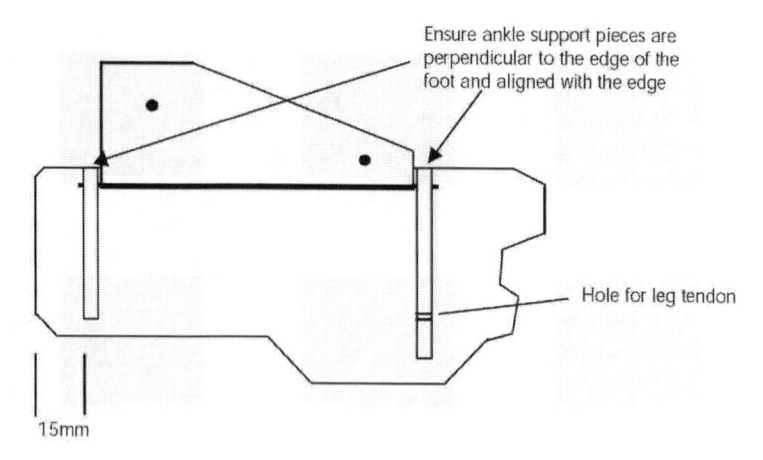

FIGURE 9.5
   Foot construction (Copyright Milford Instruments Ltd)

9.1.1.3 Foot Construction

The foot construction is shown in Figure 9.5.

9.1.1.4 Leg Construction

The leg construction is shown in Figure 9.6.

9.1.1.5 Body Board

The servomotor board is glued to the top board as shown in Figure 9.7. The servomotor block should be perpendicular to the board.

9.1.1.6 Leg-Body Connections

The legs are secured to the robot body using wires that are constructed by bending wire as shown in Figure 9.8.

9.1.1.7 Leg-Tendons

The legs are connected to the roll servomotor (the servomotor that lifts the legs up before the pitch servomotor moves the other leg forward to move forward) using piano wires that are constructed by bending wire as shown in Figure 9.9.

9.1.1.8 Toe Switches

The toe switches are constructed out of piano wire and are attached to the feet as shown in Figure 9.10.

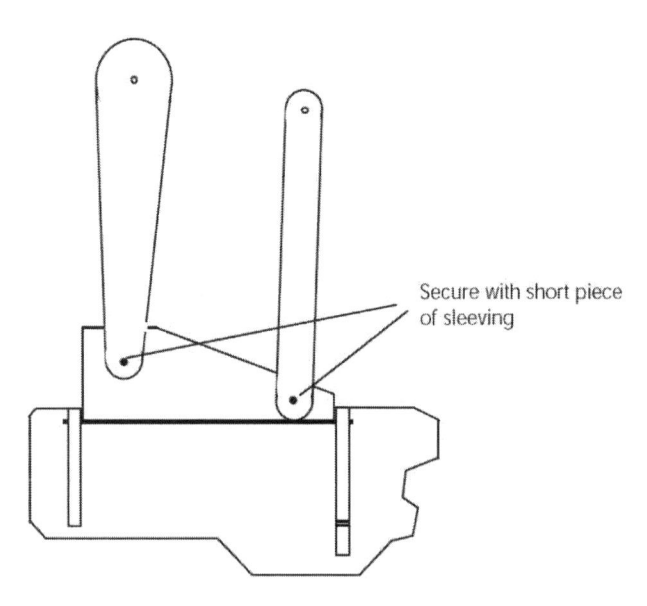

Biped Robots

FIGURE 9.6
    Leg construction (Copyright Milford Instruments Ltd)

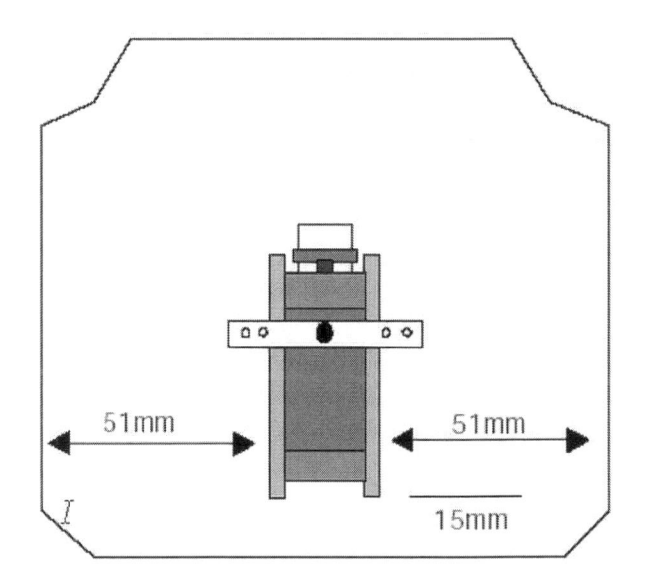

FIGURE 9.7
    Body board (Copyright Milford Instruments Ltd)

364   Practical and Experimental Robotics

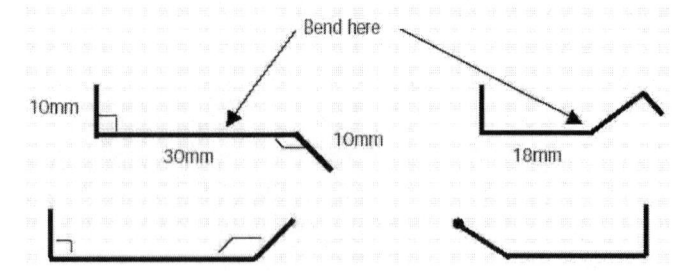

FIGURE 9.8
   Wire pieces for leg connections (Copyright Milford Instruments Ltd)

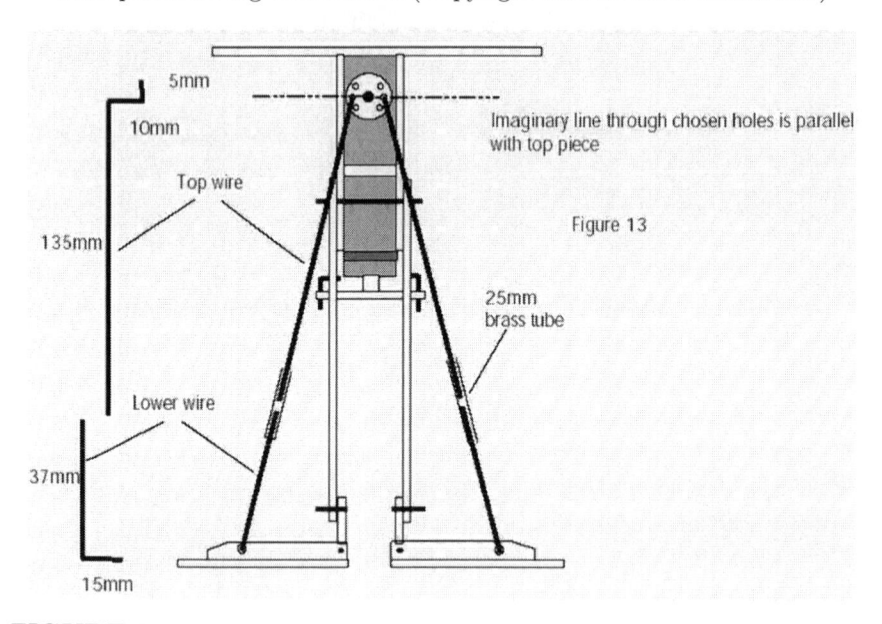

FIGURE 9.9
   Leg tendons (Copyright Milford Instruments Ltd)

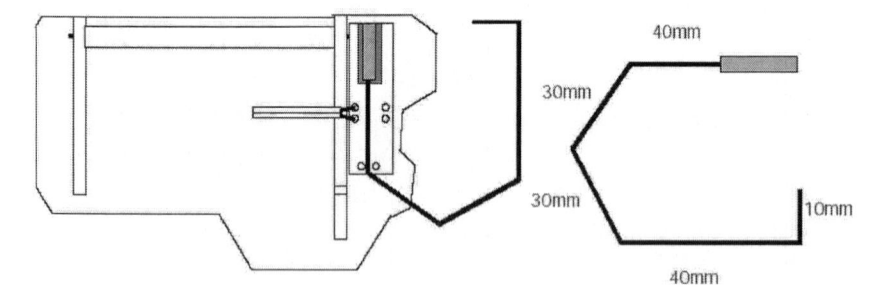

FIGURE 9.10
   Toe switch (Copyright Milford Instruments Ltd)

Biped Robots 365

FIGURE 9.11
PCB mounting (Copyright Milford Instruments Ltd)

9.1.1.9 PCB Mounting

The PCB board that has the BASIC Stamp microcontroller is glued to the main board and the wires from the battery, servomotors, and the toe switches are connected to the board as shown in Figure 9.11.

9.1.2 Programming

The robot is controlled by the Basic Stamp microcontroller from Parallax Inc. A close up of the control board is shown in Figure 9.12.

The program that comes with the robot can be modified to make the robot perform different actions from the pre-programmed actions. The listing of the symbol definitions for the Bigfoot is given in section 9.1.2.1.

9.1.2.1 Symbol Definitions Listing

```
'BigFoot.bas
'Original programme and model by D Buckley
'This version rev1.0 by Milford Instruments- 16-4-99
'
'
```

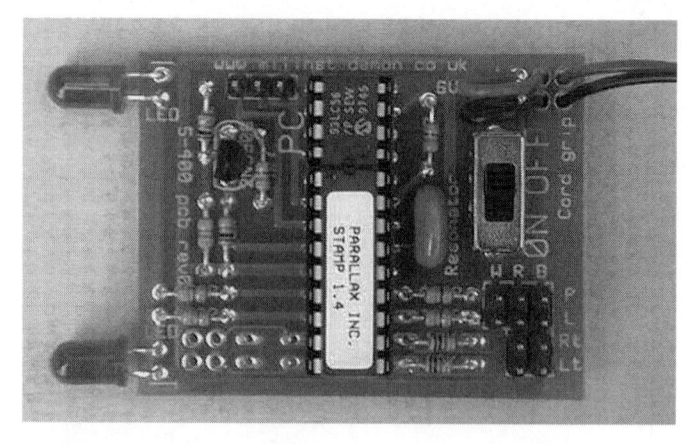

FIGURE 9.12
Control board (Copyright Milford Instruments Ltd)

```
'Define the constants and pin allocations
SYMBOL servoroll =6 'Roll servomotor connected to pin 6
SYMBOL servopace =7 'Pace servomotor connected to pin 7
SYMBOL lefteye =1 'Left LED eye
SYMBOL righteye =0 'Righ LED eye
SYMBOL righttoe =pin5 'Right Toe switch to pin 5
SYMBOL lefttoe =pin4 'Left Toe switch to pin 4
SYMBOL atroll =b1
SYMBOL roll =b2
SYMBOL atpace =b3
SYMBOL pace =b4
SYMBOL atX =b5
SYMBOL toX =b6
SYMBOL servoX =b7
SYMBOL m =b9 'current move
SYMBOL i =b9 'loop counter in init
'===========================================================
' THIS IS THE SET-UP SECTION
SYMBOL r_left =125 'roll_left hand side
SYMBOL r_stand =160 'upright
SYMBOL r_right =195 'roll to right hand side
SYMBOL p_left_fd =120 'left foot forwards
SYMBOL p_right_bk =p_left_fd
SYMBOL p_stand =140 'feet together
SYMBOL p_right_fd =160 'right foot forwards
SYMBOL p_left_bk =p_right_fd
SYMBOL speed =1 'servomotor increment =1,2(make
' to'even),3+(beware)
```

Biped Robots

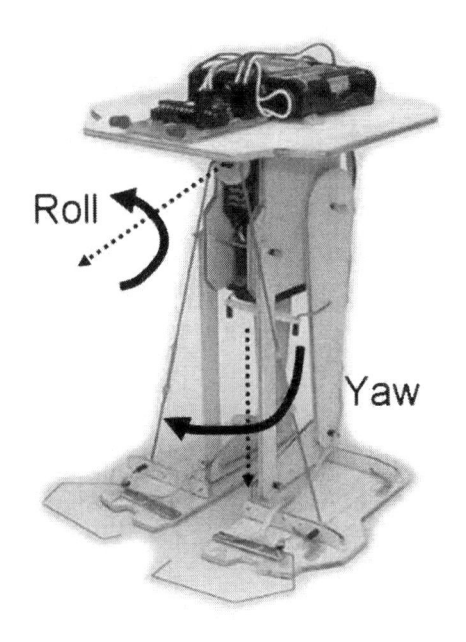

FIGURE 9.13
Roll and yaw movements (Copyright Milford Instruments Ltd)

```
SYMBOL tweenpulse =10 'delay to ensure correct pulse
' stream to servomotors
'===========================================================
SYMBOL touch_flag =bit0 'toe switches touched?
```

9.1.2.2 Software Control

We can control various aspects of the walk of Bigfoot by changing certain parameters of the program the robot comes with. The different controls are discussed here.

Upright Control We can make sure that when the robot is standing upright, it is in a good straight position. To control that, we can fine-tune the two parameters: r_stand and p_stand.

Roll Control The roll and yaw movements of the servomotors (and consequently the robot) are shown in Figure 9.13.

The amount of roll for right is controlled by $[r\_right - r\_stand]$ and for the left by $[r\_stand - r\_left]$.

Pace Control We can control the length of the pace by using the parameters: $[p_{right_f}d - p_s tand]$ and by $[p_s tand - p_{right_f}d]$.

Turn Control The robot turns by shuffling movements. These can be controlled by changing the number 5 in the Listing in section 9.1.2.3.

9.1.2.3 Code for Shuffling Movements of the Bigfoot

```
MRT: 'Shuffle to the Right
------------
------------
jumprt:
roll =r_stand+5
'===========================================================
' Adjust the "5" figure to change the turning force
'===========================================================
MLT: 'Shuffle to the Left
-----------
-----------
jumplt:
roll =r_stand-5
'===========================================================
' adjust the "5" figure to vary the turning force
'===========================================================
```

9.1.2.4 The Walking Gait

The Bigfoot robot can be made to move forward in a straight line by following the sequence below. We assume here that the robot is facing straight initially.

1. Signal the roll servomotor to lift right leg, so that the center of gravity of the robot falls on the left foot.

2. While the right leg is still up, signal the pitch servomotor to move the right leg forward.

3. Signal the roll servomotor to put the right leg back on ground, so that the center of gravity of the robot falls in the middle of the two feet.

4. Signal the roll servomotor to lift left leg, so that the center of gravity of the robot falls on the right foot.

5. While the left leg is still up, signal the pitch servomotor to move the left leg forward.

6. Signal the roll servomotor to put the left leg back on ground, so that the center of gravity of the robot falls in the middle of the two feet.

7. Go to Step 1.

We can make the robot move faster or slower by controlling the time between commands. We can also make the robot walk backwards by making the servomotor move the legs backwards instead of forward. We can make the robot make a right or left turn by making the right step bigger than left step for left turn and making the right step smaller than left step for right turn.

Biped Robots 369

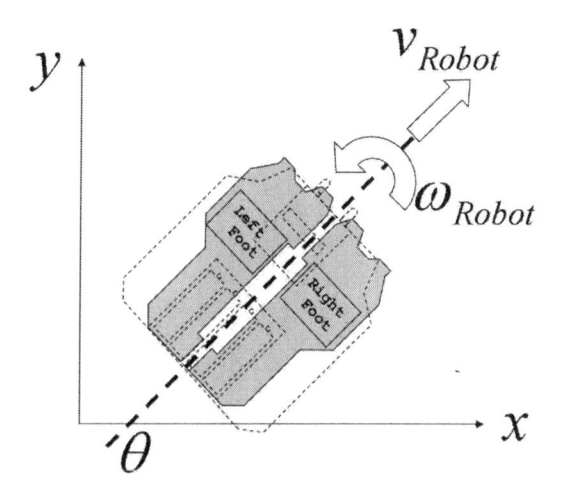

FIGURE 9.14
Robot kinematic relationships (Copyright Milford Instruments Ltd)

9.1.3 Robot Kinematics

Once we can control the robot's straight and turning movements, we can make it follow any path we want. Now, we can integrate this robot with a camera that can watch the robot from top and know its coordinates in (x, y) directions. We can use the same camera to track a soccer ball, and then give commands to move the robot based on its location and the location of the soccer ball obtained from the camera. The location of the robot and the soccer ball will be in the (x, y) coordinates and therefore we should know the relationship between the robot's linear and angular velocities (controlled by its walking gait) and the (x, y) coordinates of its position. Consider the Figure 9.14 to understand these relationships.

In Figure 9.14, we show the linear speed of the robot as v_R and its angular speed as:

$$\omega_{Robot} = \frac{\theta}{dt} \qquad (9.1)$$

Now, it is possible that we are given a path that the robot should follow. However, the path might be given in terms of the x and y coordinates and the yaw angle θ of the robot, rather than the desired wheel speeds (or robot linear and angular speed), then we will need a relationship between the world coordinates x, y, and θ, and the control variables of the robot (the linear and angular speeds, which we can control by the robot gait). This relationship can be derived as:

$$\frac{dx}{dt} = v_{Robot} \cos \theta \qquad (9.2)$$

$$\frac{dy}{dt} = v_{Robot} \sin \theta \qquad (9.3)$$

$$\frac{d\theta}{dt} = \omega_{Robot} \qquad (9.4)$$

We can make the robot follow trajectories that are given in world coordinates by designing controllers that take use of equations 9.2, 9.3, and 9.4. The trajectories can come from using a camera on top from where the robot motions can be observed and the camera can be used by a controller to decide where we want our robot to go. Instead of using the linear and angular speeds as inputs, we can also use the servomotor angle movement rates as inputs. We can come up with the relationship between the angles each servomotor moves and how it relates to forward and angular motion. Then, using the time during that movement, we can come up with the desired relationships. The advanced topics that can be studied using this robot include trajectory planning and control, among others.

9.2 The Lynxmotion Biped

The Lynxmotion Inc. robots [25] are very modular and can be built using their brackets attached to servomotors. The biped robot from Lynxmotion is shown in Figure 9.15. The robot can be built by constructing legs, arms, and torso, then attaching them together. The general technique involves creating the assembly with brackets first and then populating the assembly with servomotors.

9.2.1 Leg Assembly

The assembled leg is shown in Figure 9.16. This leg is a 5 degrees of freedom (DOF) leg, and uses five servomotors to produce the five independent rotations (one of the five servomotors is in the torso assembly for each leg). Figure 9.16 shows the five axes of rotation: foot, ankle, knee, hip pitch and hip roll.

The leg is built out of two types of brackets and a foot panel that Lynxmotion provides. The two brackets are C-bracket and multi-purpose servomotor bracket. A C-bracket can be connected to a servomotor through the rotation axis of the servomotor. To attach a servomotor to a C-bracket in other way (as when attaching the hip pitch servomotor to the topmost C-bracket in Figure 9.16), we need to attach a multi-purpose bracket to the C-bracket, and then

Biped Robots 371

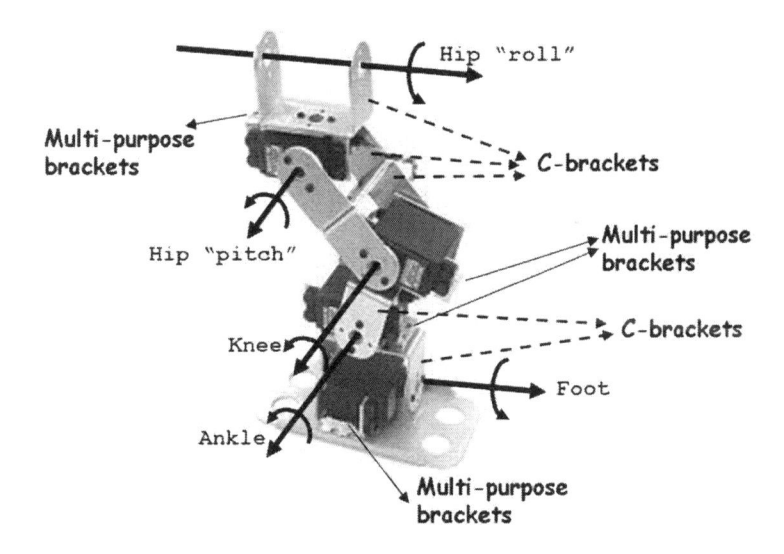

FIGURE 9.15
Lynxmotion Biped (Copyright Lynxmotion Inc.)

FIGURE 9.16
5 DOF assembled leg (Copyright Lynxmotion Inc.)

FIGURE 9.17
Some bracket assemblies in the leg (Copyright Lynxmotion Inc.)

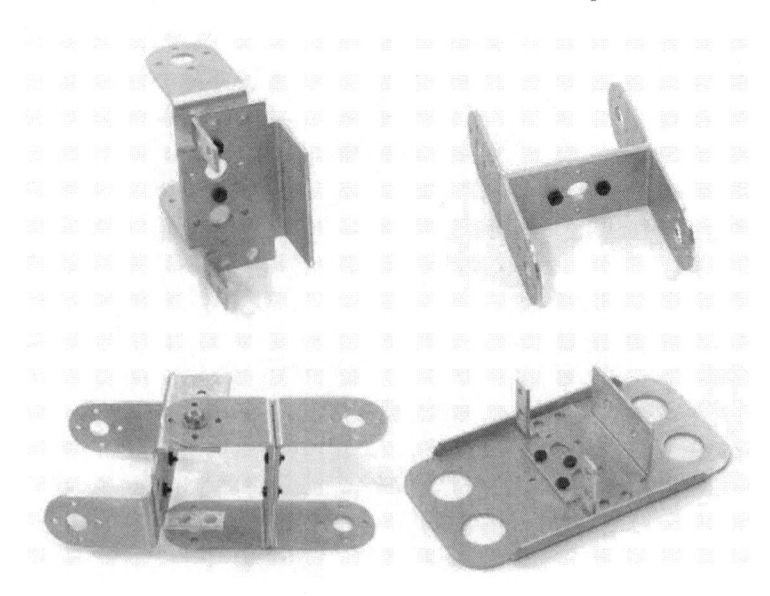

the assembly is attached to the servomotor. Some of these bracket assemblies are shown in Figure 9.17.

How the entire leg is connected to the torso assembly is shown in Figure 9.18.

9.2.2 Arm Assembly

The assembled arm is shown in Figure 9.19. This arm is a 4 DOF linkage. However, one of the servomotors is in the torso assembly for each arm. Hence the arm outside the torso has only three servomotors.

The arm is assembled by creating the bracket assembly and then populating the assembly with servomotors. The stepwise process is shown in the sequence of pictures in Figure 9.20.

9.2.3 Torso Assembly

The assembled torso is shown in Figure 9.21.

The torso has five servomotors: two for legs, two for arms, and one for the head. It is assembled by creating the bracket assembly and populating the assembly with servomotors. The stepwise process is shown in the sequence of pictures in Figure 9.22.

FIGURE 9.18
   Leg attachment to torso (Copyright Lynxmotion Inc.)

FIGURE 9.19
   The arm assembly (Copyright Lynxmotion Inc.)

374   *Practical and Experimental Robotics*

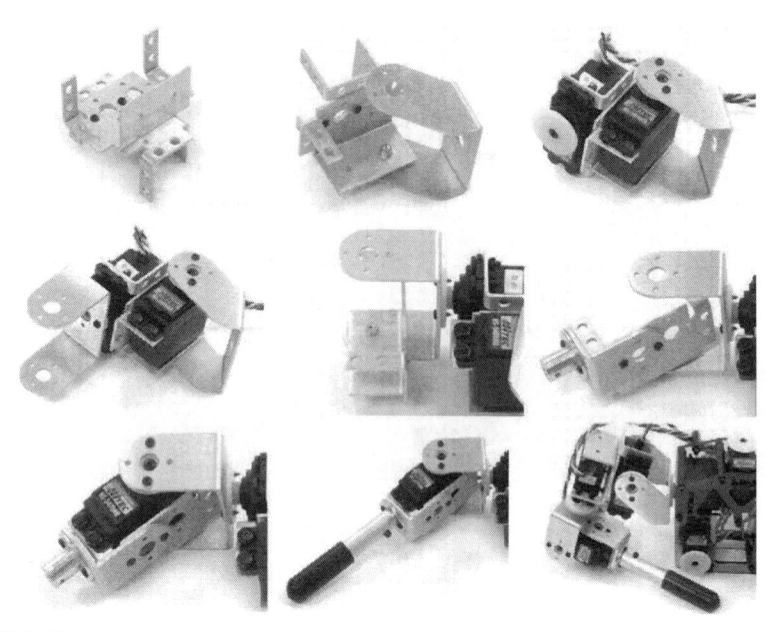

FIGURE 9.20
    The arm assembly sequence (Copyright Lynxmotion Inc.)

FIGURE 9.21
    The torso assembly (Copyright Lynxmotion Inc.)

Biped Robots 375

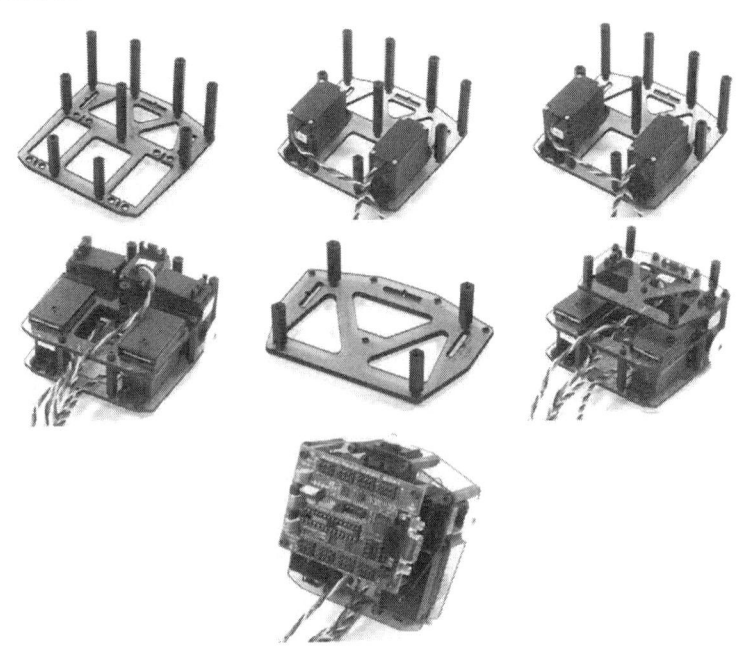

FIGURE 9.22
The torso assembly sequence (Copyright Lynxmotion Inc.)

9.2.4 Hand Assembly

The assembled hand is shown in Figure 9.23.

The hand assembly uses two servomotors. The stepwise process is shown in Figure 9.24.

9.2.5 Controller

The Lynxmotion biped uses four servomotors per leg, three servomotors per arm, five servomotors for the torso, and if hands are needed two more servomotors per hand. This means that we have twenty-three servomotors to control. Lynxmotion has a servomotor controller SSC-32 that can control 32 servomotors. The servomotor-controller is shown in Figure 9.25.

The controller has a serial connector for communication with a PC and we can connect 32 servomotors to it. The servomotor controller is based on using a microcontroller with four 8-bit serial-in-parallel-out shift register. Therefore, we can use four output pins, each one connected to a serial-to-parallel shift register. A general circuit (not the same as SSC-32) using a PIC16F84A microcontroller is shown in Figure 9.26 as an example.

In Figure 9.26 we see the output pin RB0 of PIC16F84 connected to the serial input of 74LS595 which is an 8-bit serial-in-parallel-out shift register.

FIGURE 9.23
 The hand assembly (Copyright Lynxmotion Inc.)

FIGURE 9.24
 The hand assembly sequence (Copyright Lynxmotion Inc.)

Biped Robots 377

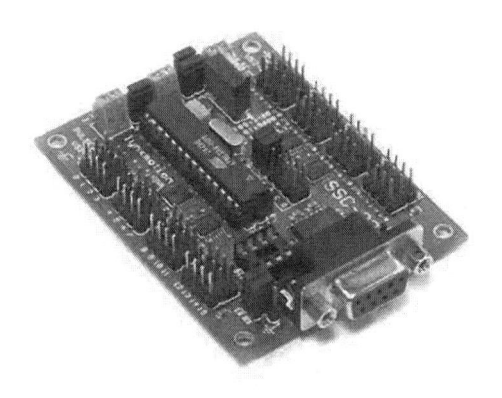

FIGURE 9.25
SSC-32 Lynxmotion servocontroller (Copyright Lynxmotion Inc.)

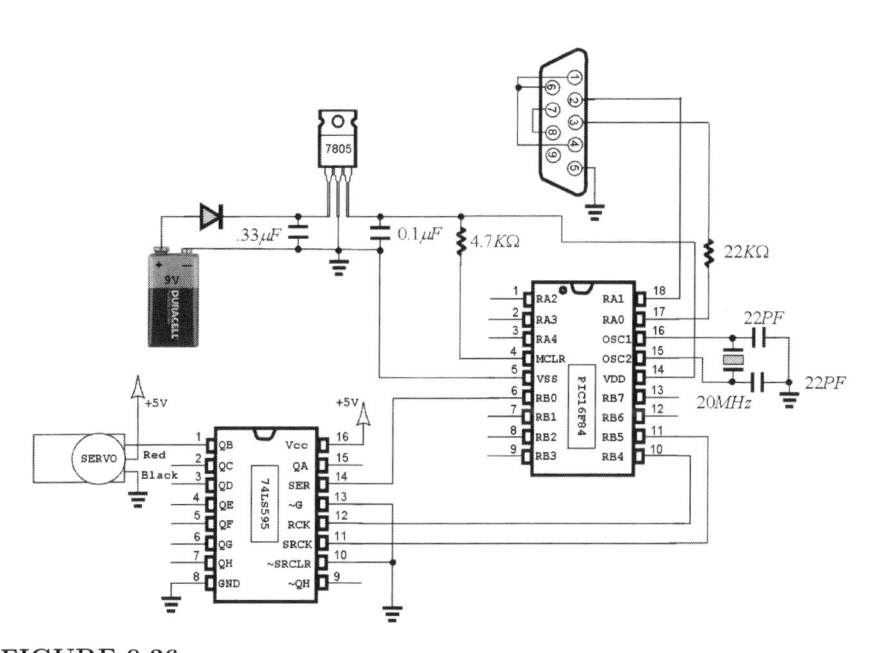

FIGURE 9.26
Servocontroller cicruit

Similarly, we can connect RB1, RB2, and RB3 to three more 74LS595 chips. All four of these 74LS595 chips have two clock inputs. These clock inputs are positive edge triggered. The serial clock SRCK is used to shift serial data in to the shift registers. When eight bits of data are in, then we can use the positive edge of the storage register clock RCK to send the signal out of the 74LS595 output pins. This way, we can control up to 32 servomotors using four 74LS595 chips. We can write a serial communication program on a PC that communicates with the PIC processor to let it know the desired motion of each servomotor, and then the PIC processor code should convert that into the actual signals to be sent to the serial-in-parallel-out chips. The internal functional representation of 74LS595 from Texas Instruments is shown in Figure 9.27.

To control each servomotor we need to use a PWM (Pulse Width Modulated) signal. The time period of the signal should be about 20 ms. The width of the on-time of the signal should be about 1ms to get the left most rotation, about 1.5 ms for center position, and about 2.0 ms for the right most rotation. Each servomotor should be tested to get its actual range and response. These servomotor commands are shown in Figure 9.28.

9.3 The Robosapien Biped

The Robosapien robot is a humanoid robot designed by Mark Tilden and commercialized by Wow Wee. Here we describe its basic functionality and how it can be sued for robotic experiments. Figure 9.29 shows the Robosapien biped.

The Robosapien has the infrared receiver on the head. It also has touch sensors on the fingers and toes, so that it knows when its hands or feet hit something. It has a sound sensor near the middle of its lower chest. It has a power switch at its back. It has the following 67 command functions.

1) Right Arm Up
2) Right Arm Down
3) Right Arm In
4) Right Arm Out
5) Tilt Body Right
6) Left Arm Up
7) Left Arm Down
8) Left Arm In
9) Left Arm Out
10) Tilt Body Right
11) Turn Right
12) Walk Forward

Biped Robots 379

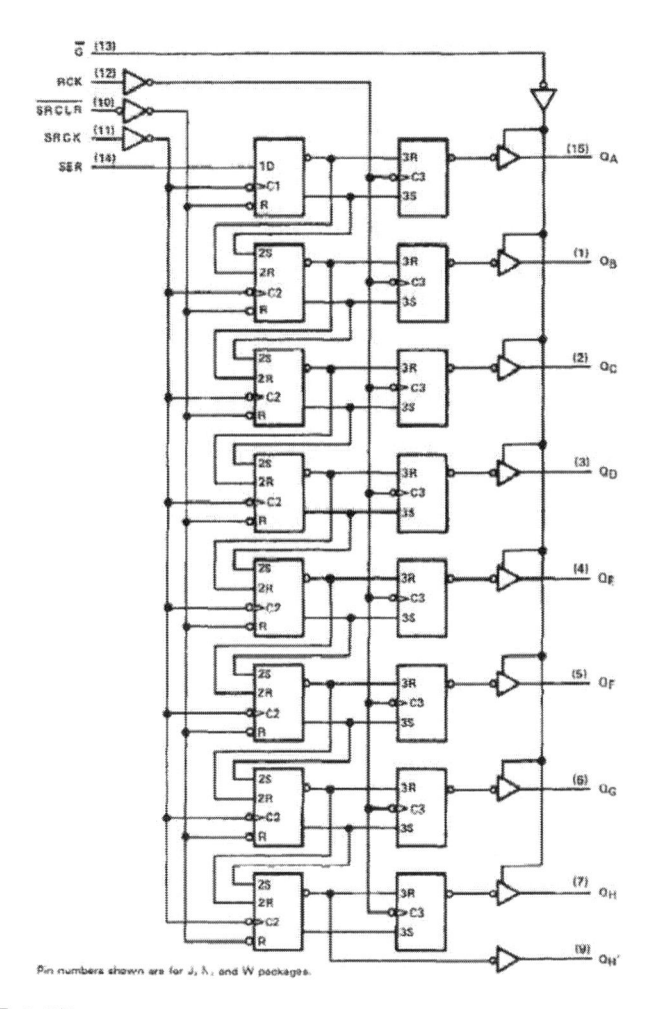

FIGURE 9.27
74LS595 functional diagram (Courtesy of Texas Instruments)

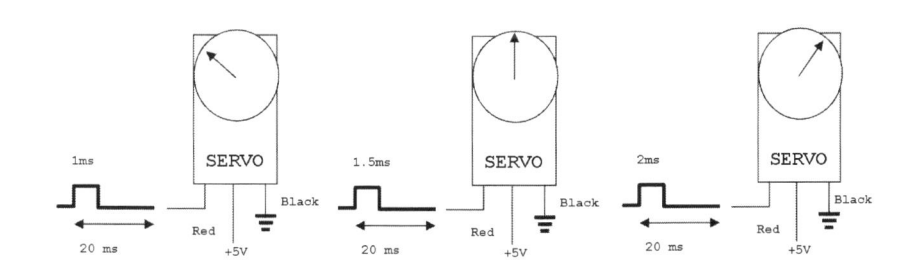

FIGURE 9.28
Servomotor commands

FIGURE 9.29

Robosapien (Copyright WowWee Ltd)

13) STOP Button
14) Turn Left
15) Walk Backward
16) (R>) Right Sensor Program
17) (S>) Sonic Program
18) (L>) Left Sensor Program
19) (R>) Right Sensor Program
20) (P) Master Command Program
21) (SELECT) Advance to GREEN Keys
22) Right Hand Thump
23) Left Hand Pickup
24) Lean Backward
25) Right Hand Throw
26) Sleep
27) Left Hand Thump
28) Left Hand Pickup
29) Lean Forward
30) Left Hand Throw
31) Listen
32) Forward Step
33) Right Turn Step
34) Backward Step

Biped Robots 381

35) Right Sensor Program Execute
36) Master Command Program Execute
37) Wake Up
38) Reset
39) Left Turn Step
40) (SELECT) Advance to ORANGE Keys
41) Left Sensor Program Execute
42) Sonic Sensor Program Execute
43) Right Hand Sweep
44) High 5
45) Right Hand Strike 1
46) Burp
47) Right Hand Strike 2
48) Left Hand Sweep
49) Talk Back
50) Left Hand Strike 1
51) Whistle
52) Left Hand Strike 2
53) Bulldozer
54) Right Hand Strike 3
55) Oops!
56) Demo1
57) All Demo
58) Power Off
59) Roar
60) Left Hand Strike 3
61) (Select) Return to RED Command Functions
62) Demo2
63) Dance Demo
64) <, < Combination "Right Walk Turn"
65) >, > Combination "Left Walk Turn"
66) Forward, Forward Combination "Slow Walk Forward"
67) Backward, Backward Combination "Slow Walk Backward"

There are also four programming modes. Three are sensor programs and one is a master program. We can program some sequence of actions in each mode. The sensor programming mode starts executing if that sensor is tripped. The master programming mode starts executing when the corresponding button is pressed. The handheld remote controller is used to control the different functions of the robot. The remote controller communicates with the robot by sending IR codes. The remote controller is shown in Figure 9.30.

FIGURE 9.30
Robosapien remote control (Copyright WowWee Ltd)

9.3.1 Robosapien Motors

The Robosapien has seven motors. Two motors for controlling the two legs, four for controlling the two arms, and one for controlling the torso. Each leg motor controls two joints, one in the hip and one in the knee. The torso motor can move the robot laterally. The shoulder motor can raise the arm up and down, while the elbow motor can twist the forearm, while opening or closing the hand grippers. These motors are shown in Figure 9.31.

9.3.2 Walking

Robosapien walking is accomplished by using the torso motor for shifting the body center of gravity (C.G.) and the two leg motors for moving forward or backward. The torso motor tilts the body laterally so that the body C.G. falls on the static foot, and then the other leg is moved by that leg's motor. Then the torso motor tilts the body to the other leg, so that the other leg can be moved. This sequence is shown in Figure 9.32.

9.3.3 PC Control

We can have PC control of the robot instead of using the hand remote controller. We can also add a camera that interfaces with a PC and also controls the Robosapien. One example of this would be to have the CMU camera identifying where the robot is and where a ball is with respect to the robot. Then, we can give commands from a PC that uses IR transmitter to make

Biped Robots 383

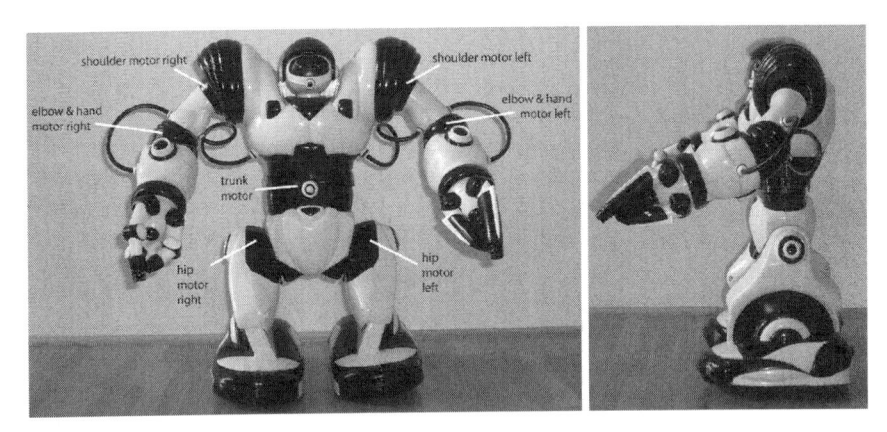

FIGURE 9.31
Motors (with permission from Dr. Sven Behnke [4])

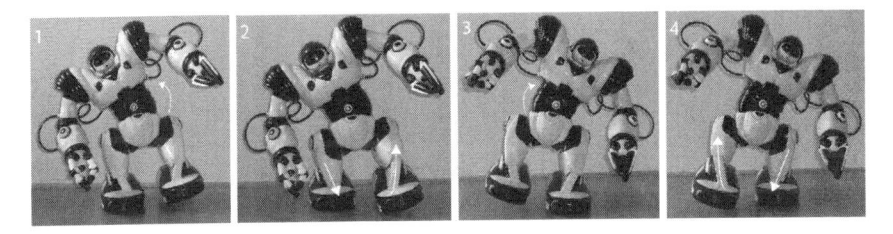

FIGURE 9.32
Robosapien walking gait (with permission from Dr. Sven Behnke [4])

FIGURE 9.33
USB-UIRT (Copyright Jon Rhees - http://www.usbuirt.com

the robot move to the ball and kick it. One PC based IR transmitter we can use is the USB-UIRT. Figure 9.33 shows the USB-UIRT.

The scenario of using the camera for sensing where the robot and another object is, USB-UIRT for sending commands to the Robosapien to walk to the object and pick it up, and a computer for interfacing, is shown in Figure 9.34.

One can obtain the software development kit for the USB-UIRT so that programs can be written to interface with it. More information on USB-UIRT can be obtained from USB-UIRT web site [34]. One alternative to using USB-UIRT is to building your own. A circuit that could be used to accomplish that is given in Figure 9.35.

The IR communication and programming details for the Robosapien can be obtained from Aibo Hack web site [35].

9.3.4 Autonomous Robosapien

Another way to interface the Robosapien is to mount the camera on its head, interface the camera with a processor and do the IR communication with the robot right there instead of using a separate PC. A similar concept is used where a pocketPC is used as a processor on the robot NimbRo RS from the University of Frieburg, Germany. The Robosapien-API is available for embedded Visual C++ for programming from Dr. Sven Behnke's NimbRo project web site [36]. This robot is shown in Figure 9.36. The NimbRo robot uses the Toshiba E755 PocketPC as the controller. Multiple modified Robosapiens playing soccer are shown in Figure 9.37.

The camera the robot uses is either Pretec CompactCam or Lifeview Flycam. Lifeview Flycam camera is shown in Figure 9.38.

The programming details of the Robosapien-API can be obtained from Dr. Sven Behnke's NimbRo project web site [36].

FIGURE 9.34
Robosapien with PC control (Copyrights WowWee Ltd, Acroname Inc, and Jon Rhees)

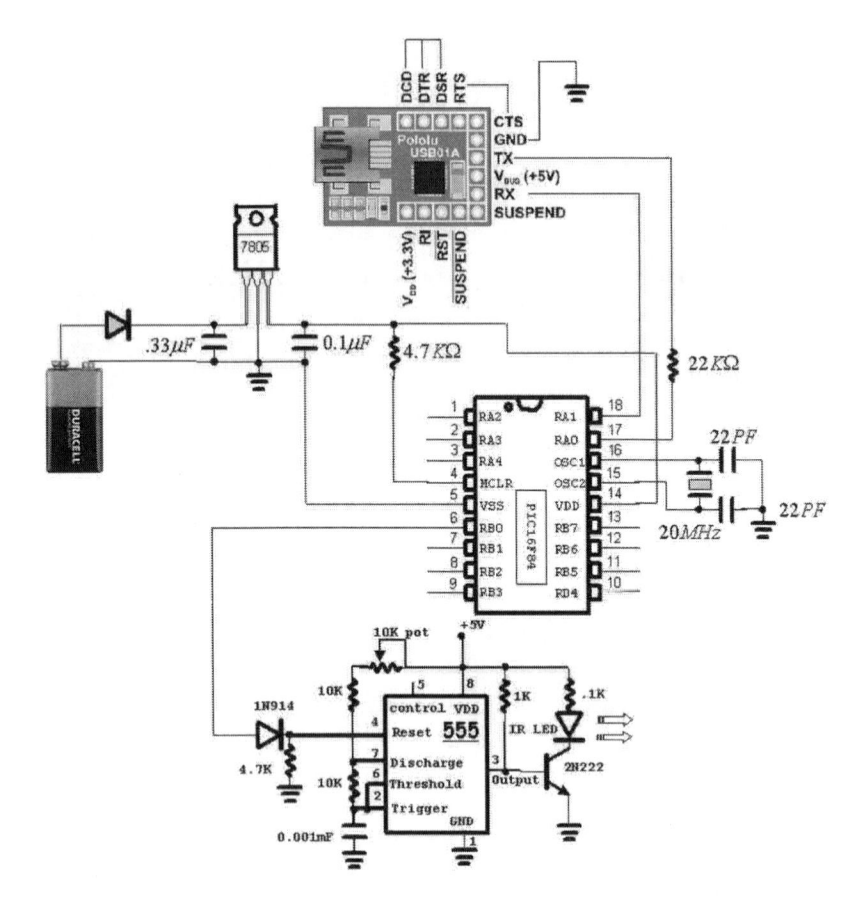

FIGURE 9.35
PC controlled IR transmitter (Copyright Pololu Inc)

FIGURE 9.36
NimbRo RS (with permission from Dr. Sven Behnke [4])

FIGURE 9.37
NimbRo RS playing soccer (with permission from Dr. Sven Behnke [4])

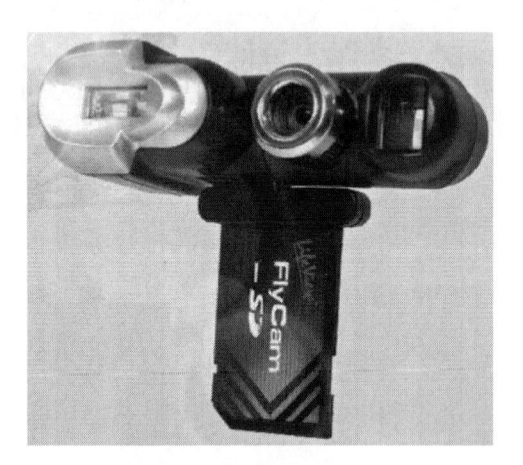

FIGURE 9.38
NimbRo RS camera by Flycam (Copyright 2006 Lifeview Animation Technologies Inc.)

10

Propeller Based Robots

In this chapter we will study robots that we have not studied till now. These are: flying robotic plane, robotic helicopters, robotic boats, and submarines. The technique followed will be essentially taking remote controlled systems (such as RC planes and RC helicopters) and using a microcontroller to control the actuators of the robots instead of the RC receivers doing the control.

10.1 Wings

To understand how the air flowing around a wing section produces lift, let us study Figure 10.1.

Figure 10.1 shows air flowing around a wing section. The streamline (a line whose tangent shows the direction of flow at every point) on the left that ends with an arrow on the wing surface is the one that reaches the stagnation point on the wing. This means that above that the fluid flows on top of the wing and below that point the fluid flows under the wing. The figure shows that the air from top and the bottom will have the same magnitude and direction at the trailing edge. This is a condition (Kutta condition) that creates lift from air flowing around a wing at some angle of attack. If we place a wing in an airflow, we would expect the streamlines to be as shown in Figure 10.2. This airflow, however, has no circulation. Without air circulation, there can be no lift. In fact the lift on a wing is proportional to the air-circulation about

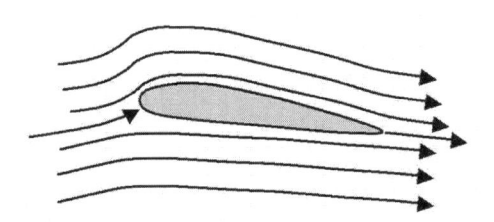

FIGURE 10.1
      Airflow around a wing section (Kutta condition)

FIGURE 10.2

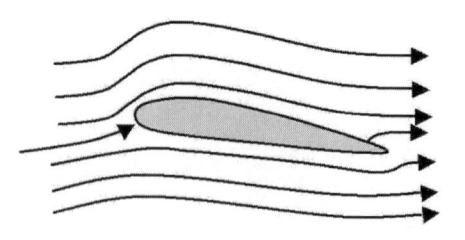

Non-circulatory airflow around a wing section

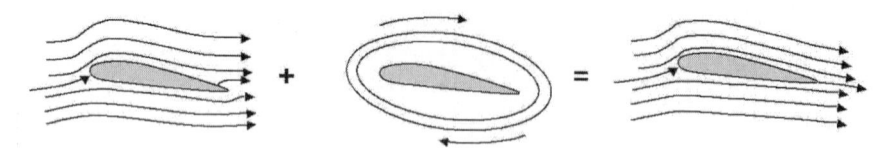

FIGURE 10.3

Total airflow around a wing section

the wing (Kutta-Joukowski theorem, [1] and [3]).

According to the Kutta-Joukowski theorem, the lift per unit length of a wing is given by:

$$Lift = airspeed \times aircirculation \times airdensity \qquad (10.1)$$

The airflow that follows the Kutta condition (the trailing edge boundary condition) can be obtained by superimposing a circulatory airflow on the non-circulatory air-flow. This is shown in Figure 10.3.

There is another theorem (Kelvin's Circulation Theorem, [3]) says that the rate of change of circulation around a closed curve that consists of the same fluid material is zero. When an airplane starts accelerating it does not follow the Kutta condition, and consequently, it does not have circulation. After the speed increases further, the Kutta condition is satisfied (creating circulation around the wings), and to make the total circulation zero, another circulating air is created behind the wing (called the wake). This is shown in Figure 10.4.

Due to the addition of clockwise circulation that is flowing in the direction of the regular airflow, the speed on top of the wing is higher than the airspeed below the wing. Bernoulli's theorem shows that where the airspeed is high, there the pressure is low, and where the airspeed is low, there the air pressure is high [9]. Since the air pressure is higher at the bottom of the wing than the pressure on the top, there is an upward force (lift) produced on the wing. This is shown in Figure 10.5.

Now when we look at a wing with a finite span, because of low pressure on top of the wing, the air from the bottom starts circulating to the top, and the

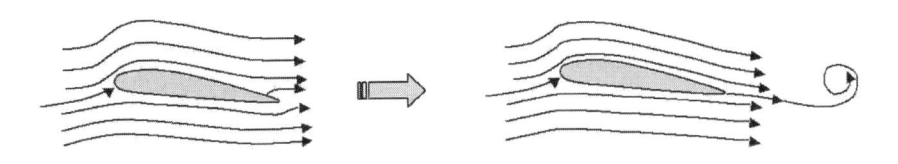

Propeller Based Robots 391

FIGURE 10.4
  Total air circulation around a wing section

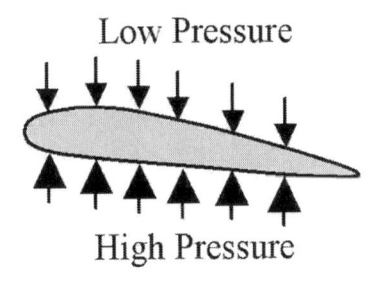

FIGURE 10.5
  Lift due to pressure difference produced on a wing section

whole circulation travels behind the plane. This is shown in Figure 10.6.

The lift force that is generated by the wings is related to the angle of attack (i.e., the angle that the air velocity at a distance in front of the wing makes with the wing chord line). When the angle of attack is at zero degrees, there is no lift produced. As the angle of attack is increased the lift produced increases, up to some angle of attack, after which further increase in the angle decreases the lift. This is shown in Figure 10.7.

10.2 Propellers

Propellers are extremely important for robotic planes, helicopters, boats, and submarines. They provide the thrust for forward motion as well as for lift (as in helicopters). Figure 10.8 shows a three blade propeller. Propellers can come in many blade configurations such as two-blade, three-blade, and four-blade.

Propellers produce forward thrust by accelerating air/fluid back. The amount of fluid forced back depends on the speed of rotation of the propeller as well as at its pitch angle. With a small pitch angle, the amount of air pushed back

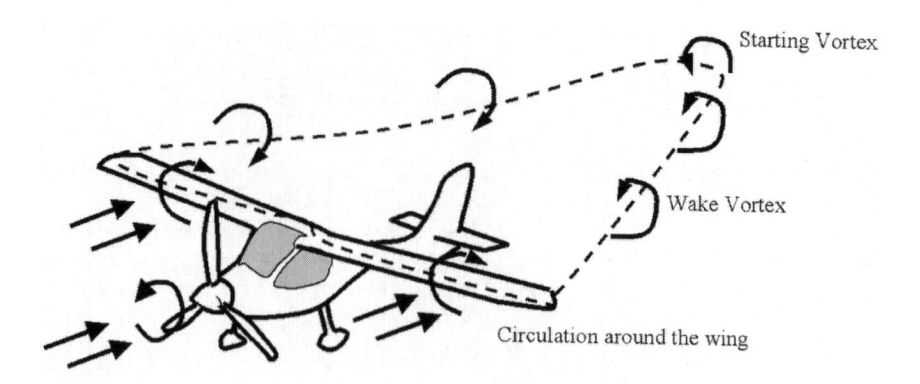

FIGURE 10.6

Vortex in flight

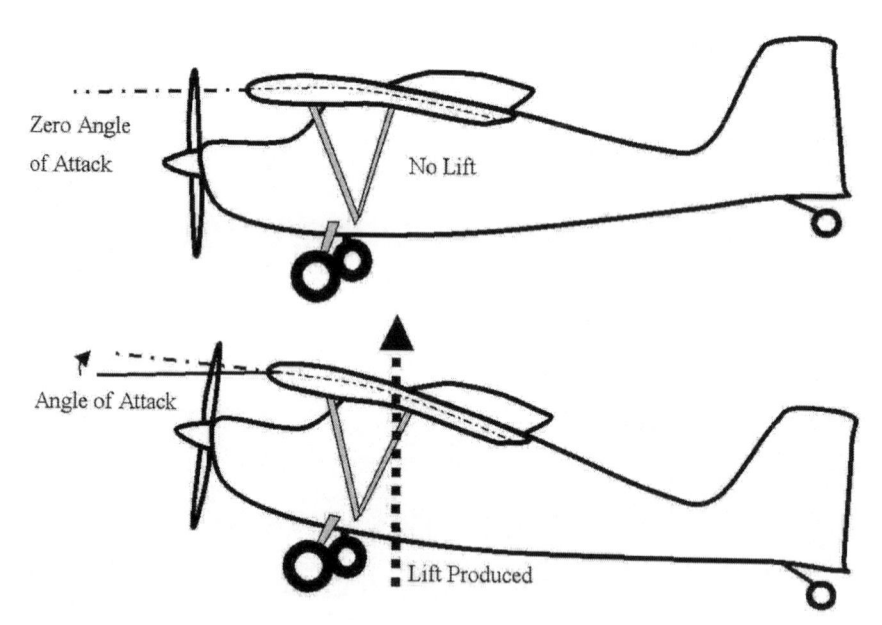

FIGURE 10.7

Angle of attack and lift

FIGURE 10.8

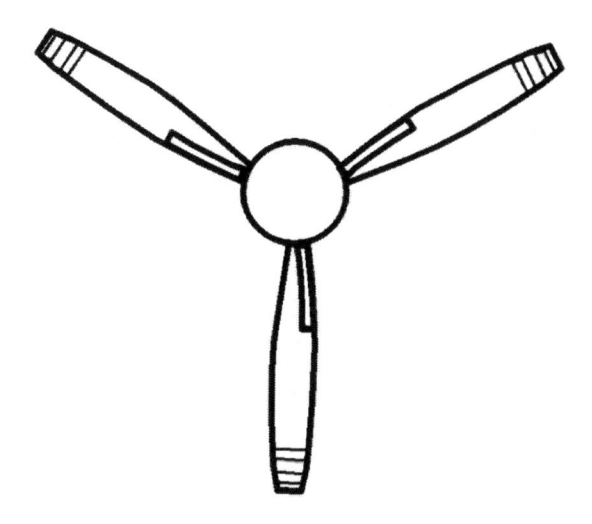

Propeller

is small and, therefore, less forward thrust is produced. However, in that case the propeller can rotate faster. With a high pitch angle, the blades rotate slower but produce more thrust by pushing more fluid back. Propellers can be designed to have fixed pitch, or can be made such that the pitch of the propeller can be changed in real-time (by an actuator). The concept of variable pitch can be compared to having gears in a motor. The pitch could be controlled automatically (using some electronics) to produce a constant speed of rotation.

To obtain quantitative measure of the thrust produced by propellers, we can use Newton's second law to relate force to change in momentum of the fluid that the propeller pushes through. Another way to get a good estimate of the propeller thrust, we can apply the theory of wing section to the propeller blade to figure out the lift force that is generated on the blade as the fluid flows across it. The airspeed that any part of the propeller blade is subjected to depends on the distance of that part from the center of rotation. Therefore, the propeller blades are twisted since at different radii, the relative speeds are different, and that makes the angle of attack different, if the blade is not designed to be twisted.

Let us look at the parts of a propeller in more details as shown in Figure 10.9.

The leading edge of the propeller is where the fluid first cuts through the blade. The trailing edge is where the fluid exits the propeller surface. The blade face is the side of the blade that pushes the fluid away creating a positive pressure on the surface. The blade back is the back side of the blade which gets a negative pressure as the fluid is moved away from that region by the

FIGURE 10.9

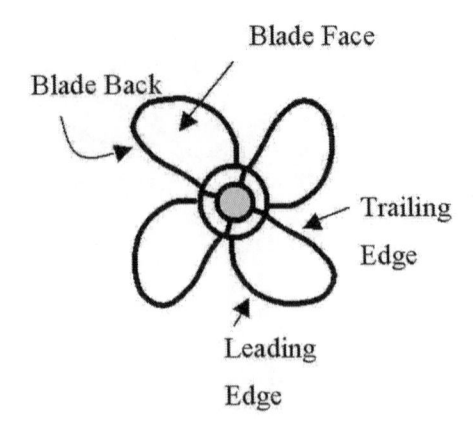

Propeller parts

rotating blade.

As the propeller rotates, the blade face pushes the fluid back and the back side starts pulling the fluid in (push-pull mechanism), similar to a fan, and this creates a forward thrust from the propeller (see Figure 10.10).

As the propeller makes one revolution, the geometric pitch is the distance the propeller "screw" would have moved through a soft material like soft wood if it were a screw. The effective pitch (smaller than the geometric pitch) is the actual distance the plane or a boat (or any other robot) moves. This difference is due to the propeller slippage through the fluid (see Figure 10.11).

10.3 Robotic Planes

Planes have essentially four actuators that can be used to control their motion. These four actuators are: propeller, aileron, elevator, and rudder. These are shown in Figure 10.12.

The forward speed of the plane is controlled by the propeller, the roll movement by the ailerons, the pitch movement by the elevators, and the yaw movement by the rudder. These controls and movements are shown in Figure 10.13.

10.3.1 RC Planes

RC planes can be used for conversion into robotic planes. These can be either gas powered or electric powered. The servos are powered by a battery, and

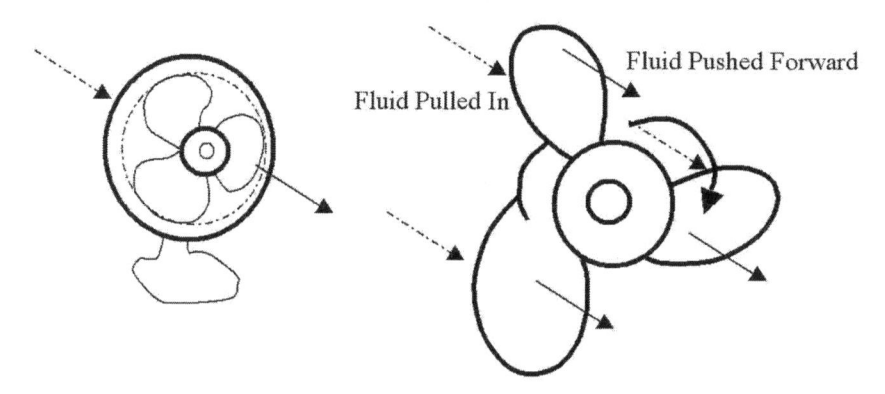

Propeller Based Robots

FIGURE 10.10 Propeller thrust

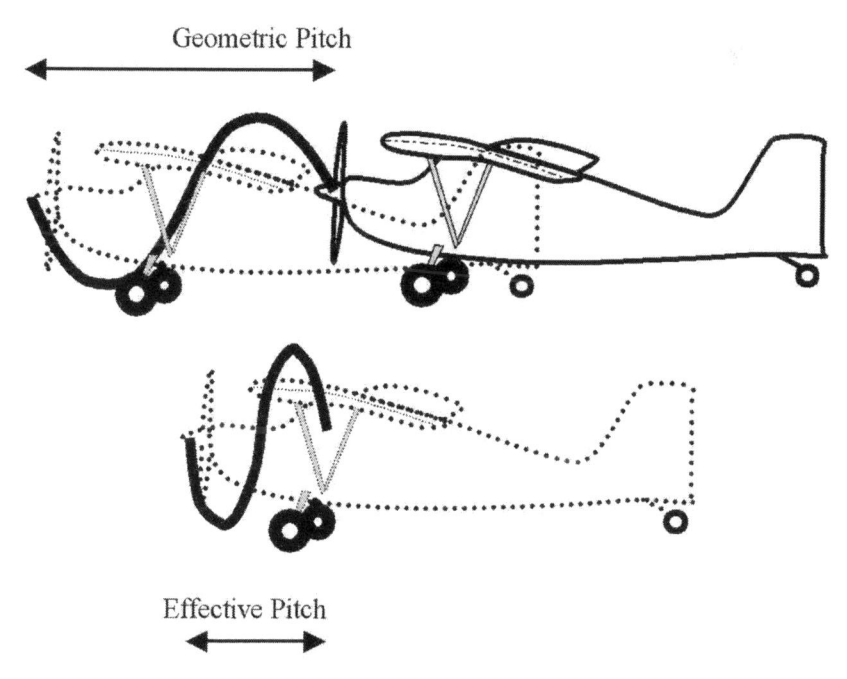

FIGURE 10.11 Propeller pitch

FIGURE 10.12

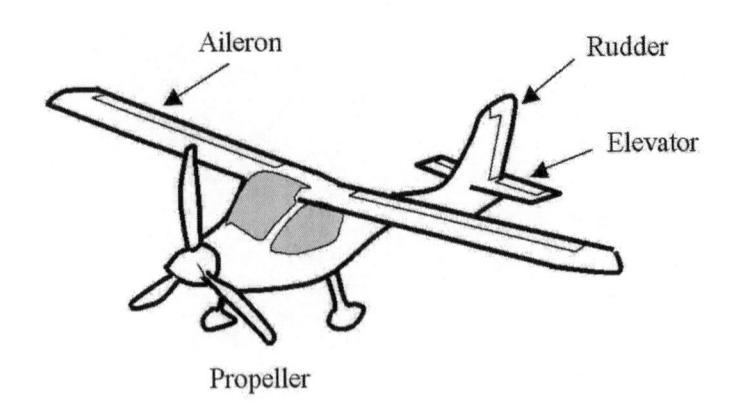

Robotic plane actuators

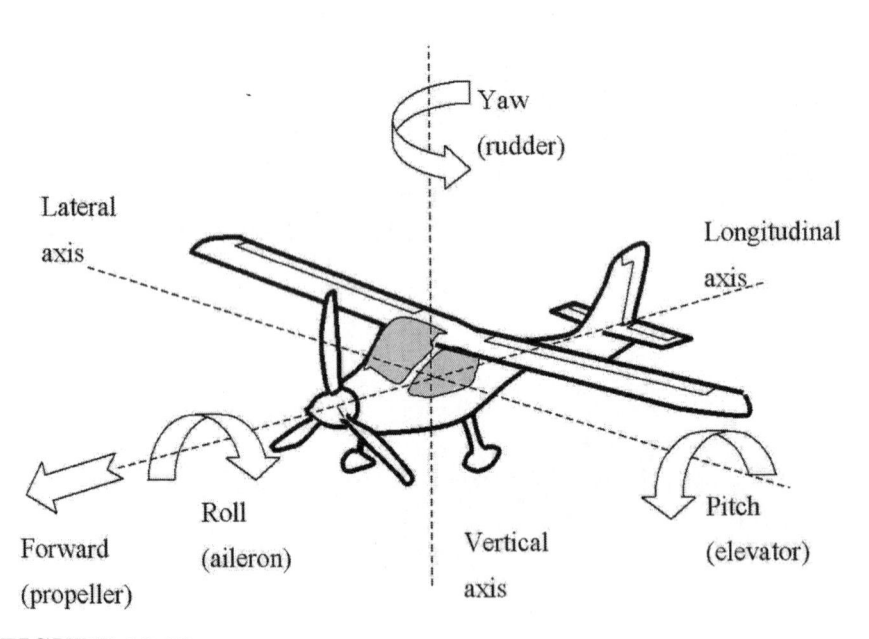

FIGURE 10.13

Robotic plane control

Propeller Based Robots   397

FIGURE 10.14
DC motor powered propeller for SuperstarEP plane with ailerons (Copyright Hobbico Inc. Reproduced with permission)

the propeller can be gas or electric powered.

10.3.1.1 Electric Powered RC Plane

The propeller of the plane can be powered by a DC motor, or by an engine. An example of a DC motor driving a propeller of an RC plane (SuperstarEP with Ailerons by Hobbico Inc.) is shown in Figure 10.14.

10.3.1.2 Gas Powered RC Plane

Figure 10.15 shows an example of a gas powered RC plane. We will be using off-the-shelf controllers and sensors and because of the sizes of those will opt for bigger RC planes.

This model is NexSTAR RTF Nitro .46 radio-controlled model airplane by Hobbico Inc. This has a 68.5" span. It comes with:

- O.S. .46 FXi Engine
- Futaba 4YBF 4 Channel Radio system
- Real Flight R/C Simulator NexSTAR Edition Windows CD-ROM

Let us start with understanding the engine. The engine is shown in Figure 10.16.

FIGURE 10.15
NexSTAR Select gas powered RC plane (Copyright Hobbico Inc. Reproduced with permission)

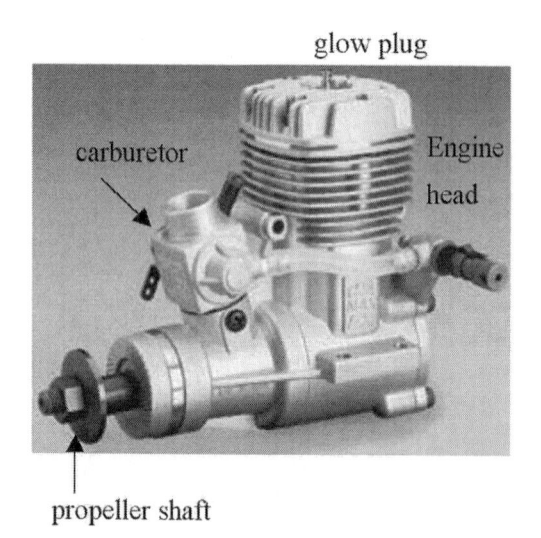

FIGURE 10.16
RC plane engine (Copyright Hobbico Inc. Reproduced with permission)

Propeller Based Robots

FIGURE 10.17

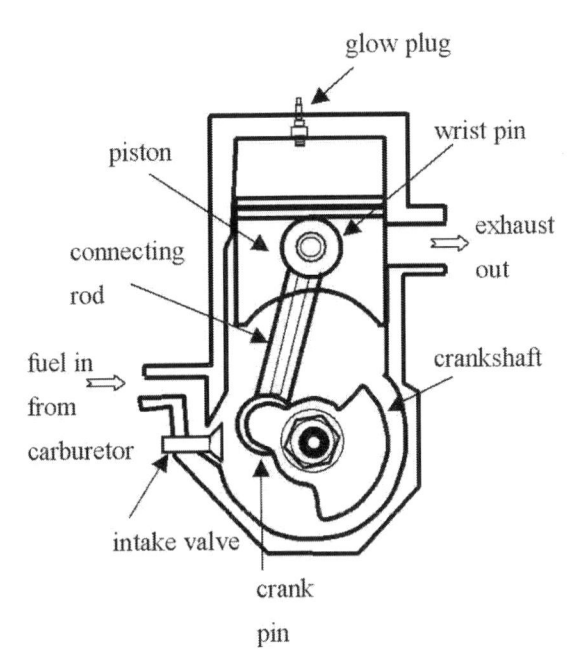

Parts of a two-stroke engine

10.3.1.2.1 Two Stroke Engine The two stroke engine that is used is built on the following specifications.

- Displacement: 0.455 cu in (7.5 cc)

- Bore: 0.866 in (22.0 mm)

- Stroke: 0.772 in (19.6 mm)

- Output: 1.65 hp @ 16,000 rpm

- RPM Range: 2,000-17,000

- Weight w/muffler: 17.2 oz (489 g)

Parts of an engine are shown in Figure 10.17.

Now, let us study how a 2-stroke engine uses fuel to convert the power into rotating the propeller. The engine operation is divided into the following parts.

Compression and Intake: This stage is shown in Figure 10.18. Here, the piston is moving up compressing the air-fuel mixture. At the same time, the intake valve opens as it gets pulled out by the expanding air.

Power: This stage is shown in Figure 10.19. Here, the piston has reached the Top Dead Center (TDC) point on the cylinder. The air-fuel mixture is

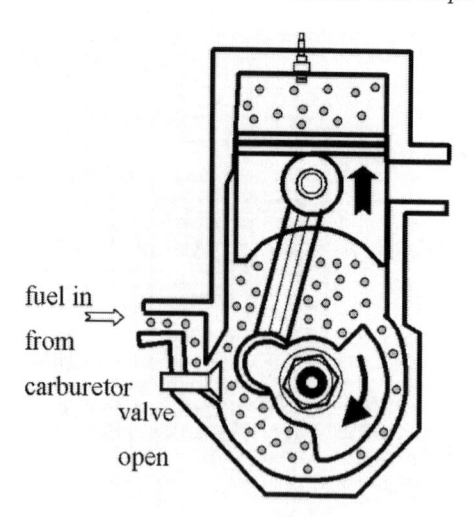

FIGURE 10.18

Compression and intake

the most compressed. At this time, the fuel mixture gets ignited by the glow plug. The explosion starts pushing the piston down, compressing the air-fuel mixture below, which in turn pushes the valve shut.

Exhaust: This stage is shown in Figure 10.20. Here, the piston has reached the Bottom Dead Center (BDC) point on the cylinder. The valve is still shut, and the air fuel mixture is going to the top chamber. At the same time, the exhaust gases are escaping through the exhaust manifold. Right after this stage the compression and intake stage will start again.

Notice that the crankshaft keeps rotating throughout all the stages, and since the crankshaft is connected to the propeller, it causes its rotation.

10.3.1.3 Servo Control

An airplane can be controlled by controlling the speed of the propeller, and by controlling the elevator, rudder, and the aileron. For the propeller in a gas engine based plane, we can use the throttle to control it using a servo motor. The throttle controls the amount of air-fuel mixture that goes in the intake manifold for combustion, and therefore controls the speed of the propeller. For a DC motor based plane, we can use a speed controller. Servo motors are used to control the elevator, rudder, and the aileron. The servos for throttle, elevator, and rudder are shown in Figure 10.21, and the servo for aileron is shown in Figure 10.22.

Figure 10.23 shows how the pushrods connected to servos control the rudder and elevator.

Figure 10.24 shows how the aileron are controlled in the opposite direction by the aileron servo to provide roll control.

Propeller Based Robots 401

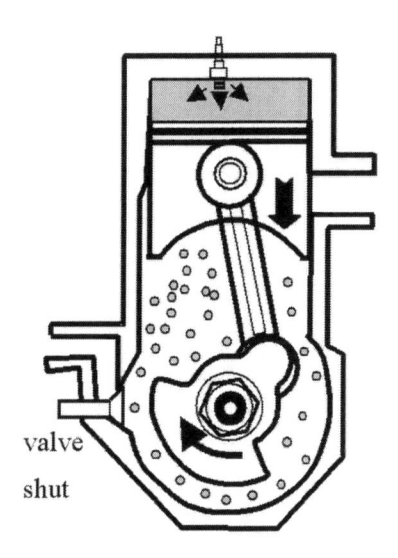

FIGURE 10.19

Power

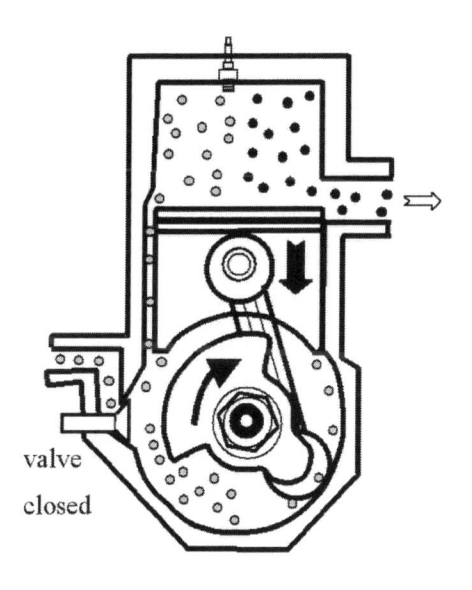

FIGURE 10.20

Exhaust

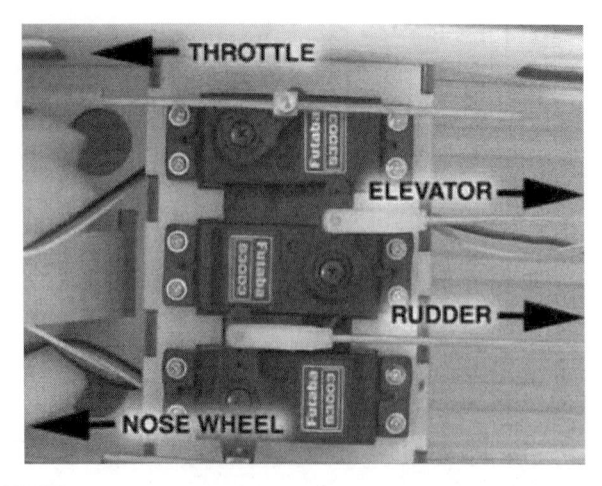

FIGURE 10.21
Servos for throttle, elevator, and rudder (Copyright Hobbico Inc. Reproduced with permission)

FIGURE 10.22
Servos for aileron (Copyright Hobbico Inc. Reproduced with permission)

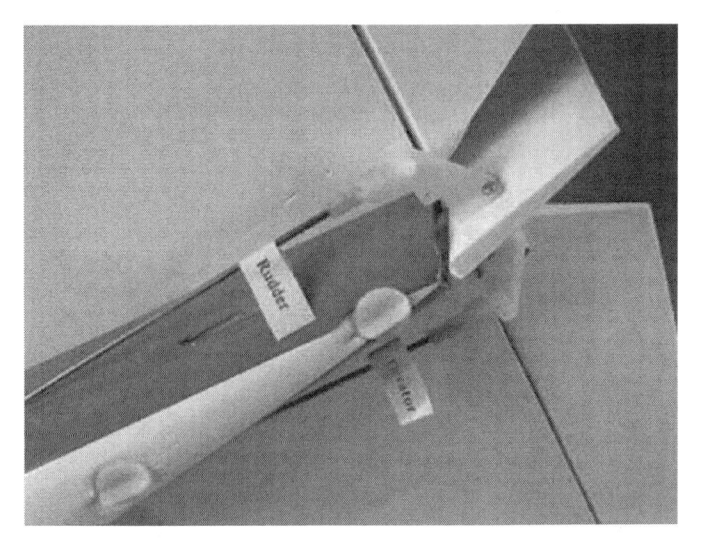

Propeller Based Robots

FIGURE 10.23
Pushrods from servos connected to the rudder and elevator (Copyright Hobbico Inc. Reproduced with permission)

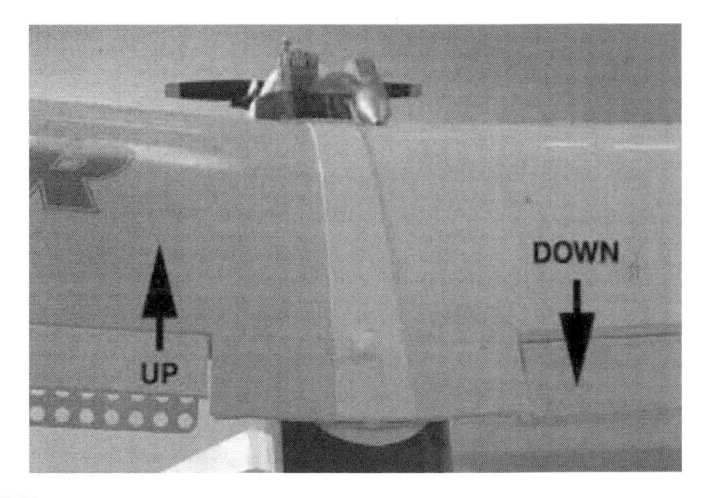

FIGURE 10.24
Aileron servo control (Copyright Hobbico Inc. Reproduced with permission)

FIGURE 10.25
Manual control (Copyright Hobbico Inc. Reproduced with permission)

10.3.2 Manual Control for RC Planes

Manual control of RC planes is accomplished by using a hand held controller, where different levers provide control over the various servo movements. The manual handheld controller (transmitter) for the NexSTAR RTF Nitro plane is shown in Figure 10.25.

10.3.3 Automatic Controllers for RC Planes

There are some off the shelf controllers available for RC planes that can be integrated with sensors to make the planes perform different tasks. For example, the planes can be programmed to go to certain points, or follow some trajectory etc. using the controllers. One such controller is the MP2028 by MicroPilot Inc., and another one is AP50 Autopilot made by UAV Flight Systems, Inc.

10.3.3.1 MP2028 Controller for RC Planes

MicroPilot is the world leading manufacturer of miniature autopilots for unmanned aerial vehicles (UAV) and micro aerial vehicles (MAV). Weighing 28 grams, the MP2028g is the world's smallest full functional autopilot. Capabilities include airspeed hold, altitude hold, turn coordination, GPS navigation, plus autonomous operation from launch to recovery. Included with the MP2028g autopilot is the HORIZONmp ground control software that offers a user friendly point and click interface for mission planning, parameter adjustment, flight monitoring, and mission simulation.

For versatility, MicroPilot has developed a full complement of autopilot accessories including a Compass Module, Analog to Digital Converter Module, Ultrasonic Altitude Sensor, Configuration Wizard, and XTENDERmp Software Developers Kit. For example, XTENDERmp gives the systems integrator the ability to fully customize the MP2028g autopilot and the HORIZONmp ground control software to match their end user's requirements. Using the XTENDERmp it is possible to implement alternate control laws, collect and display data from custom sensors, and cameras.

For dependability, MicroPilot offers a comprehensive selection of training, integration, flight testing, along with custom software and hardware development services. As an added convenience, clients can choose to have a MicroPilot expert come to their site or they can send a person or team to the MicroPilot UAV Facility. Situated on 40 acres of flat prairie, the MicroPilot UAV Facility includes head office, production, R&D, and flight testing all at one site.

The MP2028g features the following:

- 1,000 programmable waypoints or commands

- Powerful command set allowing flexibility when describing your mission

- Sensors required for complete airframe stabilization fully integrated into a single circuit board

- Controls up to 24 servos or relays

- Complete autonomous operation from launch to recovery

- Autonomous launch methods including runway takeoff, hand launch, bungee launch, and catapult launch

- Autonomous recovery methods including runway landing, parachute recovery, and deep stall landing

- Supports manually directed and autonomous flight modes, as well as an integrated RC override

- Supports flaps, flaperons, split rudders, split ailerons, flap/aileron mixing, elevons, v-tail, and x-tail

- Extensive user programmable feedback gains and flight parameters allowing tailoring of the MP2028g to your airframe and requirements

- Extensive data log capability simplifying post flight diagnostics and analysis

- Integrated POST ensures reliability and repeatability

- Low battery warnings, both on the ground and in flight

- User-programmable error handlers for loss of GPS signal, loss of RC signal, engine failure, loss of data link, and low battery voltage

- Extremely low 28 g weight suitable for micro UAVs

- Includes HORIZONmp ground control software

- Feedback loop gains adjustable while in flight

- Easy to use setup wizard to simplify system integration

Block diagram of MP Micro Pilot is shown in Figure 10.26. MP 2028 interfaces with RC receiver, RF modem, ultrasonic altimeter, servomotor multiplexers, and on-board GPS unit. On MP 2028, there are sensors to detect pressure, pitch rate, roll rate, yaw rate, and airspeed. Through RC receiver, the MP 2028 can control an aircraft using RC joystick. This makes the system suitable for manual landing and takeoff. MP 2028 can also be programmed to control the aircraft without any user interaction through its programming interface.

Some sensors on the board are shown in Figure 10.27.

10.3.3.2 AP50 Controller for RC Planes

The UAV Flight Systems website, [39], provides the detailed information on the AP50 controller. The AP50 has two main functions: navigation and flight stability control. Mission-related tasks such as controlling three mission servos, three TTL-level digital outputs, engine shutdown, and data logging are handled by the navigation function. The stability control is managed by a 6-state user-defined control law table for each flight control. This control law is based on a modified proportional integral derivative (PID) control algorithm.

The block diagram of the AP50 controller is shown in Figure 10.28, and picture in Figure 10.29.

10.3.4 Kinematics

In order to design algorithms for automatic control for robotic plane flights, we need to come up with the mathematical model of the planes. We can use models that utilize the mass, moment of inertia, forces etc. in the model. These models are based on the dynamics of the system. However, they tend to be more complex to deal with for control law design. We can use kinematics* based models for trajectory control if we can provide good speed control of the planes. This section presents the kinematics model of the plane. It assumes that we can control the linear speed (using the propeller), and the three angular rotations of the plane using servos (aileron, elevator, and rudder). Using

---

*The "kinematics" section should be omitted by readers who have not studied advanced robotics or mathematics courses

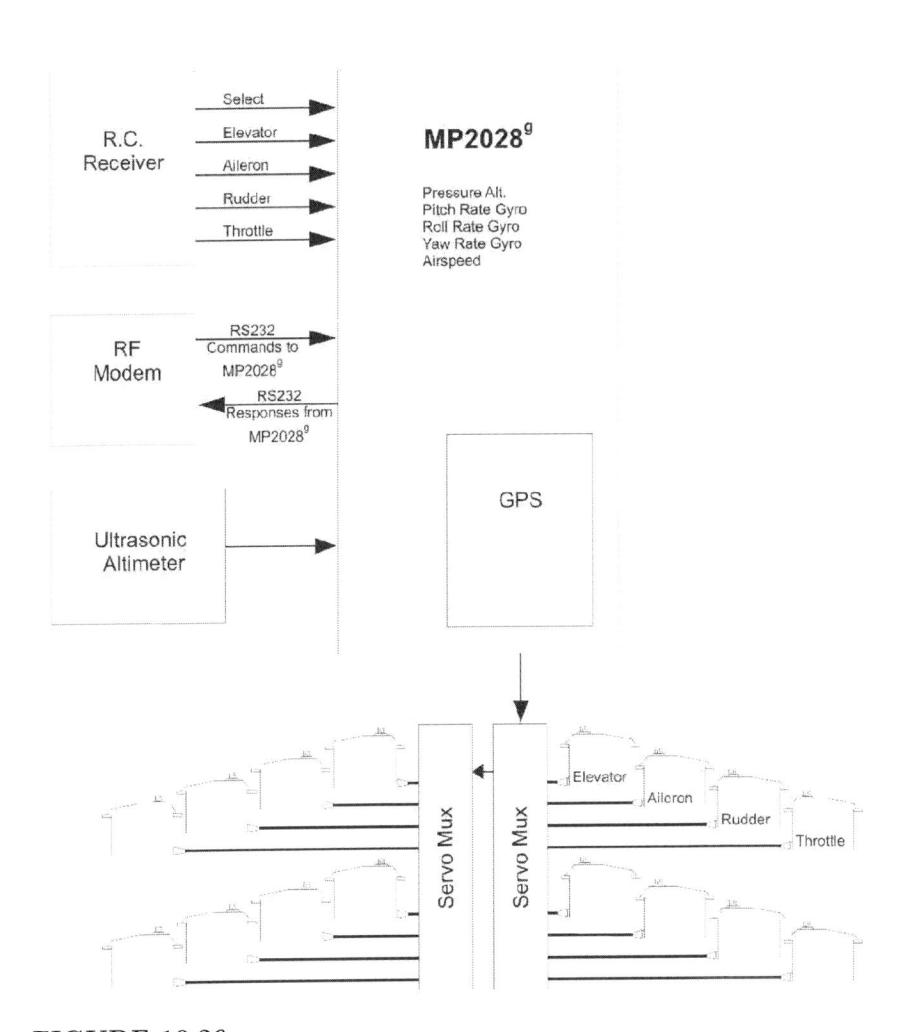

Propeller Based Robots

FIGURE 10.26
Block diagram of MP2028 (Copyright 2006 MicroPilot Inc.)

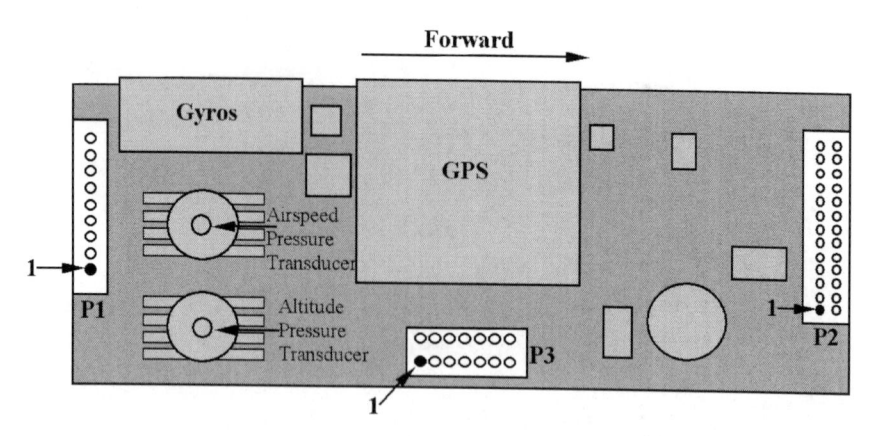

FIGURE 10.27
Board sensors (Copyright 2006 MicroPilot Inc.)

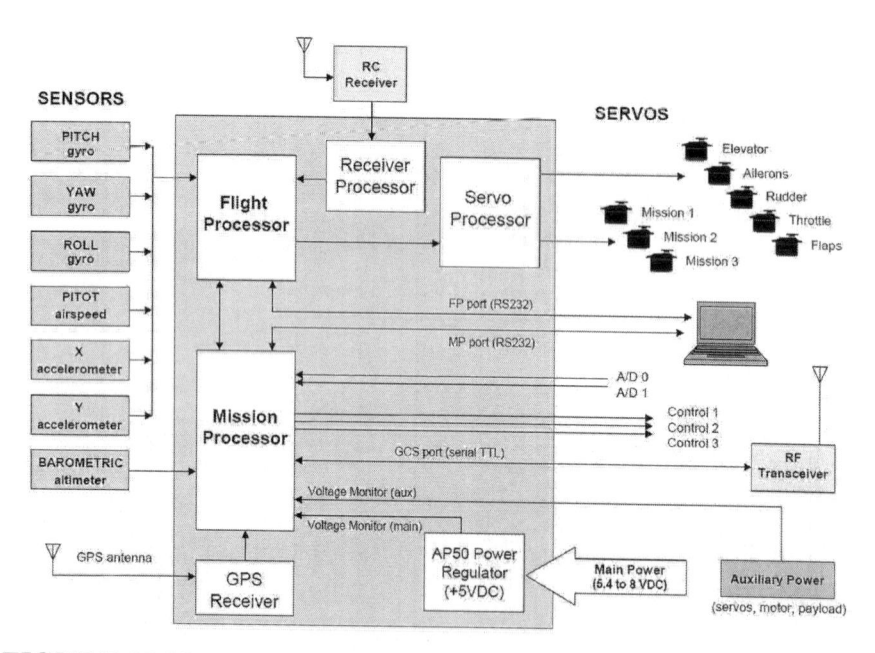

FIGURE 10.28
AP50 block diagram (Copyright 2006 UAV Flight Systems, Incorporated)

FIGURE 10.29
AP50 board (Copyright 2006 UAV Flight Systems, Incorporated)

these four controls, we need to control the path of the robotic plane. This section developed the kinematics model that shows how the global position and orientation of the plane change with the four controls. This section does not develop the control laws. However, the kinematics based control laws can be designed by using the control laws developed in [18].

Consider Figure 10.30 for developing the kinematics equations.

Global coordinates: The global or the inertial frame coordinates are denoted by (P, X, Y, Z). The frame remains fixed at the ocean surface with origin P. The unit vector in the Z direction points down into the water while the unit vectors along X and Y direction complete the right handed system.

Local coordinates: The local or the body frame coordinates are denoted by (p, x, y, z). The frame remains fixed on the vehicle with origin surface with origin p.

The kinematics of the vehicle is described by six state variables and four input variables. The kinematics relationships describing the transformations between the two coordinate systems can have a number of parameterizations. The one used here is the Euler angle parameterization. In the Euler angle representation the orientation between the inertial and the local coordinate frame is expressed in terms of a sequence of three rotations: roll(ϕ), pitch(θ), and yaw(ψ) about the axes x, y, and z respectively.

Let q be the vector of six generalized coordinates required to specify the kinematics of the vehicle. The six coordinates are the Cartesian coordinate vector $p = [x, y, z]^T$ of the vehicle in the local frame and the orientation coordinate vector $\eta = [\phi, \theta, \psi]^T$. The orientation vector is the vector of Euler angles which give the orientation of the body frame with respect to the inertial frame. The transformation from the local coordinate frame to the global

FIGURE 10.30

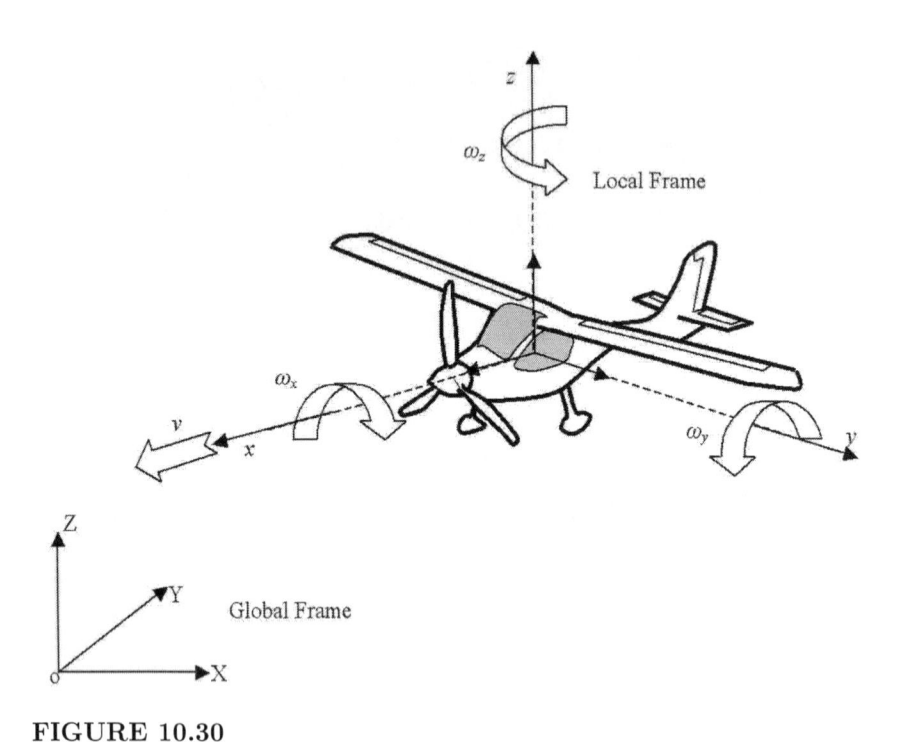

Global and local frames

Propeller Based Robots

coordinate frame is given by means of a transformation matrix called Rotation matrix $R \in S(O3)$. R satisfies the relation $R^T R = I$, i.e., $R^T = R^{-1}$ or R is an orthogonal matrix and $det(R) = 1$. The matrix R is given below as

$$R = \begin{bmatrix} r_{11} & r_{12} & r_{13} \\ r_{21} & r_{22} & r_{23} \\ r_{31} & r_{32} & r_{33} \end{bmatrix} = \begin{bmatrix} n \\ s \\ a \end{bmatrix} \quad (10.2)$$

with

$$\begin{aligned}
r_{11} &= \cos\theta \cos\psi \\
r_{12} &= \sin\theta \sin\phi \cos\psi - \cos\phi \sin\psi \\
r_{21} &= \cos\theta \sin\psi \\
r_{22} &= \sin\theta \sin\phi \sin\psi + \cos\phi \cos\psi \\
r_{23} &= \sin\theta \cos\phi \sin\psi - \sin\phi \cos\psi \\
r_{31} &= -\sin\theta \\
r_{32} &= \cos\theta \sin\phi \\
r_{32} &= \cos\theta \cos\phi
\end{aligned}$$

Let $v = [v_x, 0, 0]^T$ be the linear velocity of the vehicle, i.e., the vehicle has linear velocity along the x-axis only and $\omega = [\omega_x, \omega_y, \omega_z]^T$ be the angular velocity components along x, y, and z directions respectively in the body frame. The velocity vector along three coordinate axes and the time derivative of the Euler angles are obtained from the following relations:

$$\dot{p} = Rv = \begin{bmatrix} n \\ s \\ a \end{bmatrix} v \quad (10.3)$$

$$\dot{R} = RS(\omega) \quad (10.4)$$

where $S(\omega)$ is the skew-symmetric matrix given as:

$$S(\omega) = \begin{bmatrix} 0 & -\omega_z & \omega_y \\ \omega_z & 0 & -\omega_x \\ -\omega_y & \omega_x & 0 \end{bmatrix} \quad (10.5)$$

The above equations give the following equations on solving:

$$\dot{p} = J_1(\eta)v \dot{\eta} = J_2(\eta)\omega \quad (10.6)$$

with

$$J_1(\eta) = \begin{bmatrix} \cos\theta\cos\psi & \cos\theta\sin\psi & -\sin\theta \end{bmatrix}^T$$

$$J_2(\eta) = \begin{bmatrix} 1 & \sin\phi\tan\theta & \cos\phi\tan\theta \\ 0 & \cos\phi & -\sin\phi \\ 0 & \sin\phi\sec\theta & -\cos\phi\sec\theta \end{bmatrix}$$

The above set of equations can be written as the following equations.

$$\dot{x} = r_{11}v = \cos\phi\cos\theta v \quad (10.7)$$
$$\dot{y} = r_{21}v = \sin\phi\cos\theta v \quad (10.8)$$
$$\dot{z} = r_{31}v = -\sin\theta v \quad (10.9)$$
$$\dot{\phi} = \omega_x + \sin\phi\tan\theta\omega_y + \cos\phi\tan\theta\omega_z \quad (10.10)$$
$$\dot{\theta} = \cos\phi\omega_y - \sin\phi\omega_z \quad (10.11)$$
$$\dot{\psi} = \sin\phi\sec\theta\omega_y + \cos\phi\sec\theta\omega_z \quad (10.12)$$

This can be written in the matrix form as:

$$\begin{bmatrix} \dot{x} \\ \dot{y} \\ \dot{z} \\ \dot{\phi} \\ \dot{\theta} \\ \dot{\psi} \end{bmatrix} = \begin{bmatrix} \cos\theta\cos\psi & 0 & 0 & 0 \\ \cos\theta\sin\psi & 0 & 0 & 0 \\ -\sin\theta & 0 & 0 & 0 \\ 0 & 1 & \sin\phi\tan\theta & \cos\phi\tan\theta \\ 0 & 0 & \cos\phi & -\sin\phi \\ 0 & 0 & \sin\phi\sec\theta & \cos\phi\sec\theta \end{bmatrix} \begin{bmatrix} v \\ \omega_x \\ \omega_y \\ \omega_z \end{bmatrix} \quad (10.13)$$

The equations can be written in the generalized vector form as

$$\begin{bmatrix} \dot{x} \\ \dot{y} \\ \dot{z} \\ \dot{\phi} \\ \dot{\theta} \\ \dot{\psi} \end{bmatrix} = \begin{bmatrix} \cos\theta\cos\psi \\ \cos\theta\sin\psi \\ -\sin\theta \\ 0 \\ 0 \\ 0 \end{bmatrix} v + \begin{bmatrix} 0 \\ 0 \\ 0 \\ 1 \\ 0 \\ 0 \end{bmatrix} \omega_x + \begin{bmatrix} 0 \\ 0 \\ 0 \\ \sin\phi\tan\phi \\ \cos\phi \\ \sin\phi\sec\theta \end{bmatrix} \omega_y + \begin{bmatrix} 0 \\ 0 \\ 0 \\ \cos\phi\tan\phi \\ -\sin\phi \\ \cos\phi\sec\theta \end{bmatrix} \omega_z$$

(10.14)

The system here is subject to two nonholonomic constraints. The constraints are on the linear velocities along y and z directions. The velocities along these directions are zero. The two constraints are

$$s^T \dot{p} = 0 \quad (10.15)$$
$$a^T \dot{p} = 0 \quad (10.16)$$

These can also be written as:

Propeller Based Robots

$$r_{12}\dot{x} + r_{22}\dot{y} + r_{32}\dot{z} = 0 \tag{10.17}$$
$$r_{13}\dot{x} + r_{23}\dot{y} + r_{33}\dot{z} = 0 \tag{10.18}$$

or as:

$$(\cos\psi \sin\theta \sin\phi - \sin\psi \cos\phi)\dot{x} + $$
$$(\sin\theta \sin\phi \sin\psi + \cos\phi \cos\psi)\dot{y} + (\cos\theta \sin\phi)\dot{z} = 0 \tag{10.19}$$

$$(\sin\psi \sin\theta \cos\phi - \sin\psi \sin\phi)\dot{x} + $$
$$(\sin\theta \sin\phi \cos\psi - \cos\phi \cos\psi)\dot{y} + (\cos\theta \cos\phi)\dot{z} = 0 \tag{10.20}$$

The above equations are of the form:

$$A(q)\dot{q} = 0 \tag{10.21}$$

with

$$A(q) = \begin{bmatrix} r_{12} & r_{22} & r_{32} & 0 & 0 & 0 \\ r_{13} & r_{23} & r_{33} & 0 & 0 & 0 \end{bmatrix} \tag{10.22}$$

Expressing the feasible velocities as the linear combination of vector fields $g_1(q)$, $g_2(q)$, $g_3(q)$, and $g_4(q)$ spanning the null space of matrix $A(q)$ we have the following kinematic model

$$\dot{q} = g_1(q)v_1 + g_2(q)v_2 + g_3(q)v_3 + g_4(q)v_4 \tag{10.23}$$

which in the matrix notation is:

$$\dot{q} = \begin{bmatrix} g_1(q) & g_2(q) & g_3(q) & g_4(q) \end{bmatrix} \begin{bmatrix} v_1 \\ v_2 \\ v_3 \\ v_4 \end{bmatrix} \tag{10.24}$$

where

$$g_1(q) = \begin{bmatrix} \cos\theta \cos\psi \\ \cos\theta \sin\psi \\ -\sin\theta \\ 0 \\ 0 \\ 0 \end{bmatrix} \tag{10.25}$$

$$g_2(q) = \begin{bmatrix} 0 \\ 0 \\ 0 \\ 1 \\ 0 \\ 0 \end{bmatrix} \tag{10.26}$$

$$g_3(q) = \begin{bmatrix} 0 \\ 0 \\ 0 \\ \sin\phi \tan\theta \\ \cos\phi \\ \sin\phi \sec\theta \end{bmatrix} \quad (10.27)$$

$$g_4(q) = \begin{bmatrix} 0 \\ 0 \\ 0 \\ \cos\phi \tan\theta \\ -\sin\phi \\ \cos\phi \sec\theta \end{bmatrix} \quad (10.28)$$

and $v_1 = v = v_x$; $v_2 = \omega_x$; $v_3 = \omega_y$; and $v_4 = \omega_z$. More generally, we can write

$$\dot{q} = G(q)v \quad (10.29)$$

The above equations are the kinematics model of the system. The system is nonlinear and under actuated, which means that the number of inputs to the system is less than its states. The generalized velocity vector $\dot{q}$ cannot assume any independent value unless it satisfies the *nonholonomic constraints*. The constraints are the examples of the Pfaffian Constraints which are linear in velocities. The admissible generalized velocities are contained in the null space of the constraint matrix $A(a)$.[†]

10.3.5 Robotic Experiments

Some experiments that can be performed using the robotic planes can include many of the following:

1. Add a wireless camera to the plane for surveillance.

2. Follow a target on the ground or another flying target (another plane).

3. Carry some load (we can design a servo based gripper for that) and drop it at some known location.

4. Perform a formation flight of multiple planes.

[†]Nonholonomic constraints are those constraints on velocity that do not end up constraining the places the robot can go to. For example, the robotic plane can move only forward (and of course can have three rotations). It can not move sideways. However, it can make maneuvers to reach a point that is adjacent to it by making flying acrobatic loops. The flying robot has no constraints on where it can fly to although it has constraints on its velocities. This means the constraints on the velocities can not be integrated to get constraints on positions (since it has no position constraints). In other words, the robotic plane is "controllable."

Propeller Based Robots

FIGURE 10.31

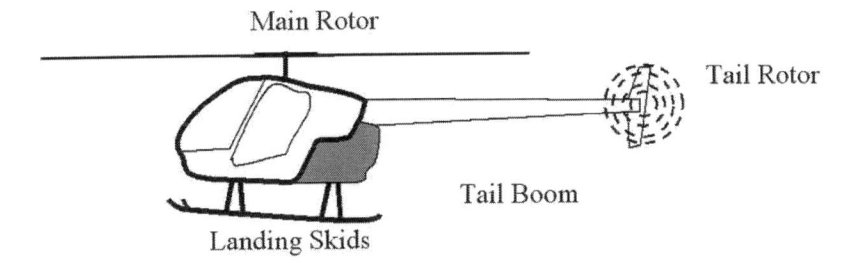

Helicopter parts

10.4 Robotic Helicopter

Helicopters are more versatile than planes because they can do maneuvers that planes cannot do, such as hovering at a place, fly backwards, rotate at a place, and move sideways. These are accomplished by some added complexity in the design. The main parts of a helicopter are shown in Figure 10.31: main rotor, tail rotor, landing skids, and the tail boom. The main rotor system and the tail rotor system provide the mechanisms for all movements of the helicopter.

10.4.1 Controlling Movements

There are many movements that can be accomplished by a helicopter. Let us understand them one by one. To get an idea of overall movements, refer to Figure 10.32.

10.4.1.1 Yaw Control

When the main rotor rotates, it creates a torque on the entire helicopter. To prevent the helicopter from spinning due to that torque, the tail rotor is used that provides a counter torque to balance the rotating torque. If we need to create a yaw motion of the helicopter, we can change the thrust from the tail rotor. This is shown in Figure 10.33 below. A gyroscope is used to sense the rotation, so that it can be balanced by counter-torque. Without a gyro, the RC helicopter might spin out of control.

The helicopter tail rotor can get its power from the main rotor power (gas engine or a DC motor), or can have its own power (usually a DC motor). If the power is obtained from the main rotor, then the change in the rotor thrust is accomplished by rotating the rotor blades so that the angle of attack on the tail rotor blades is changed. The change in the blade angles is shown in Figure 10.34. The servo that accomplishes this is called rudder servo.

416   *Practical and Experimental Robotics*

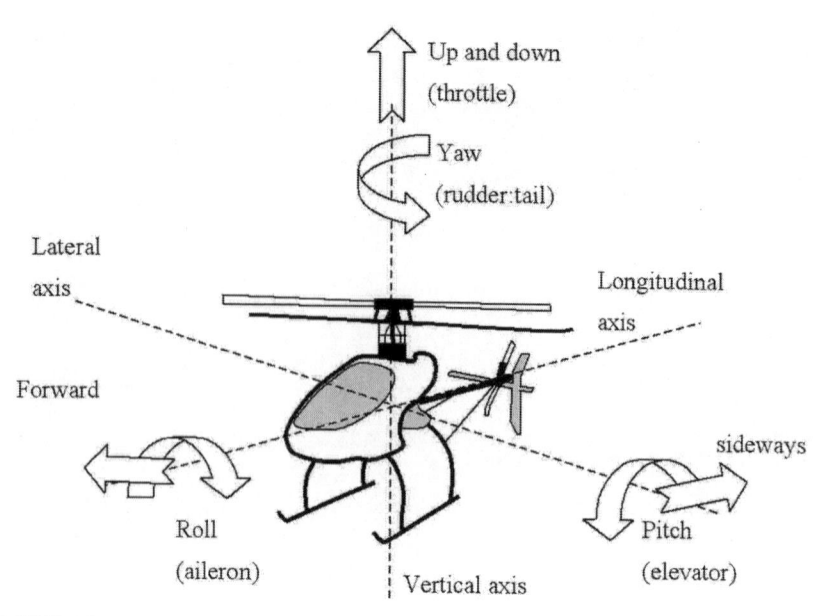

FIGURE 10.32

Helicopter movements

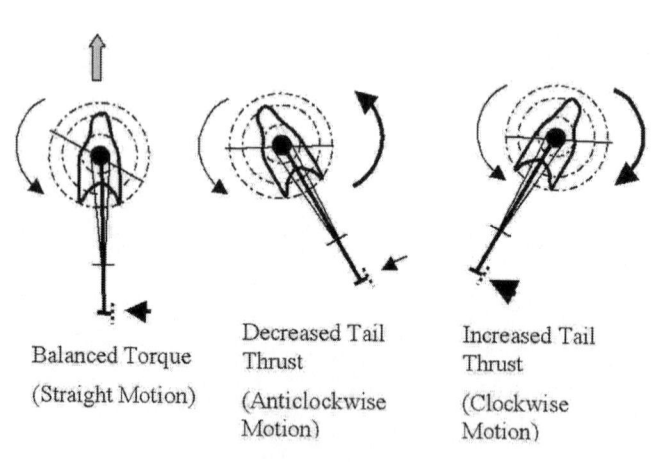

FIGURE 10.33

Yaw control

Propeller Based Robots 417

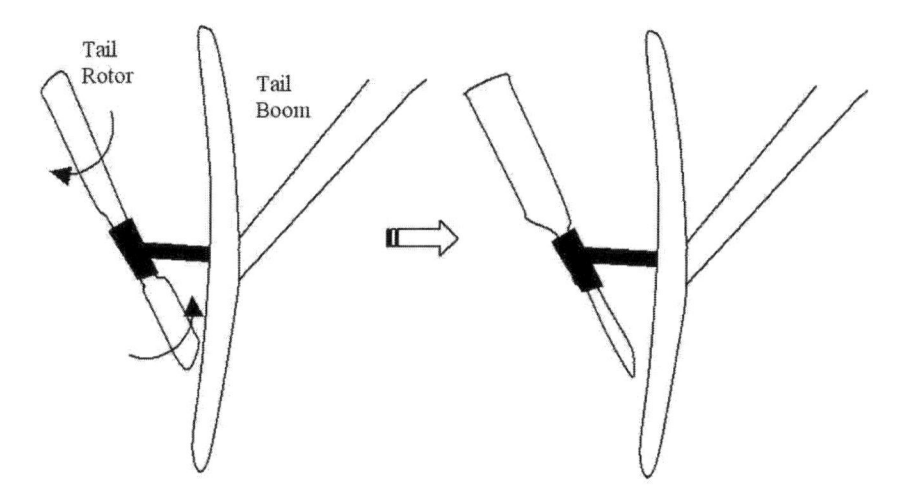

FIGURE 10.34
Rotor blade angle of attack change

An example of a rudder servo changing the angle of attack for rotor blades is shown in the Hummingbird Elite RC helicopter in Figure 10.35.

An example of a DC motor based independent tail rotor is a Helimax Rotofly Submicro Helicopter. The rotor motor can be seen in Figure 10.36. The figure shows the internal and external view of the helicopter [28].

The components of this helicopter are:

- FLYBAR - Adds stability to the rotor head to increase steadiness in flight.

- MICRO SERVOS

FIGURE 10.35
Rudder servo for rotor blades in Hummingbird Elite RC helicopter (Copyright 2006 Century Helicopter Products)

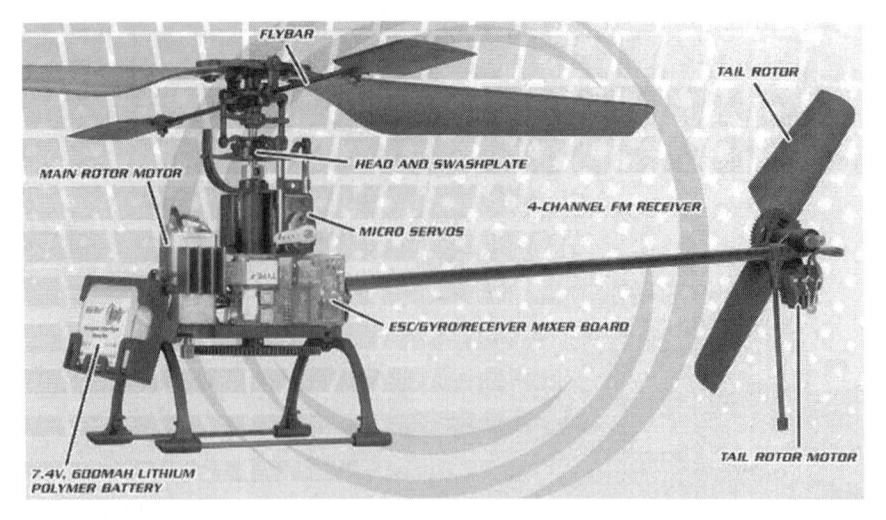

FIGURE 10.36
Helimax Rotofly Submicro helicopter (Copyright 2006 Hobbico Inc. Reproduced with permission)

- 4-CHANNEL FM RECEIVER

- TAIL ROTOR MOTOR - An N20 motor powers the tail rotor blades.

- TAIL ROTOR - With skid to protect tail rotor blades.

- ESC/GYRO MIXER BOARD - Space and weight saving one-piece unit.

- MAIN ROTOR MOTOR - A 180-size motor powers the main rotor blades

10.4.1.1.1 Swash Plate (Pitch and Roll) Swash plate is a mechanism that allows using the rotor blades in such a way that we can get different rotations. We will present a very simple swash plate here that can provide the pitch and roll rotations using two servos (one for each rotation). The swash plate used in the Hummingbird V3 [29] is shown in Figure 10.37.

The swash plate has two plates. The one at the bottom is the static one, and one on the top is the one that rotates with the main rotor. The bottom one has two protruding balls at two tips (as shown in the figure). This is where two servo push rods are connected so that the two points can be independently moved up and down. The top plate has four such balls. These are where one end of rods is connected that connects these points to the four blades (two of the main rotor and two for the flybar). The up or down movement of the bottom plate makes the upper rod move also when it reaches that fixed part of the plate, and that changes the angle of attack for the blade. That changes the upward thrust of the main rotor on one side of rotation. This

FIGURE 10.37
Swash plate used in Hummingbird V3 (Copyright 2006 Century Helicopter Products)

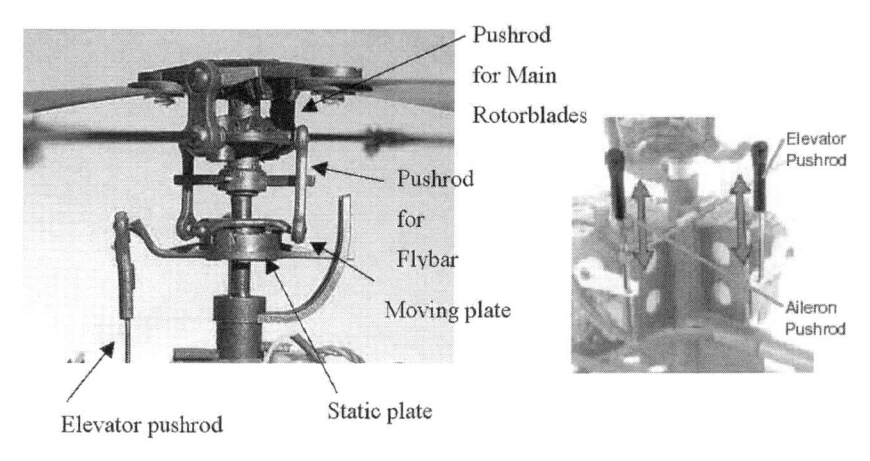

FIGURE 10.38
Swash plate details for Hummingbird V3 (Copyright 2006 Century Helicopter Products)

allows for the helicopter to pitch or roll. More details of the swash plate and its connections to servo pushrods and the main rotor blades and flybar are shown in Figure 10.38.

An alternative scheme uses three servos instead of two for swash plate control. That allows for roll, pitch, and collective control. Collective control means that we can change the angle of attack for all blades on the main rotor together, instead of changing only when the blade gets close to where the pushrod has pushed that part of the plate. This scheme is used in Hummingbird 3D Pro [29]. Figure 10.39 shows clearly how the push rod changes the blade angle, when the rod is moved.

FIGURE 10.39
Swash plate details for Hummingbird 3D Pro (Copyright 2006 Century Helicopter Products)

10.4.1.2 Pitch Control

The elevator servo changes the angle of attack of the main rotor so that the helicopter achieves a pitch rotation. The pitch control is achieved by the elevator as shown in Figure 10.40.

10.4.1.3 Roll Control

The aileron servo changes the angle of attack of the main rotor so that the helicopter achieves a roll rotation. The roll control is achieved by the aileron as shown in Figure 10.41.

10.4.1.4 Up-down Control

The throttle servo in a gas engine based RC helicopter controls the air-fuel mixture that powers the main rotor. By increasing the throttle, the rotor gets more power and moves the helicopter vertically upwards. Decreasing the throttle makes the helicopter move vertically down. The same can be achieved by an electric RC helicopter by controlling the speed of the DC motor. The vertical movement control by the throttle or DC motor is shown in Figure 10.42.

Propeller Based Robots 421

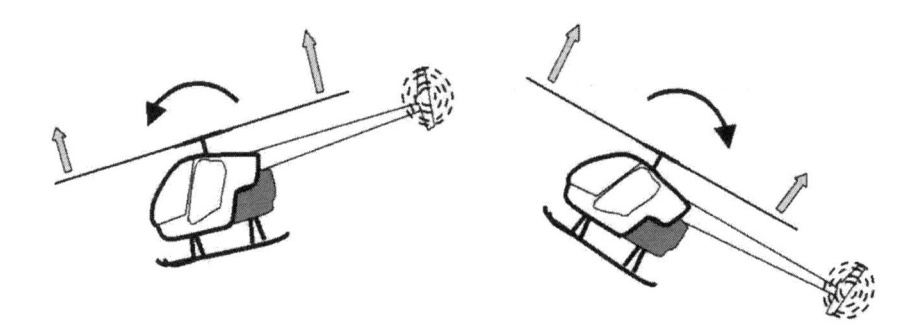

FIGURE 10.40
Elevator pitch control

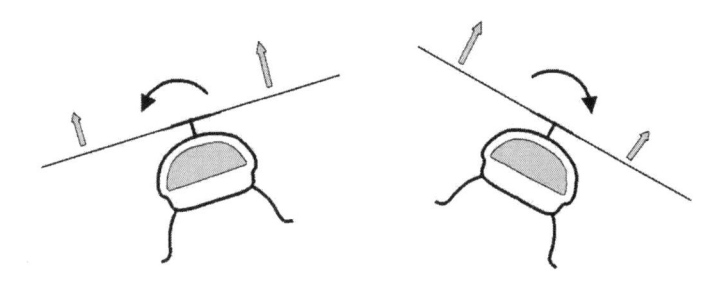

FIGURE 10.41
Aileron roll control

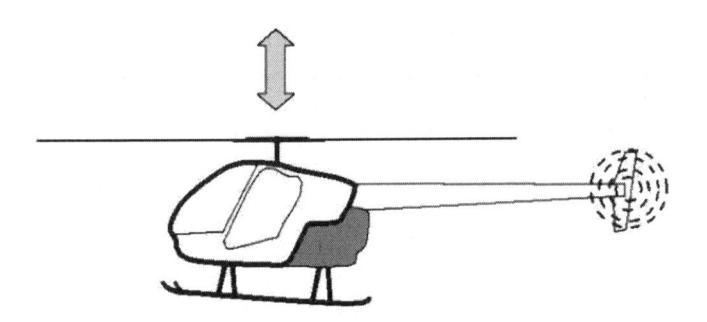

FIGURE 10.42
Throttle/DC motor vertical control

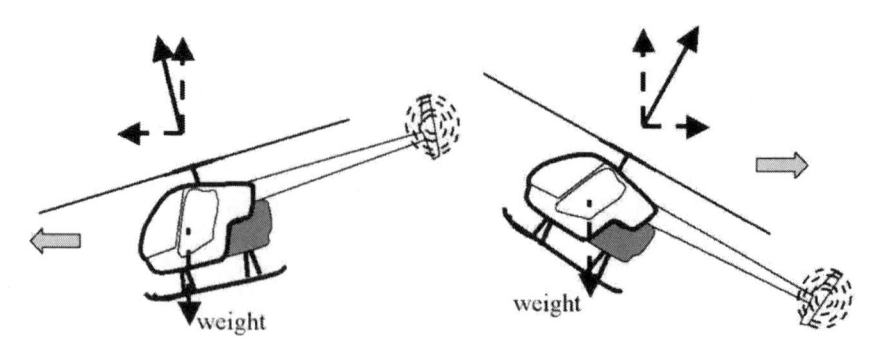

FIGURE 10.43

Forward/Backward movement

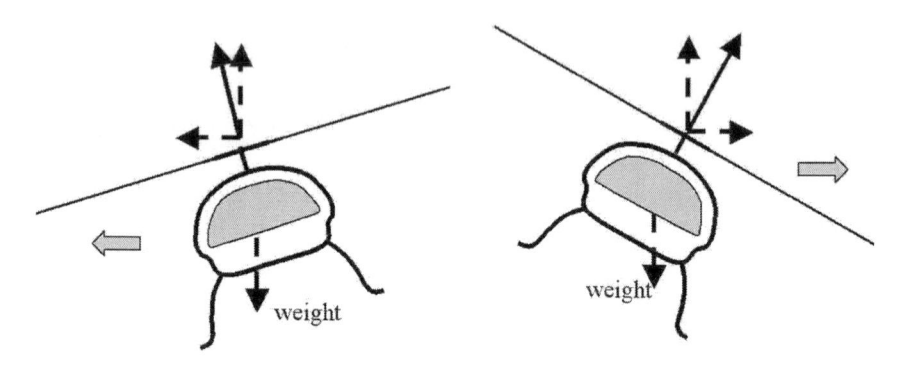

FIGURE 10.44

Sideways movement

10.4.1.5 Forward-Backward Movement Control

Forward movement of the helicopter is accomplished by first tilting the helicopter forward and then using enough thrust on the main rotor so that the vertical component of the thrust cancels the helicopter weight, and the horizontal component moves it forward (as shown in Figure 10.43). Backward movement of the helicopter is accomplished by first tilting the helicopter backward and then using enough thrust on the main rotor so that the vertical component of the thrust cancels the helicopter weight, and the horizontal component moves it backward (as shown in Figure 10.43).

10.4.1.6 Sideways Movement Control

Sideways movement of the helicopter is accomplished by first tilting the helicopter sideways and then using enough thrust on the main rotor so that the vertical component of the thrust cancels the helicopter weight, and the horizontal component moves it sideways (as shown in Figure 10.44).

Propeller Based Robots 423

FIGURE 10.45
The UAV helicopter controller by Rotomotion LLC (Copyright 2003, 2004 by Rotomotion, LLC.)

10.4.2 Automatic Controllers for RC Helicopters

In this section we explore a automatic controller for RC helicopters for autonomous flight. Rotomotion LLC has a product called UAV Helicopter Controller. This UAV system is an integrated package design to control and guide an RC helicopter. The box is shown in Figure 10.45, to be mounted to the helicopter.

With this system, the user has to take-off and land the helicopter manually. Once the helicopter is successfully taken-off and in the hovering state, the helicopter can go into autonomous mode by flipping a switch on the transmitter. The autonomous flight control system has an advanced stable-hover control system with multiple modes of operation. The system has four modes of operation: velocity command mode (VC-Mode), position command mode (PC=Mode), waypoint route mode (WAY-Mode), and fast forward flight mode (FFF-Mode).

In the VC-Mode, the helicopter position is controlled by the transmitter using proportional velocity commands. In the PC-Mode, the helicopter position is changed by the transmitter proportional to the stick command. The helicopter flies a preprogrammed series of waypoints in the WAY-Mode. Finally, the helicopter can move fast forward similar to fixed wing aircraft in the FFF-Mode and orbit an area in this mode.

10.5 Robotic Boats

It is very easy to convert an RC boat into a robotic boat. Boats usually have only two actuators: throttle (or electric motor) and steering (servo). These can be easily controlled using a microprocessor, and we can interface sensors to

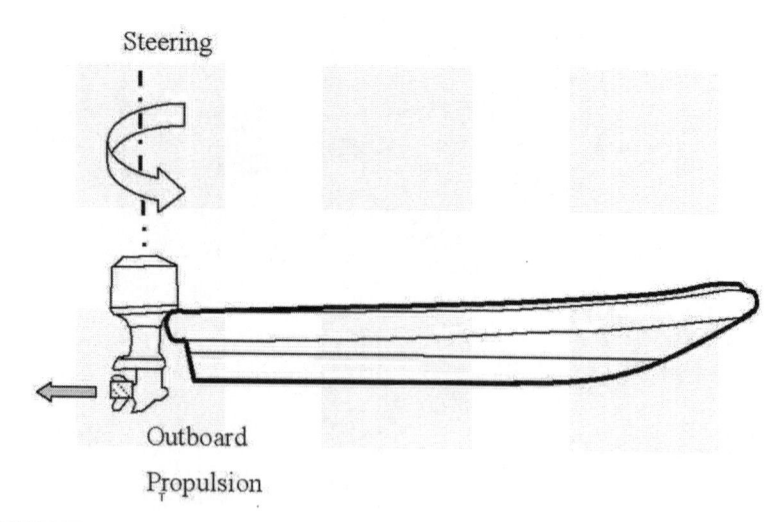

FIGURE 10.46 Outboard propulsion

the microprocessor. We can interface sensors, such as a GPS sensor, cameras, and ultrasonic sensors.

Boats are of two types depending on what their bottom looks like: displacement hulls (designed for stable slow motion) and planing hulls (designed to move fast above water). The planing hull can be a monohull (flat bottom, deep V, or a shallow V), or a multi-hull, or a hydroplaning hull.

10.5.1 Propulsion

The propulsion system can be outboard, inboard, or outboard/inboard. The outboard propulsion system has everything outside the boat: the power generation (gas or electric), the propeller, and the steering system. Figure 10.46 shows an outboard propulsion system. Another way to control the boat is to have two motors, so that by controlling the direction of each, we can have forward/reverse motion, and steering (right or left).

Most electric RC boats have one DC motor for the propeller and a servo for the rudder. As an example of this consider the Tower Turbo Vee II from Tower Hobbies [30] as shown in Figure 10.47.

There is a direct drive system from the DC motor to the propeller, and a rudder for steering controlled by a servo. These are shown in Figure 10.48.

Most recent electric boats have multiple DC motors for the propeller and a separate servo motor for steering. A boat with such a design (Villain EX by Traxxas Corporation) is shown in Figure 10.49.

The anatomy of the Villain EX twin motor electric boat is shown in Figure 10.50. Wiring diagram of the boat is given in Figure 10.51.

Propeller Based Robots

FIGURE 10.47
Tower Turbo Vee II (Copyright Hobbico Inc. Reproduced with permission)

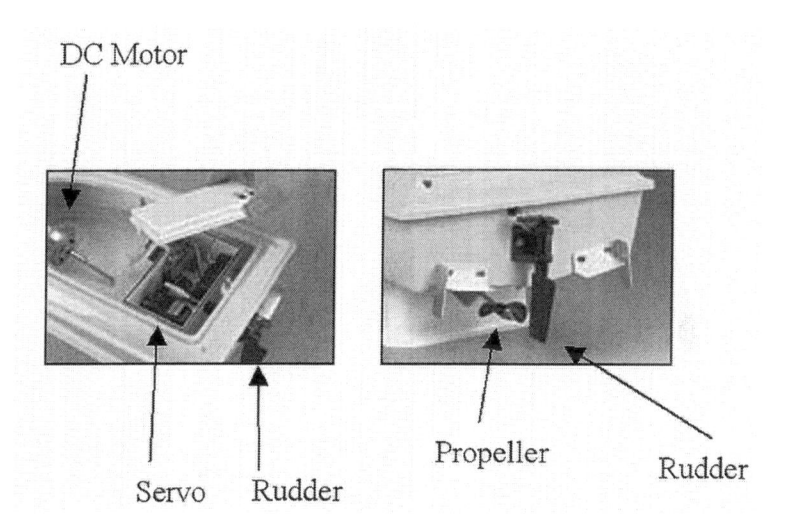

FIGURE 10.48
Rudder and propeller based boat (Copyright Hobbico Inc. Reproduced with permission)

FIGURE 10.49
Twin motor propulsion (Copyright 2003 Traxxas Corporation)

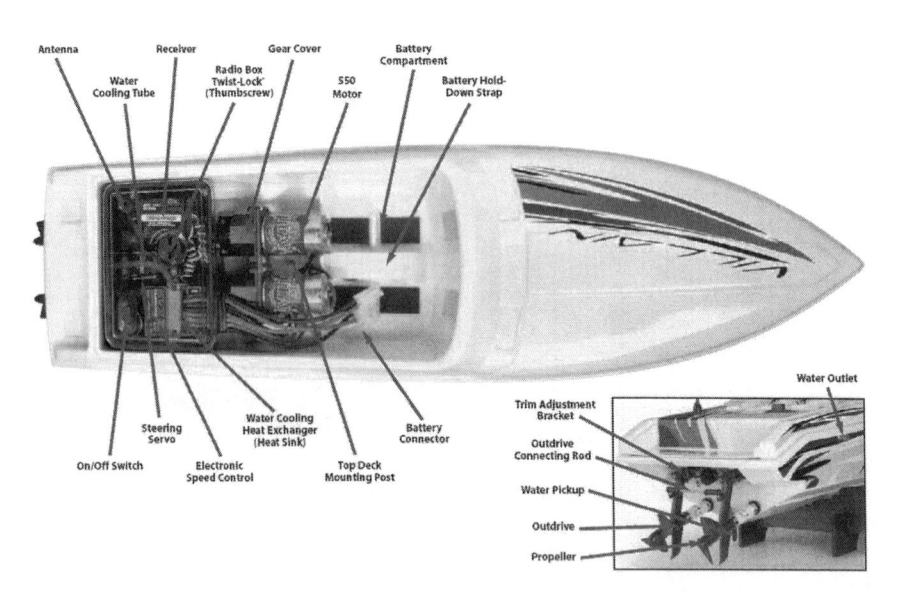

FIGURE 10.50
Anatomy of Villain EX (Copyright 2003 Traxxas Corporation)

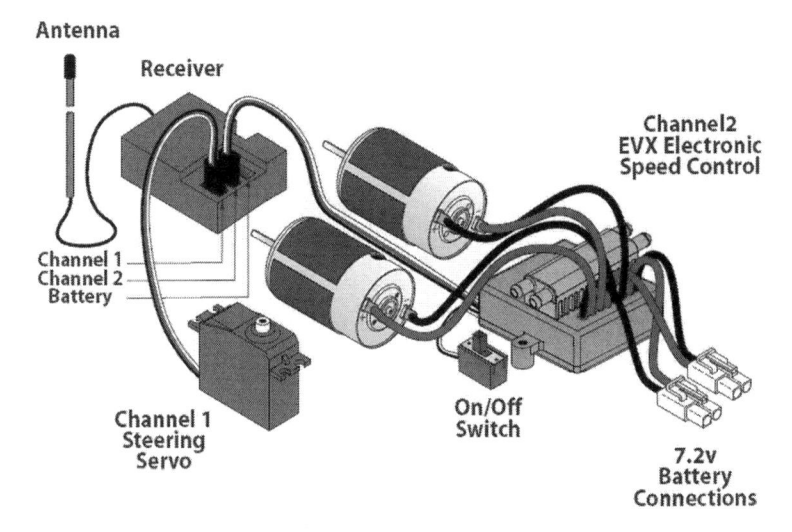

Propeller Based Robots 427

FIGURE 10.51
Wiring diagram of Villain EX (Copyright 2003 Traxxas Corporation)

Another propulsion system uses a propeller like an airplane (completely out of the water). The steering is done using a rudder (by a servo), and speed is controlled via a throttle servo. This system is used by an aquacraft from Tower Hobbies [30] and Hobbico Inc. [31] as shown in Figure 10.52.

10.6 Robotic Submarines

Submarines can be dynamic or static. Dynamic submarines usually will float on water and are actively pushed down by propellers to keep them underwater. Static submarines on the other hand add water into a submarine chamber (called ballast tank) to add weight to it, so the submarine can get heavier. On the other hand when it wants to come up, it uses compressed gas to push the water out, so that the submarine becomes light again.

RC submarines have a motor that drives a propeller at the back that pushes the vehicle forward, and it has rudder and stern (and sail) planes to enable it to produce yaw motion and pitch motion. Rudders enable the yaw motion by their rotation, and stern planes produce the pitch motion just like elevators do in RC planes. These actuators are shown in Figure 10.53.

A more detailed view of rudder and stern are shown in Figure 10.54.

The weight distribution keeps the submarine facing the right side up by keeping heavier weight at the bottom compared to the top. The sail plane

FIGURE 10.52
Aquacraft by AirForce (Copyright Hobbico Inc. Reproduced with permission)

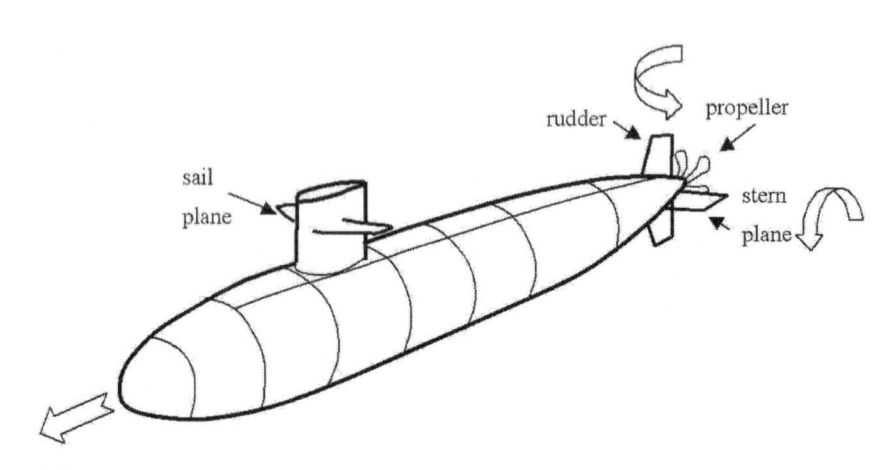

FIGURE 10.53
Submarine actuators

Propeller Based Robots 429

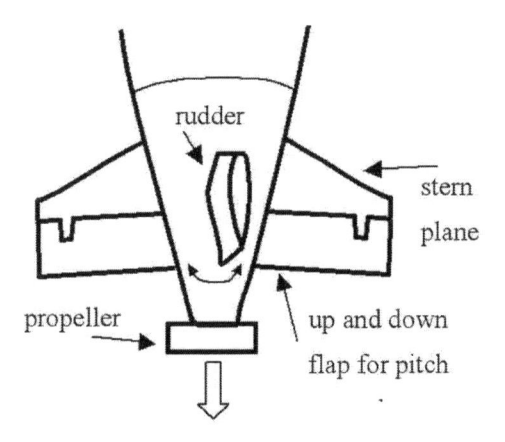

FIGURE 10.54
Submarine rudder and stern

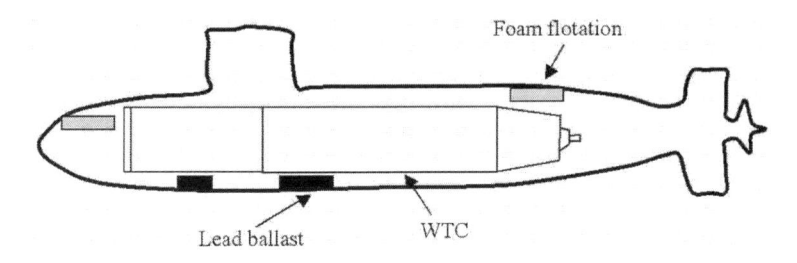

FIGURE 10.55
WTC location and weight Distribution

with the stern plane keeps the submarine balanced. The electronics is kept inside the WTC (water tight containers) and contains two servos and a DC motor. One servo controls the rudder angle, the other controls the stern. The location of the WTC and the weight distribution is shown in Figure 10.55.

The WTC has generally three compartments. The first compartment will have battery and speed controller, the second one will have a ballast system, and the third one the servos. Typically, the ballast system will have a tank with pressurized air and valves. The valves are used to allow water to flood in to increase the weight of the submarine to enable it to go down, or to let water come out when the pressurized air is let into the chamber, so that the weight of the submarine is reduced to enable it to rise. The other electronic components that are usually present are speed controller connected to the radio signal receiver (which receives the signal from the remote control to control the motor and servos), pitch control (using an accelerometer to control the pitch angle), and a microsafe (a device which turns the ballast system off if the radio signal is lost for some time, so that the submarine will float on top in case of lost radio signal).

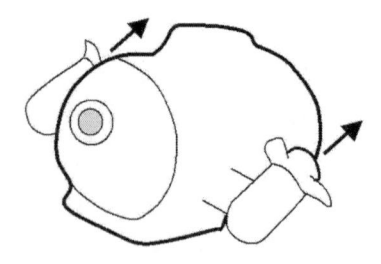

FIGURE 10.56
Twin propeller based dynamic RC submarine

FIGURE 10.57
Sea Scout twin propeller RC submarine (Copyright Hobbico Inc. Reproduced with permission)

Submarines communicate under water using sonar signals using audio frequencies. Submarines also use GPS for navigation (however that signal is lost underwater), and therefore underwater INS (Inertial Navigation System) is used that is composed of accelerometers and gyroscopes to measure how the submarine is moving so that it can keep track of its position and orientation.

Dynamic submarines do not have the ballast system and therefore they have to be actively forced down in the water, otherwise they float up on the surface. Many inexpensive dynamic RC submarines are built using twin motors that are used to dive, surface, turn etc. The angle of the propellers can be changed before putting the submarine into water to get different movements. One such submarine is illustrated in Figure 10.56. Figure 10.57 presents an example of a twin propeller submarine, Sea Scout by Hobbico Inc. [31].

The kinematics of a submarine can be derived using a similar mathematical technique that was used for deriving the kinematics for the airplane. Many experiments can be designed for robotic submarines for performing underwater surveillance (using wireless cameras in WTC (Water Tight Containers)), using various other sensors, performing formation "flight", etc.

References

[1] Abbott Ira H. and Von Doenhoff Albert E. *Theory of Wing Sections*. Dover Publications; New York (June 1, 1980).

[2] Alexander Charles and Sadiku Matthew. *Fundamentals of Electric Circuits*. McGraw-Hill Science/Engineering/Math; Second edition (May 26, 2004).

[3] Anderson John D. *Fundamentals of Aerodynamics*. McGraw-Hill Science/Engineering/Math; Third edition (January 2, 2001).

[4] Behnke Sven. "Playing Soccer with Humanoid Robots," *KI - Zeitschrift Knstliche Intelligenz*. No. 3, pages 51-56, (2006).

[5] Brown Henry T. *507 Mechanical Movements: Mechanisms and Devices*. Dover Publications (August 15, 2005).

[6] Craig John J. *Introduction to Robotics: Mechanics and Control*. Prentice Hall; Third edition (October 12, 2003).

[7] Kachroo Pushkin and Mellodge Patricia *Mobile Robotic Car Design*. McGraw-Hill/TAB Electronics; First edition (August 12, 2004).

[8] Lynxmotion *ssc-32v2.pdf, SSC-32 User's Manual Version 2*. (2005).

[9] Munson Bruce R., Young Donald F., and Okiishi Theodore H. *Fundamentals of Fluid Mechanics*. John Wiley & Sons; 5 edition (March 11, 2005).

[10] Nilsson James W. and Riedel Susan. *Electric Circuits*. Prentice Hall; Seventh edition (May 20, 2004).

[11] Ulaby Fawwaz T. *Fundamentals of Applied Electromagnetics*. Prentice Hall; Media edition (2004).

[12] Raibert Marc H. *Legged Robots That Balance*. MIT Press; Reprint edition (April 2000).

[13] Nise Norman S. *Control Systems Engineering*. Wiley; Fourth edition (2004).

[14] Iovine J. *PIC Microcontroller Project Book : For PIC Basic and PIC Basic Pro Complilers*. McGraw-Hill/TAB Electronics; Second edition (2004).

[15] Sclater Neil and Chironis Nicholas. *Mechanisms and Mechanical Devices Sourcebook*. McGraw-Hill Professional; Third edition (June 13, 2001).

[16] Spong Mark W., Hutchinson Seth, and Vidyasagar M. *Robot Modeling and Control*. Wiley; (2006).

[17] Texas Instruments *sn54ls595.pdf, 74LS595 Datasheet*. (1988).

[18] Wadoo Sabiha *Feedback Control and Nonlinear Controllability of Nonholonomic Systems*. http://scholar.lib.vt.edu/theses/available/etd-01162003-101432. M.S. Thesis, Bradley Dept. of Electrical and Computer Engineering, Virginia Tech, (2003).

[19] Norton Robert L. *Machine Design: An Integrated Approach, 2/E*. Prentice Hall; (2000).

[20] Auslander David M. and Kempf Carl J. *Mechatronics: Mechanical System Interfacing*. Prentice Hall; (1996).

[21] Histand Michael B. and Alciatore David G. *Introduction to Mechatronics and Measurement Systems*. McGraw-Hill (1999).

[22] Gajski Daniel D. *Principles of Digital Design*. Prentice Hall (January 1997).

[23] Bolton W. *Mechatronics: Electronic Control Systems in Mechanical and Electrical Engineering*. Longman, Second edition (1999).

[24] The Oject Oriented PIC Web Site. *http://www.oopic.com*.

[25] The Lynxmotion Inc. Web Site. *http://www.lynxmotion.com*.

[26] Parallax Inc. Web Site. *http://www.parallax.com*.

[27] Chaney Electronics Web Site. *http://www.chaneyelectronics.com*.

[28] Heli-Max Website. *http://www.helimax-rc.com*.

[29] Century Helicopter Products Web Site. *http://www.centuryheli.com*.

[30] Tower Hobbies Web Site. *http://www.towerhobbies.com*.

[31] Hobbico Inc. Web Site. *http://www.hobbico.com*.

[32] Fairchild Semiconductor Web Site. *http://www.fairchildsemi.com*.

[33] National Semiconductor Web Site. *http://www.national.com*.

[34] USB-UIRT Web Site. *http://www.usbuirt.com*.

[35] Aibo Hack Web Site. *http://www.aibohack.com/robosap/ir_codes.htm*.

[36] Dr. Sven Behnke's Nimbro Project Web Site. *http://www.nimbro.net/robots.html*.

[37] Internals Web Site. *http://www.internals.com*.

References

[38] Texas Instruments Web Site. *http://www.ti.com*.

[39] UAV Flight Systems, Inc. Web Site. *http://www.uavflight.com*.

[40] Geek Hideout Web Site. *http://www.geekhideout.com/iodll.shtml*.

[41] Programmers Heaven Web Site. *http://www.programmersheaven.com/zone15/cat610/16353.htm*.

[42] Logix4u Web Site. *http://www.logix4u.net*.

[43] Borland Compiler Web Site. *http://www.borland.com/downloads/download_cbuilder.html*.

[44] MCII Corporation Web Site. *http://www.mciirobot.com/download/download.htm*.

[45] Pololu Inc. Web Site. *http://www.pololu.com/products/pololu/0391/#adapterspecs*.

[46] The Code Project Web Site. *http://www.codeproject.com/csharp/SerialCommunication.asp*.

[47] National Semiconductor *LM555.pdf, LM555 Datasheet*. (February 2000).

[48] National Semiconductor *LM386.pdf, LM386 Datasheet*. (August 2000).

Index

12-servo hexapod, 339
3-state buffer, 168
3-to-8 decoder, 166
4WD, 269
555 Timer, 236
555 Timer as a One-Shot, 253
74LS, 32
74LS126, 168
74LS244, 115, 123
7805 Regulator, 164
78xx, 35, 36
9-pin serial cable, 76

7404, 110

AC, 32
AC adapter, 125
Address for LPT, 105
Adjusting Servomotors, 353
aileron, 420
aluminum risers, 273
aluminum spacer, 318
angle gussets, 269
angular speed, 223
angular speeds, 268
anode, 8
applied torque, 226
Arithmetic Operations, 88
ASCII, 92, 136
ATAN, 206
audio amplifier, 252
audio IC, 253
automatic control, 173

back emf, 224
back emf constant, 224
Backward Motion, 231
Backward Right Turn, 232

Backward Timed Motion, 242
ball links and nuts, 309
ball socket assembly, 311
Bar Coil, 5
Basic Atom, 355
Basic Atom Pro, 301, 355
Basic Robot Movements, 235
BASIC Stamp, 67, 269
BASIC Stamp 1, 67
BASIC Stamp 2, 67
Basic Stamp 2, 301, 357
BASIC Stamp 2 Carrier Board, 74
BASIC Stamp 2e, 71
BASIC Stamp 2p, 69
BASIC Stamp 2sx, 69
BASIC Stamp Activity Board, 75
BASIC Stamp Editor, 76
BASIC Stamp editor, 87
BASIC Stamp math, 88
BASIC Stamp Super Carrier board, 75
Belt System, 222
binary logic, 25
BIOS, 103
Board of Education, 75
Body Construction, 315, 333
Boolean, 26
breadboard, 221
breakdown voltage, 8, 10
BS2-IC, 67, 74
BS2e, 339
BS2e-IC, 74
BS2e-IC pins, 71
BS2p24-IC, 69, 75
BS2p40-IC, 69
BS2sx-IC, 69, 74
buffers, 115

Index

camera
    CMUcam2+, 183
capacitor, 5
    electrolytic and nonelectrolytic, 6
cathode, 8
CD4013BC, 258
Centering a servomotor, 306
circulation, 389
clutch, 147
CMOS, 19, 31
coding
    binary, 179
    Gray, 179
CON, 87
Connecting the BS2e, 344
connection scheme, 234
constant, 87
constraints
    nonholonomic, 412
Construction and Mechanics, 221
control panel, 140
coordinate
    Cartesian, 409
    orientation, 409
coordinates
    joint, 192
    world, 192
current limiting resistor, 136

D Flip-Flop, 258
Dancer Robot, 260
Day-Runner Robot, 249
DB-25, 71
DB-9, 71
DB-9 connector, 76
DB25, 103, 125
DB25 connector, 166
DB25F, 125
DB9, 130
DC, 32
DC Motor Dynamics, 224
DC motors, 222
DC signal, 260
DEBUG, 92

DEC, 98
DEC4, 99
decimal system, 105
Denavit-Hartenberg, 188
Depletion Zone, 17
desired torque, 226
desired wheel speeds, 268
DH table, 194
dielectric, 17
differential drive robot, 219
Differentiator, 21
diode, 8
    laser, 8
    LED, 11
    light-emitting, 8
    photo, 8
    zener, 8, 9, 35
DIP, 67
DOF, 301
dog bone, 311
Double Timer Circuit, 245
Download, 346
DPDT, 229
driver
    ULN2803, 158
dynamic equations, 225

EEPROM, 81
electret microphone, 252
Electrical Control, 339
Emotional Robot, 252
encoder
    absolute, 174
    incremental, 174
encoders, 174
END, 94
experiments, 219
Extreme Hexapod, 301

Feedback Control, 266
flight control, 423
flip-flop, 27
    clocked, 28
    D, 28
    JK, 28

SR, 28
flybar, 418
FOR...NEXT, 93
force, 223
formatters, 99
forward kinematics, 158
Forward Motion, 231
Forward Right Turn, 232
four-link mechanism, 147
four-wheeled robot, 269
FSM, 27

gait
    dynamic, 359
    static, 359
Gate potential, 17
gear ratio, 145
gear train, 145
gearbox, 139
GOSUB, 94
gripper, 147
GWS S03N servo, 269

H-bridge, 262
handshaking, 130
HCT, 32
HEDS, 177
helicopter, 417
hex nuts, 272
hexadecimal system, 105
Hi-Z, 171
high impedance, 171
hinge, 303
Hitec HS-422, 301, 303
hull
    displacement, 424
    planing, 424

ideal pulley system, 224
image processing, 156
index marker, 177
inductance, 5
infrared, 14
inpout.dll, 128
inpout32.dll, 125

Input-Output Interfacing, 123
INS, 430
Instrumentation Amplifier, 23
INT, 97
Integrator, 21
internal coil, 224
internal resistance, 224
inverse kinematics, 158
inverted TTL, 136
inverter, 19
IO.dll, 123
IOLineP, 293
IR LED, 249
IR light, 252
IR receiver, 250
IR system, 250
IR-transmitter, 251
IRPD, 293, 315, 335

JM-SSC16, 198
Joint
    Base, 139
    Elbow, 139
    Shoulder, 139
    Wrist, 139
joint angles, 158

Karnaugh maps, 26
kinematics
    forward, 192, 201
    inverse, 192, 205
Kirchhoff's voltage law, 224
Kirchoff's laws, 14
Kutta condition, 389

L brackets, 303
Lagrangian, 193
LCD display, 346
LED, 175
linear speed, 223
LM386, 253
load torque, 226
logic gates, 26
LPT, 103
LPT ports, 105

Index

matrix
    rotation, 411
    skew-symmetric, 411
MAX232, 186
mediation gear, 149
Mini SSC II, 301, 353
minimization, 27
minimum torque, 226
motor shaft, 222

Newton's law, 225
Next Step Carrier Board, 340
Night-Runner Robot, 245
NiMh battery, 278
NPN transistor, 112, 229
null modem, 130
numbering systems, 88
nylon rivet fasteners, 303
nylon spacer bars, 306

Obstacle Avoidance, 252
oButton, 285
OEMBS2e, 71
OEMBS2sx, 69
Ohm's law, 23
Ohms, 1
oIRPD1, 293
OOP, 282
OOPic, 301
OOPic-R, 269, 315
operational amplifiers, 20
optical switch, 174
oscillators, 20
oServo, 286
oSonarPL, 291
OWI-007, 156

p-n junction, 11
Parallax Board of Education, 269
parallel port, 103, 171
PAUSE, 94
PBASIC programming fundamentals, 84
PC Control, 265
PC to PC serial communication, 130

peak detectors, 20
permittivity, 20
photodiode, 14
phototransistor, 18, 175
photovoltaic effect, 14
PIC Basic, 265
PIC controller, 265
PIC16F84, 164
PICBasic, 133
PICmicro MCU, 136
pinchoff, 18
pitch, 394
plate
    swash, 418
Plexiglas, 221
PNP transistor, 112
Polaroid sonar, 291
Port Access Library, 106
potentiometer, 182, 253
potentiometers, 1
Power System, 227
Programming the Hexapod, 346
propeller, 391
pulse width modulated, 344
push-button switch, 227
PVC, 269
PWM, 165, 166, 265, 286, 344

quadrature, 175

R-S Flip-Flop, 236
rack and pinion, 147
radius ratio, 223
RC, 20
RC servo, 286
register, 28, 31
registers, 103
Relay Board, 229
Relay Interface, 124
relays, 158
remote controls, 249
resistors, 1
RETURN, 94
Rev. A, 74
Rev. B, 74

Rev. C, 74
reverse biased, 14
Robot Base, 221
Robot Eyes, 249
Robot Kinematics, 266
Robot Speed Control, 262
robotic arm, 139
rotation span, 344
rotational velocity, 224
rotor, 417
RS-232, 135
RS-232 connector, 266
RS-232 wireless modem, 266
RS232 interface, 138
Rubber belts, 222

Sample Code, 346
screw, 394
sequential circuits, 27
Serial port, 130
serial port class, 131
SEROUT, 96
servo, 394
servo brackets, 269
servo horn, 273, 329
Servo Operation, 344
servo speed values, 343
servomotor driver, 301
servomotors, 194, 269, 301
Sharp Right Turn, 234
short-tolerant, 136
Signed modifiers, 99
SIL package, 5
SIP, 67
six-legged robot, 301
slippage, 394
socket head cap screws, 306
solderless breadboard, 221
Sound-Activated Robot, 260
spacer bar, 306
SPDT, 229
speed, 222
SSC-12, 301, 355
SSC-12 Servo Controller, 342
STAMP memory map, 79

static friction coefficient, 225
Steady State Analysis, 225
streamline, 389
submarine, 430
summing amplifier, 24
surveillance, 430

teach pendant, 156
theorem
    Bernoulli, 390
    Kelvin's Circulation, 390
    Kutta-Joukowski, 390
Time-Controlled Sharp Left Turn, 243
Time-Controlled Sharp Right Turn, 242
Timed Movements, 234, 239
top deck, 273
torque, 141, 222, 223, 225
torque constant, 225
torque limiting, 149
traction force, 226
traction system, 221, 222
trajectory, 193
trajectory planning, 156
transformations
    homogeneous, 204
transformer, 32
transistor, 14
    BJT, 14
    FET, 14
    MOSFET, 14
    NPN, 15
    photo, 28
    PNP, 15
transmitter circuit, 250
transmitter-receiver, 175
TTL, 186
TTL Current, 111
TTL Interfacing, 112
TTL inverter, 110
TTL logic, 109
TTL signals, 109
TTL Voltage, 110

ULN2803, 121

Index

unity gain, 23
USB interfacing, 137
USB ports, 137
USB to serial converter, 137

viscous friction coefficient, 225
Visual Basic Form, 130
voltage comparators, 20
voltage regulator, 10

WAIT, 101
wake, 390
Walking Scheme, 346
wheeled robots, 219
wheels, 269
wing section, 389
WinIo, 106
world coordinates, 268
WTC, 430

zener diode, 120

An environmentally friendly book printed and bound in England by www.printondemand-worldwide.com

PEFC Certified
This product is from sustainably managed forests and controlled sources
www.pefc.org
PEFC/16-33-415

MIX
Paper from responsible sources
FSC® C004959
www.fsc.org

This book is made entirely of sustainable materials; FSC paper for the cover and PEFC paper for the text pages.

#0141 - 300514 - C0 - 234/156/25 [27] - CB